AF899481

Kant's Revolutionary Theory of Modality

Kant's Revolutionary Theory of Modality

Uygar Abacı

OXFORD
UNIVERSITY PRESS

OXFORD
UNIVERSITY PRESS

Great Clarendon Street, Oxford, OX2 6DP,
United Kingdom

Oxford University Press is a department of the University of Oxford.
It furthers the University's objective of excellence in research, scholarship,
and education by publishing worldwide. Oxford is a registered trade mark of
Oxford University Press in the UK and in certain other countries

© Uygar Abacı 2019

The moral rights of the author have been asserted

First Edition published in 2019
Impression: 1

All rights reserved. No part of this publication may be reproduced, stored in
a retrieval system, or transmitted, in any form or by any means, without the
prior permission in writing of Oxford University Press, or as expressly permitted
by law, by licence or under terms agreed with the appropriate reprographics
rights organization. Enquiries concerning reproduction outside the scope of the
above should be sent to the Rights Department, Oxford University Press, at the
address above

You must not circulate this work in any other form
and you must impose this same condition on any acquirer

Published in the United States of America by Oxford University Press
198 Madison Avenue, New York, NY 10016, United States of America

British Library Cataloguing in Publication Data
Data available

Library of Congress Control Number: 2018962720

ISBN 978–0–19–883155–6

Printed and bound in Great Britain by
Clays Ltd, Elcograf S.p.A.

Links to third party websites are provided by Oxford in good faith and
for information only. Oxford disclaims any responsibility for the materials
contained in any third party website referenced in this work.

To Bedriye and Kathleen

Contents

Preface and Acknowledgments

The last decade has witnessed an explosion of interest in Kant's views on modality, which, valuable exceptions in German such as Guido Schneeberger (1952) and Bernward Grünewald (1986) notwithstanding, had not previously been intensively studied in the vast literature on Kant. Thanks to a new generation of Kant scholars including Ian Blecher, Andrew Chignell, Toni Kannisto, Jessica Leech, Tobias Rosefeldt, Timothy Rosenkoetter, Nicholas Stang, and Reed Winegar, different aspects of Kant's theory of modality have recently been brought to the attention of the Kant community as well as the wider philosophical audience. Stang's excellent book (2016) was the first book-length study dedicated solely to Kantian modality in English. I intend this book to complement my own work on the subject and contribute to the ongoing efforts of this dynamic group.

I have been working on Kant's treatment of modal notions for more than a decade. My first fascination with the subject goes back to my graduate studies at Boğaziçi University, İstanbul. İlhan İnan was first to direct my attention to the intriguing question of what it means to exist, and Stephen Voss and Lucas Thorpe helped me refine my initial thoughts on Kant's theses on existence that result in my first publication (2008). My gradual realization that Kant's theses on existence constitute the crux of a much more comprehensive theory of modality came to motivate my doctoral dissertation at the University of Pennsylvania. An important portion of my ideas in this book are rooted in my dissertation in one way or another. I am forever grateful to my dissertation supervisor Paul Guyer. He masterfully guided my voyage through the stormy oceans of Kant's philosophy to the safe shores of interpretive clarity and truth. My other advisors, Karen Detlefsen and Charles Kahn, provided me with generous support and insightful feedback on my narrative regarding the broader history of modal thought in Western metaphysics. Andrew Chignell helped me immensely as my external reader. Without Andrew's criticisms and suggestions, I would not have recognized some of the important intricacies of Kant's account of real modality in my dissertation. I would also like to thank The Woodrow Wilson National Fellowship Foundation for generously awarding me the Newcombe Fellowship in support of my dissertation project.

I first conceived the idea of developing my dissertation into a book during my two years of teaching at the University of British Columbia. The writing process took place during my three years at the University of Richmond and two years at the Pennsylvania State University. I benefited from the generous support of these three institutions, the input of my colleagues, and the insightful questions of my students who took my graduate and undergraduate seminars on Kant's theoretical philosophy. I am especially thankful to my current department at Penn State for hosting and funding a manuscript review workshop in October 2017. The participants, Amy Allen, Brady Bowman, Christopher Moore, Emily Grosholz, Mark Sentesy, and Timothy Rosenkoetter, provided me with extremely helpful substantial, organizational, and stylistic feedback on a complete draft of this book. Ben Randolph, Reed

Winegar, and Mike Nance have also been generous enough to read and comment on the drafts of various individual chapters.

I am indebted to Peter Momtchiloff of Oxford University Press for believing in this project from the very beginning and navigating me through a smooth review and publication process. Thanks also to two anonymous readers for their meticulous notes and constructive suggestions on the whole of the manuscript. I believe the revisions made as a result of the readers' reports substantially improved the manuscript. I would also like to thank the following publishers for permission to reuse some material from my previously published papers: thanks to Cambridge University Press for 'Kant's Only Possible Argument and Chignell's Real Harmony' (*Kantian Review* 19(1):1–25, 2014) used in chapter 4; thanks to John Wiley and Son for 'The Coextensiveness Thesis and Kan's Modal Agnosticism in the "Postulates"' (*European Journal of Philosophy* 24(1): 129–58, 2016), used in chapter 8; and thanks to John Hopkins University Press for 'Kant, The Actualist Principle, and The Fate of the Only Possible Proof' (*Journal of the History of Philosophy* 55(2): 261–91, 2017), parts of which appeared in chapters 4 and 8.

Finally, I would like to express my gratitude to the very special people in my life. I am so lucky to have the parents I have, Bedriye and Ali Abacı, who understood and supported my rather drastic and risky career shift from engineering to philosophy. This book would not have been really possible without the emotional and intellectual inspiration of Kathleen Harbin. She has given me her constant and loving patience at every single stage of the development of this project, from an early dissertation draft to a complete book manuscript, and kept me going even at times of deep frustration with my own writing. The entire process of writing in the last five years has also made me realize once again that I have such great friends as Sanem Soyarslan, Gaye Çankaya Eksen, Kerem Eksen, Aslı Silahdaroğlu Bekmen, and Ahmet Bekmen. Though each was deeply engaged in their own scholarly projects, they have been so kind as to put up with my ceaseless preoccupation with this project throughout.

Introduction

Although interest in Kant's views on modality has surged only recently, Kant had a great deal to say about modal notions throughout his long philosophic career, from his early works of the 1750s and 60s to his critical works. While there may also be various reasons to be interested in Kant's recurrent discussions of modality from the viewpoint of contemporary epistemology and metaphysics of modality, as Jessica Leech and, to some extent, Nick Stang demonstrate in their works, they deserve particularly special attention from both broader historical and Kant scholarship points of view. For not only do these discussions constitute a genuine turning point in the history of modal thought, but they also provide a framework for a novel interpretation of Kant's philosophical trajectory.

This book will approach the subject of Kantian modality from these broad and narrow historical angles. I aim to offer a comprehensive study of Kant's views on modality by i) locating these views in their broader historical context; ii) establishing their continuity and transformation across Kant's precritical and critical texts; iii) determining their role in the substance as well as the development of Kant's philosophical project. I make two overarching claims. First, Kant's precritical views on modality, which are critical of the tradition only from within its prevailing paradigm of modality and are thus revisionist in character, develop into a historically revolutionary theory of modality in his critical period. Second, this revolutionary theory of modality is not only a crucial component of Kant's critical epistemology, simply as one among its other major doctrines, but it is in fact directly constitutive of the critical turn itself. Thus, tracing the development of Kant's conception of modality provides us with an alternative reading of Kant's overall philosophical development.

Kant presents his precritical views on modal notions mostly in the context of his critique of the ontological argument for the existence of God. Western metaphysics in general, and the ontotheological tradition in particular, with its different versions of the ontological argument, construed existence and modal notions as fundamental ontological predicates expressing different modes or ways of being of things. The Kant of the early 1760s shows some signs of breaking with the tradition, for instance, when he famously claims in *The Only Possible Argument* (1763) that "existence is not a predicate or determination of a thing" (Ak. 2:72), and, even more strongly, when he defines existence as a predicate "not so much of the thing itself as of the thought which one has of it" (Ak. 2:72). Yet, these reflections on existence are oriented toward revising the ontological argument and thus toward reviving rather than dismantling the ontotheological project of proving God's existence from mere concepts. Therefore, despite his immensely important discovery that existence should be reinterpreted as a feature of our representational relation to objects, the precritical

Kant mostly neglects the groundbreaking implications of this discovery for modality in general. Ultimately he remains within the traditional paradigm, conceiving modal notions in ontological terms.

However, in the mid to late 1760s, Kant starts realizing the truly novel character of his discovery and its radical implications for all modal notions. He begins to lay the ground for a revolutionary theory of modality that will find its fullest and most systematic articulation only in the *Critique of Pure Reason* (CPR). This theory of modality primarily consists in breaking with the traditional paradigm by redefining modal notions as features of our conceptual representations of objects rather than as features of objects themselves. Thus, on Kant's revolutionary paradigm, the modality of an object involves a certain reference or relation to the cognitive subject. Possibility, actuality, and necessity all express different modes or manners in which our conceptual representations of objects are related to our cognitive faculty. Accordingly, the modal assertion of an object does not specify a predicate of that object but rather asserts or 'posits' its representation in relation to the conditions of our cognition of objects in general. The modal differences between possibility, actuality, and necessity therefore amount to different ways in which this relation holds, and not to the differences in the contents of our representations of objects.

This revolutionary theory of modality is indeed central to Kant's overall theory of knowledge in the *CPR*, despite the tendency among classical commentators to diminish or altogether ignore its importance. More crucial and even more neglected, however, is that Kant's modal breakthrough is indispensable to the origination and development of the critical project itself. The revolutionary shift in Kant's conception of modality begins to unfold earlier than, and independently of, his earliest formulation of the very idea of a critical turn in philosophy in his famous letter to Markus Herz of 1772. Thus, the former cannot be explained as a logical consequence of the latter. On the contrary, the shift in Kant's conception of modality is constitutive of the critical turn. Kant's radical idea that modal notions pertain to our representations of things and thus involve an ineliminable reference to the cognitive subject is what forces him to transform the guiding question of his philosophy from the ontological question, 'what does it mean to be possible?', into the transcendental question, 'under what conditions can objects be related to our cognition?', as he articulates it in the letter to Herz. Moreover, by the late 1760s, the shift in Kant's conception of modality has already initiated the critical transformation in his understanding of rational theology as well as metaphysics in general—at a point in time, therefore, before any clear announcement of the critical turn itself. The radical critique and reconstruction of metaphysics and theology in the Transcendental Ideal, by which Kant replaces the more revisionist and immanent critique of ontotheology he espoused in the early 1760s, extends from this transformation and turns on his revolutionary conception of modality. The latter, then, can be read as a motor force of Kant's overall critical project.

I.1 Breakdown of Chapters

The book is composed of three parts, devoted, respectively, to the history of modality before Kant, Kant's precritical views on modality, and his critical and revolutionary theory of modality.

Part I presents an account of modal thought in Western metaphysics prior to Kant, with particular emphasis on the early modern period. The 'ontotheological tradition,' that is, the history of the various versions of the ontological argument plays the central role in this account. This tradition has a direct impact on the development of modality in that it treats questions about the meanings and interrelations of modal notions as subsequent to the question of God's existence. Moreover, especially in his precritical period, Kant himself often situates his discussions of modal notions in the context of his critique of the ontological argument. In Part I, I therefore aim to tease out the conceptions of modality underlying the various versions of the argument in order to attain a better understanding of the novelty of Kant's own views on modality.

Chapter 1 first offers a general framework for reading ontotheology, according to which any version of the ontological argument consists of two logical steps. First, it introduces existence into the concept of God in one way or another; second, it infers the existence of God from the concept of God and asserts identity between two distinct notions of God, viz. as the most real being and as the necessary being. With this framework in place, the chapter then examines the classical version of the ontological argument, introduced by St Anselm and popularized by Descartes. I will demonstrate that while Kant's primary objection, namely that existence is not a real predicate, applies equally to both Anselm's and Descartes' arguments, Descartes importantly anticipates what I will call the 'actualist principle', i.e., facts about possibility must be grounded on facts about actuality. This will come to be a major insight and turning point in Kant's theory of modality.

Chapter 2 primarily examines the modal version of the argument, propounded by Kant's more immediate predecessors in the German rationalist school such as Leibniz, Wolff, and Baumgarten. Yet I also look at these figures' accounts of modality in other metaphysical contexts with a view to presenting a more accurate and comprehensive picture of where Kant's views stand in relation to the prevalent conception of modality in the school tradition. I take issue with two claims concerning the school metaphysicians put forth by a historical narrative that is favored in the literature: (i) they were committed to a logicist account of modality, according to which claims about possibility and necessity can be exhaustively explained through formal-logical principles, while Kant introduced a real or metaphysical account of modality, involving extra-logical truth-makers of modal claims; (ii) they were committed to a view of existence to which Kant vehemently objected, namely that existence is a real predicate or determination. I argue that especially Leibniz and Wolff had robust conceptions of real possibility and necessity, irreducible to logical principles, and in their mature metaphysical works, they carefully avoided committing to the conception of existence as a distinct determination of things and even anticipated Kant's position on existence in significant ways. This, of course, raises important questions about where to locate the historical novelty (such as it is) of Kant's views on existence and modality in general.

Part II addresses Kant's precritical views on modality, with a focus on the question of their novelty vis-à-vis the background provided in Part I. Chapter 3 examines major historical objections to the ontological argument. There are two main lines of objection. The first, pursued by Gaunilo and Aquinas against Anselm, Caterus against Descartes, and Crusius against Wolff, aims to block the argument's second logical step, inferring the existence of God *qua* object from the concept of God. The second line of objection, originated by Kant himself in *The Only Possible Argument* (OPA) (1763) (and not by Gassendi as is often claimed), aims to block the argument's first step, by arguing that since existence is not a predicate or determination, it is fallacious to introduce existence into the concept of God in the first place. On one prominent interpretation, this thesis means that any object that instantiates a concept necessarily also instantiates the predicate "exists," presumably because existence is a precondition of being an object at all, implying thereby that existence is a first-order predicate that universally or unrestrictedly applies to all (and not a subset of) objects. This, I argue, is exactly Gassendi's view, but not at all Kant's. The upshot of Kant's negative thesis is rather that existence is not a predicate of any object and thus could not be contained in the intension or content of any concept of an object. This separation of existence from the intensions of conceptual representations of things is the most consistently recurring aspect of Kant's reflections on existence. However, given Leibniz's and Wolff's efforts to define existence as an extrinsic denomination, I argue that the historical novelty of Kant's conception of existence does not lie in this negative thesis. Instead, the novelty is to be found in his two positive theses, "Existence is a predicate not so much of the thing itself as of the thought which one has of the thing" (*OPA*, Ak. 2:72), and "Existence is the absolute positing of a thing" (*OPA*, Ak. 2:73). These theses point to a radical discovery: existence is to be reinterpreted as a feature of conceptual representations of things, i.e., the feature of being instantiated by an object outside the representation; even more importantly, existence should be reinterpreted in terms of a cognitive act, i.e., the cognitive act through which a representation is asserted by the cognitive subject as instantiated by an object outside or distinct from that representation. Unfortunately, however, the Kant of 1763 does not pursue the groundbreaking implications of his positive account of existence for modality in general. Instead, his early criticisms of the traditional understanding of possibility and necessity remain revisionist, for they are primarily oriented toward revising the ontological argument, rather than toward putting a definitive end to the ontotheological project of proving God's existence *a priori*. Consequently, even though Kant strongly commits himself to the negative thesis that 'existence is not a predicate or determination of a thing,' he still remains within the broad conception of modal notions as expressing fundamental ontological features or modes of being of things.

Chapter 4 offers a reconstruction and analysis of Kant's reformulated ontological argument, which moves from the 'actualist principle' (AP), that every real possibility must be grounded in actuality, to the conclusion that there exists a unique really necessary being, i.e., the *ens realissimum*, which grounds all real possibility. This argument turns on Kant's rigorous distinction between real modality, i.e., possibility and necessity of existence, on the one hand, and logical modality, i.e., possibility and necessity of thought, on the other. The literature on this argument usually focuses on the fact that the argument's premises do not warrant the singularity of the ground of all real possibility but allow a plurality of grounds, a problem Kant seems to fail to address. While I too address this problem of singularity of the ground, I raise a further question: what grounds the AP itself? The AP can be interpreted as an ontological principle, expressing the conditions of real possibility *per se*, or as an epistemological principle, expressing the conditions of our cognition of real possibility. I argue that Kant ends up adopting the ontological interpretation of the AP despite flirting with the epistemological interpretation, and yet does not provide a justification for the former.

Part III illustrates how, in Kant's critical period, a revolutionary theory of modality emerges from the radical but initially unexplored core idea in his early positive theses on existence—a theory that redefines all modal notions (possibility, existence, and necessity) as various modes or ways in which the conceptual representations of things are related to the cognitive subject. This theory marks a historic break from the general conception of modalities as basic, genuine, and absolute ontological features of things, a conception held not only by Kant's immediate rationalist predecessors but also by the greater metaphysical tradition.

Chapter 5 examines the trajectory of this development in the transitional period between the publication of the *OPA* (1763) and that of the *CPR* (1781). From the mid-1760s on, we observe a consistent trend in Kant's reflections on modality: he interprets his core radical idea that existence involves a relation to the cognitive faculty more broadly, applying it to the concepts of possibility and necessity. In the late 1760s, we also observe a clear shift in his conception of the AP, which he ceases to treat as an ontological principle concerning the conditions of real possibility, and begins to understand as an epistemological principle concerning the conditions of our cognition of real possibility. This very shift plays an essential role in Kant's realization of the need for a 'critical turn' in philosophy, explicitly stated first in his 1772 letter to Herz, where Kant formulates it in terms of a problem that will prove fundamental to his critical project: 'How do we cognize that our conceptual representations, especially the pure ones that do not derive from our experience of objects, do indeed represent really possible objects or that they are objectively valid?' For what problematizes the objective validity of pure concepts is the epistemological interpretation of the AP, stating that cognition of actuality is a prerequisite of any cognition of real possibility. This strongly suggests that Kant's emerging revolution in modality should be construed as constitutive of his critical turn rather than as a mere logical consequence of it.

Chapters 6, 7, and 8 reconstruct Kant's theory of modality as presented in the *CPR*. Here at least four steps are to be distinguished. (i) The first systematic discussion of modality appears in the Metaphysical Deduction, where Kant presents the 'modality of judgments' as one of the four classes of logical functions of judgment from which he then derives the 'categories of modality' (A74–6/B100–1). (ii) In the Schematism, Kant provides an initial account of real modality, defining the temporal conditions under which the categories of modality can be empirically applied (A144–5/B184). (iii) In the Postulates of Empirical Thinking in General, he provides the full account of real modality by going on to specify the complete set of such conditions (A218/B266). (iv) In the Ideal of Pure Reason, Kant utilizes his critical theory of modality to reframe his 'only possible argument' as demonstrating merely the subjective necessity of the idea of God and to construct his systematic refutation of the traditional variants of the ontological argument. I discuss (i) in chapter 6, (ii) and (iii) in chapter 7, and (iv) in chapter 8.

Chapter 6 offers an alternative to two current strands in the reading of Kant's account of the modal functions of judgments in the literature. The first understands the modality of a given judgment in terms of the judger's attitude toward its content based on their epistemic or psychological states. The second understands the modality of a judgment solely in terms of its location in a formal syllogism or rational inference. I argue for a third alternative: Kant construes the modal functions of judgments as instantiating relative logical modalities and expressing the logical coherence relations of a judgment with another background judgment or set of judgments, i.e., whether a judgment is logically compatible with the background, logically grounded by it, or logically grounded by it through laws of logic. This interpretation not only fits very well with Kant's overall program of redefining modality in terms of the relations between the conceptual representations of things and the subject's cognitive faculties, but also captures the formal-logical infrastructure of Kant's account of real modality in the rest of the *CPR*.

Chapter 7 comprises a detailed discussion of Kant's account of real modality relative to the domain of experience, that is, relative to the background of the conditions of our empirical cognition of objects. This account, which unfolds in the Schematism and the Postulates, marks the culmination of Kant's longstanding revolutionary program in modality. Here we find his precritical theses on existence, both negative and positive, transform into a strong 'peculiarity' thesis about modal categories in general: "as a determination of the object they do not add to the concept to which they are ascribed in the least, but rather express only the relation to the faculty of cognition" (A219). Accordingly, possibility, actuality, and necessity are all instances of absolute positing. Each posits the conceptual representation of an object as a whole in a different relation to the background of the conditions of our empirical cognition or experience of objects, either as logically compatible with them, or as logically grounded by them, or as logically grounded by them through the law of causality. Each such act of positing constitutes a peculiar, i.e., 'subjective,' type of synthetic judgment, where the intension of the subject-concept is not at all enlarged, but a relation with a distinct cognitive faculty (i.e., with understanding, perception, and reason) is added to it. Kant's emphatic rejection of the rationalist contention that the extension of possibility is greater than that of actuality, which in turn is greater than that of necessity, is in fact an expression of his refusal to define modal differences in terms of the intensions of concepts of objects and his corresponding redefinition of modal differences in terms of how concepts of objects are posited in relation to the cognitive subject.

Chapter 8 shows how Kant's revolutionary theory of modality radicalizes his critique of ontotheology in the Transcendental Dialectic. What makes this critique radical, as opposed to Kant's precritical revisionist critique, is that it claims to demonstrate the impossibility of ontotheology as such and reframes it in terms of a natural, but only subjectively valid, procedure of pure reason. I examine Kant's radical critique of ontotheology in two parts. First, I focus on sections 2 and 3 of the Ideal of Pure Reason, where Kant provides a subtle critique of his own precritical 'only possible argument.' I argue that what leads Kant to downgrade his precritical argument from an objectively valid demonstration of the real necessity of the existence of God as the ground of all real possibility to a subjectively valid demonstration of the necessity of assuming the idea of such a being is his aforementioned shift in his conception of the AP from an ontological to an epistemological principle, a shift that starts in the late 1760s but is only fully articulated in the Postulates. Second, I discuss his refutation of the traditional ontological argument in section 4 of the Ideal. I argue that Kant follows a multilayered strategy against the ontological argument, consisting of a combination of two historical lines of objection, only the second of which presupposes his negative thesis that existence is not a real predicate, as well as an additional, third objection based on his further thesis that all existential judgments are synthetic, albeit in a peculiar sense.

Finally, Chapter 9 focuses on the question of the absolute modality of things as they are in themselves in light of the two striking modal commitments Kant makes in §76 of the *Critique of the Power of Judgment*. First, Kant states the epistemic thesis that while it is a necessary feature of our discursive understanding to distinguish between the merely possible and the actual, an intuitive understanding would not make modal distinctions and cognize only actual objects. Entailing a Spinozistic notion of God, who could not represent mere possibilities and could not have brought about a world other than the actual world, the epistemic thesis seems to undermine divine freedom. Second, Kant states the metaphysical thesis that the modal categories are merely subjectively valid for human discursive understanding and thus noumena do not have modal properties. The metaphysical thesis seems to undermine human freedom, a central commitment of Kant's practical philosophy, for if our noumenal selves do not have modal properties, our noumenal volitions could not have been otherwise. I argue that both the metaphysical and epistemic theses are rooted in Kant's revolutionary theory of modality as reconstructed in earlier chapters of the book. The mere subjectivity of modal categories directly follows from the peculiar status that Kant consistently assigns to the modal categories throughout the *CPR*. Accordingly, modal categories are distinct from other categories in that instead of purporting to express the most fundamental ways things are, they express the various ways in which the representations of objects are related to the cognitive subject. This peculiarity is what makes modalization an exclusive feature of a discursive mind to which representations of objects can be related in multiple ways and whose cognition therefore displays a progressive structure of gradual incorporation of individual representations into a whole. This brings us to the essence of Kant's revolutionary theory of modality: modality is irreducibly relational, subjective, and discursive by its very nature. Finally, I show that this way of understanding §76 as the ultimate articulation of Kant's revolutionary theory of modality presents a framework for the resolution of the tensions between the epistemic and metaphysical theses, on the one hand, and divine and human freedom, on the other.

PART I

Modal Thought Prior to Kant

1

Ontotheology and Modality I

The Classical Version of the Ontological Argument

1.1. Ontotheology as the Context of Modal Thought

While the subject of modality recurs in a wide range of contexts across Kant's corpus, the single context in which Kant first introduces a systematic approach to modal notions and continues to provide his most lucid expositions of these notions is that of the ontological argument for God's existence. From the precritical writings such as the *New Elucidation* (1755), *The Only Possible Argument* (1763), and *Inquiry* (1764), to the *Critique of Pure Reason* (1781/1787) and the *Lectures on the Philosophical Doctrine of Religion* (1783–84?), Kant remains engaged with a critique of the classical version of the ontological argument as well as with the idea of developing an alternative version while also nourishing skepticism with regard to the very possibility of the ontotheological project. As I hope to show in this book, Kant's theory of modality is much more than a by-product of his longstanding engagement with the ontological argument, as it goes far beyond Kant's concern with the question of God's existence and comes to play a substantive role in the development of his overall mature philosophy. Yet Kant's choice of the ontological argument as the primary context for his discussion of modality is by no means trivial; indeed, it is suggestive in two historical respects.

First, the particular context of the ontological argument connects Kant's discussions of modality with those of his immediate predecessors in the rationalist school in eighteenth-century Germany. For it is indeed a characteristic of the Leibnizian-Wolffian school metaphysics to discuss the significance of modal concepts in the context of the ontological argument, some version of which is a standard component of any work in rational theology in this tradition. That Kant is in direct conversation with the school's metaphysics of modality is obvious, if not always from his explicit references, from the modal terminology he uses in Latin and German, the specific examples he provides, the specific distinctions and alleged corrections he makes, and his polemical tone, all of which seem to presuppose an awareness in his audience of what was previously said on the subject. This is partly why Kant's precritical account of modality remains an immanent or revisionist critique of, or perhaps better, a contribution to, an ongoing discussion of modality at the time. Kant's revolutionary break, or his paradigm shift, as it were, from the preceding conceptions of modality in his critical period can only be fully understood by recognizing the exact place of Kant's precritical and revisionist critique of these conceptions from within the said ongoing discussion. For, as I will argue in this book, Kant's revolution in modality springs from a radical idea that he discovers through his precritical efforts to revise the ontological argument but fails to develop, perhaps also due to his revisionist agenda.

Second, the ontological argument connects Kant's treatment of modality to a much broader historical context. This context is primarily what Kant himself calls 'ontotheology,' a tradition represented in the eighteenth century by the Leibniz-Wolffian school. On Kant's account, ontotheology is the kind of speculative theology that "considers God merely in terms of concepts" (*Th.Pölitz*, Ak. 28:1003), proceeding from the merely possible to the existence of God as the principle of all possibility.[1] Ontotheology is thus a purely 'transcendental theology', differing from 'cosmotheology' and 'physicotheology,' both of which proceed from experienced existence to the existence of a highest being as the ground of all existence. Since Kant calls the generic form of arguments aiming to infer God's existence from his mere concept 'ontological,' the ontotheological tradition consists for him in the history of the various versions of the ontological argument, which can be traced as far back as Anselm of Canterbury (1033–1109). In fact, this broader historical context comprises also a counter-tradition of objections to the ontological argument. This historical back and forth revolves around alternative ways of understanding modal notions, and thus comes to form a fertile ground for the flourishing of modal thought in Western metaphysics. The major philosophical appeal of this debate is that it provides us with a chance to take a closer look at the historical evolution of the notions of possibility, existence, and necessity and their interrelations.[2]

Kant holds a unique place in the history of ontotheology. First of all, though less widely known than his critique of the argument, Kant makes a positive contribution to the tradition as the originator of a version of an ontotheological argument. Introduced and developed in his precritical works, *NE* (1755) and *OPA* (1763), this version of the argument presents an alternative that is immune to the objection Kant levels against the classical version, namely the thesis that 'existence is not a real predicate.' Yet, in the Ideal of Pure Reason of the *CPR* (1781), Kant will come to acknowledge the impossibility of demonstrating God's existence. Alongside his explicit refutations of the three classical kinds of arguments (ontological, cosmological, and physico-theological), he provides a subtle critique of the modal foundations of his own version—ironically, the very version he once presented as the only possible argument with a chance of success. Thus, Kant's understanding of modality undergoes a critical shift, leading him to abandon the ontotheological project altogether. One of my central claims in this book will be that this shift marks both an important point in the development of Kant's critique of speculative theology and a revolutionary break with traditional conception of modality in

[1] See also *OPA* (Ak. 2:156).

[2] For God, conceived as the most perfect of all beings, would be the exemplary being, the one that best exemplifies the modes of being and thus most suitable for such an investigation into modal concepts. As Heidegger suggests, the historically significance of ontotheology was purely philosophical and consisted in "the orientation of ontology towards the idea of God." See Heidegger (1982), 29.

general. Therefore, a careful scrutiny of the history of the ontological argument is crucial both to understanding Kant's modal theory and recognizing its historically revolutionary character. My aim in Part I will be to provide an account of the chief historical versions of the argument with a view to fleshing out how their proponents conceived modal notions in general.

However, before focusing on the logical mechanics of the ontological argument, it is important to recognize that the argument presupposes a certain conception of deity, philosophical reflections on which play an indispensable role not only in the evolution of modal thought but also in the very emergence of modality as a distinct philosophical theme. The deity in question is the 'God of Abraham,' a god who created everything else *ex nihilo*, and whose very name is 'He Who Is' (Yahweh). This conception of deity entails an ontological difference between God and the World, the Creator, and the Created: while God's essence is identical to his existence, for all other beings existence is a novelty, a gift, an addition bestowed upon them by God at creation, and is thus in some important sense distinct from their essences. This ontological difference requires a metaphysics that could account for the identity and distinction between essence and existence in terms of different modes of being. Such modal metaphysics is absent from Ancient Greek ontology. For since the Ancient Greeks have to accommodate neither the doctrine of creation *ex nihilo* nor, consequently, the notion of a world that might not have come into existence, they take the essence or 'what-is-it' of an existent thing as the starting point of their metaphysical inquiry without problematizing either the notion of existence as such or its identity with or distinction from essence.[3]

The ontological difference and its modal exposition becomes a significant theme later in scholastic ontology. The earliest modal formulation of the essence–existence relation is found in Avicenna's *Metaphysics*.[4] Avicenna argues that God, since his essence or 'quiddity' (*mahiyya*) is no other than existence (*aniyya/wujud*), is the "Necessary Existent" (*wajib al-wujud*).[5] Only God has the privilege of having his existence follow from His essence. In the case of all other things, essence is distinct from existence; in other words, they are "contingent" (*mumkin*). Therefore, existence, in the case of contingent beings, is something "occurring to them externally," and is thus "accidental" (*aradi*) to essence.[6] The essences of contingent beings are 'possible in themselves' (*bi-dhatihi*), but they become actual (*maujud*) only by receiving existence from the 'Necessary Existent.'

[3] See Kahn (1976), 326; (2003), x.

[4] The *Metaphysics* of Avicenna that was better known by the scholastics and referred to by Thomas Aquinas in his *De Ente et Essentia*, is the *Metaphysics* (or *Al-Ilahiyat*, the *Theology*) of his *Al-Shifa* (*The Healing*).

However, Avicenna's formulations of the essence/existence distinction and the modalities issuing from this distinction are clearer in the *Metaphysics* (*Ilahiyyat*) of his historically less significant work, *Danish nama-i ala'i* (*The Book of Scientific Knowledge*). I will thus be referring to both texts here, respectively as *Shifa* and *Danish nama*. There are important scholarly objections to the claim that Avicenna was the first to introduce the essence/existence distinction. Most notably, see Goichon (1969), 34. My claim here is rather that Avicenna was the first to explicitly thematize this distinction as a central doctrine in his philosophy.

[5] *Shifa*, 8.4. (3); *Danish nama*, §24.

[6] *Shifa*, 8.4. (12); *Danish nama*, §38.

While Avicenna's modal account of the ontological difference between God and created things has wide impact on late scholastic ontology, it leaves open questions regarding how to understand the nature of the relationship between essence and existence in created things. Is the distinction between essence and existence a real, mind-independent distinction, or is it a merely conceptual, mind-dependent distinction? Are they separable in reality or only in the mind? Either way, the further question remains, what exactly does existence add to the mere essence of a thing? Most importantly, should this addition itself be construed at the level of the essence or 'quiddity' of the thing, as though it enlarges the latter?

For instance, according to the view commonly attributed to Thomas Aquinas, the distinction between essence and existence is real in that they are mutually separable, like a distinction between a thing (*res*) and another thing (*res*), where existence has its own quidditative content.[7] This, of course, entails that the actual (essence) contains more (reality or quidditative content) than the merely possible (essence). The difficulty of this strong view of the real distinction is rather obvious: if existence has its own essence, then the distinction applies once again to existence itself, opening the door to an infinite regress. The Thomistic view therefore provokes two major reactions.

Duns Scotus takes the moderate position that while essence and existence are not really separable in any concrete individual and that the distinction is therefore not comparable to that between a thing (*res*) and another thing (*res*), where either can occur without the other, the distinction is still mind-independent and not merely conceptual. According to Scotus, the distinction between essence and existence should be understood as a 'modal distinction.' Certain forms or natures in reality come in degrees that are inseparably attached to what they really are. Scotus calls such a degree of intensity the 'intrinsic mode' of the entity to which it belongs. A color, for instance, is necessarily instantiated as having a certain shade or degree of intensity, a mode. Thus, although a color cannot be ontologically separated from its degree of intensity, the two are still 'modally' distinct in the sense that the former can be conceived or defined without the latter (but not vice versa). Scotus suggests that "One can say that the essence and existence in creatures are like quiddity and its mode. Therefore they are distinct."[8]

The more extreme reaction to the Thomistic view is to demote the distinction between essence and existence to a mere distinction of reason, imposed on things through intellectual activity. This is the view developed by Francis Suárez. Suárez insists that the distinction in question is to apply only to created and thus actual entities. The essence of an actual entity (*ens*) is then an actual essence that is already in act as opposed to an essence in mere potency.[9] Introducing a distinction between an actual entity's actual essence and actual existence is not merely metaphysically superfluous, but downright impossible, because the two express one and the same

[7] Wippel (1982a, 1982b) argues that this attribution to Aquinas is not accurate, and that this historical conflation is based on the misinterpretation of the real distinction by an early Thomist, Giles of Rome, as between two things (*duae res*), which unfortunately came to be viewed as the official Thomist position on distinction from the late thirteenth century. On the same point see also Gilson (2005), 99.

[8] *QuodQs*, q.1, add. 1:11, 485.

[9] *Disputationes*, d. 31, sec. 1.13, 52.

thing: "Existent man and a man are the same thing," as Aristotle states.[10] On this view, then, any talk of existence 'adding' anything new to a created entity is entirely misguided. According to Suárez, it makes more sense to metaphysically distinguish between the actual essence and the potential essence, which is tantamount to a distinction between being and non-being. For the essences of created beings, although they are known by God from eternity, have neither reality (*res*) nor being (*ens*), but are absolutely nothing (*nihil*) prior to God's act of creation.[11] A merely possible essence in potency is not to be construed as something real or positive in itself. Therefore, there is no such thing as actualization of once potential essences, but only creation of actual beings out of absolute nothingness. Neither existence nor essence is an ontologically fundamental item in its own right. Instead, what is ontologically fundamental is the actual individual created being (*ens*) itself. The distinction between essence and existence is therefore merely conceptual and pertains to our understanding of an actual being under different significations. Nonetheless, this conceptual distinction is not arbitrary but has some basis in reality: the fact that we can conceive of the essences of created beings in abstraction from their actualized being or non-being, while we cannot similarly abstract God's essence from his existence, reflects the contingent existence of the former and the necessary existence of the latter.[12]

It would therefore be fair to say that one central locus of the late Scholastic debate is the question whether existence should be construed as amounting to a genuine constituent of an actual individual's quidditative content, or in other words, whether the actual contains more reality or quidditative content than the merely possible. This question will assume immense importance in Kant's critique of the ontological argument based on the thesis that existence is not a real predicate. For one major implication of this thesis is that existence does not add anything quidditative to the merely possible. Of course, Kant's critique is valid only to the extent that the ontotheological tradition really employs a conception of existence as having a quiddity of its own. Whether the latter is the case is the focus of this and the following chapter.

1.2. The Framework of Ontotheology

Despite significant differences in the interpretation of the essence/existence distinction in non-divine beings, the prevalent contention regarding the metaphysical constitution of God remains the same from Avicenna through Suárez: essence and existence are identical in God. This divine privilege is what is implied by the Biblical characterization of God as one whose very name is "I AM WHO I AM" (Yahweh), a god whose 'essence' is nothing but existence.[13] The ontotheological tradition is inspired by this fundamental insight. The identity between God's essence and existence ought to warrant his existence and reduce 'God exists' to a mere statement of that identity or logical necessity. However, since the ontological argument

[10] *Metaphysics* 1003b 27. [11] *Disputationes*, d. 31, sec. 2.1, 57.
[12] *Disputationes*, d. 31, sec. 6.23, 102. [13] Exodus 3:14 (New Revised Standard Version).

purports to prove rather than merely presuppose this identity, it cannot serve as its starting point. I thus propose that the argument, throughout its historical development, can be seen as consisting in two logical steps: (i) introducing existence into the concept of God, by means of an implicit or explicit premise, and then (ii) inferring the existence of God from this existentially loaded concept of God. The historical versions of the argument amount to different ways of carrying out these two steps.

The history of objections to the argument, on the other hand, consists in different attempts to block one or the other of the argument's two steps. One line of criticism, propounded by Gaunilo and Aquinas against Anselm, Caterus against Descartes, and Crusius against Wolff, questions the legitimacy of the inference in the second step from the conceived existence of God to the actual existence of God. The other line of criticism questions the first step's introduction of existence into the concept of God as a predicate. I take Kant's objection to the argument to be the original and most systematic expression of this second line of criticism.

I also propose that, in addition to the two logical steps, each variant of the argument moves between two distinct notions of God, *ens perfectissimum* (the most perfect being) and *ens necessarium* (the necessary being), aiming to establish an identity between these two notions.[14] While these two notions are captured in the broader notion of divinity fleshed out by medieval metaphysics, each is constructed logically independent of the other, forming distinct notions of the ultimate ground of explanation and responding to distinct forms of inquiry into the world of non-divine beings. As Henrich notes, the notion of the most perfect being, i.e., a being that exemplifies all perfections, satisfies the Platonic-Augustinian pursuit of explaining the limited degrees of perfections existing in the world; in turn, the notion of the necessary being emerges to satisfy the Aristotelian-Thomistic pursuit of explaining the existence of all contingent beings in the world.[15]

Now, though the respective possible rational constructions of these two distinct notions of God do not play any direct role in the ontological argument itself, the argument's two basic historical versions differ from one another with respect to the way in which they shift from one notion of God to the other in carrying out the aforementioned logical steps. What I will henceforth call the 'classical version' of the argument, defended by Anselm and Descartes, starts from the notion of God as the *ens perfectissimum* in the first step; then, having concluded that the *ens perfectissimum* exists *simpliciter* in the second step, goes on to establish that the *ens perfectissimum* is in fact the *ens necessarium* as its stronger conclusion. What I will call the 'modal version' of the argument, defended by Leibniz, Wolff, and Baumgarten, starts from the notion of God as the *ens necessarium*, thereby carrying out the first step through such an existentially loaded concept, and appeals to the identity of the *ens necessarium* with the *ens perfectissimum* in order to prove the possibility of the former through the latter.

[14] Henrich (1960), 63 calls this "*das Problem der Verbindung*." However, I do not share Henrich's conclusion that these two distinct notions of God correspond to the two forms of the ontological argument. What differentiates the two basic versions of the ontological argument is not, as Henrich suggests, which of the two notions of God they depart from, but the way they shift from one notion to the other.

[15] Henrich (1960), 4.

These two observations form the basis of the framework in which to examine and narrate the history of the ontological argument and the objections to it. My more specific aim is to lay out the prevalent conceptions of modality employed by the proponents and the opponents of the argument. For these are the source of the modal terminology as well as the particular problems that orient Kant in his engagement with the ontological argument, and to which he contrasts his own views of existence and modality. I utilize this background in Part II, where I examine whether Kant's criticisms of the argument are fair and decisive, what conception of modality they entail, and the extent to which that conception is historically novel. The present chapter focuses on the 'classical version' of the argument, propounded by Anselm and Descartes. Chapter 2 is devoted to the 'modal version,' propounded by Leibniz, Wolff, and Baumgarten. My account of the historical objections to the argument, including Kant's, follows in Chapter 3.

1.3. Anselm's Argument(s)

Anselm is historically the first to formulate an ontological argument in his *Proslogion*. His emphasis in the Prologue on the principle 'faith seeking understanding' (*fides quaerens intellectum*) lead some readers to take his argument to be designed exclusively for Christian believers to elevate their faith to the understanding of the truth of God.[16] However, Anselm's repeated appeal to 'the fool' as the main interlocutor in the dialectic of the argument and the direct role that 'the fool's' understanding of the nominal definition of God plays in the logical mechanics of the argument give stronger credence to the alternative view that the argument is intended also to persuade unbelievers and skeptics, who, Anselm states in the *Monologion*, could be persuaded by reason alone, if they are "even moderately intelligent."[17] In any case, Anselm's argument moves from the claim that there is an intentional object in the understanding of the believers and unbelievers alike, and aims to prove that this same object exists also in reality, as the believers believe it to do. Here is Anselm's own statement of the argument in chapter 2 of the *Proslogion*:

> Now we believe that you are something than which nothing greater can be thought. So can it be that no such nature exists, since "The fool has said in his heart, 'There is no God'"? But when this same fool hears me say "something than which nothing greater can be thought," he surely understands what he hears; and what he understands exists in his understanding, even if he does not understand that it exist [in reality]. For it is one thing for an object to exist in the understanding and quite another thing to understand that the object exists [in reality] ... so even the fool must admit that something than which nothing greater can be thought exists at least in his understanding, since he understands this when he hears it, and whatever is understood exists in the understanding. And surely that than which nothing greater cannot be thought cannot exists only in the understanding. For if it exists only in the understanding, it can be thought to exist in reality as well, which is greater. So if that than which a greater cannot be thought exists only in the understanding, then the very thing than which a greater cannot be thought is something than which a greater can be thought. But that is clearly impossible.

[16] See, for instance, Goodman (1996), 52. [17] See Williams (2016).

Therefore, there is no doubt that something than which a greater cannot be thought exists both in the understanding and in reality.[18]

Anselm's starting premise is a definition of God as He is believed to be:

(1) "Now we believe that you are something than which nothing greater can be thought (*aliquid quo nihil maius cogitari posit*)."

There are two points to note here. First, the definition of God provided in this premise is not intended to restrict God's greatness to the capacity of human thought. Instead, it implies that God's greatness exhausts and indeed exceeds our capacity of thought, and that He is the greatest thing we can possibly think of. For later in chapter 15, Anselm writes: "Therefore Lord, you are not merely that than which a greater cannot be thought; *you are something greater than can be thought.*"[19] Second, even though Anselm uses an atypical expression for it, he still has in mind the notion of God as the *ens perfectissimum*, the most perfect being. Whatever qualification Anselm might mean by 'greater,' the Platonic-Augustinian operator in his definition attributes God the maximal degree of 'greatness' compared to which everything else is less great. Moreover, in chapter 5, Anselm presents a rule for determining the attributes of God: "God is whatever it is better to be than not to be." Given this explicit rule and the maximizability criterion in the definition of God, the divine attributes for Anselm must be nothing but perfections, i.e., maximal degrees of fundamental positive predicates.

What I take to be not only the first step but also the *nervus probandi* of the entire argument is not put forth by Anselm as an official premise, but is clearly employed in the argument:

(2) An object is greater if it exists in reality (as well as in the understanding) than if it exists in the understanding alone.

Anselm grounds this premise on a very important distinction that he introduces between two modes of existence, namely, between 'existence in the understanding' (*esse intellectu*) and 'existence in reality' (*esse reale*): (2a) "It is one thing for an object to exist in the understanding and quite another thing to understand that the object exists [in reality]." This distinction is supported by two principles that Anselm appeals to both in the *Proslogion* proper and in the *Replies*: (2b) Understanding what is signified by a term or word is distinct from understanding that the signified object exists in reality; and (2c) "Whatever is understood exists in the understanding (*intellectus*)." And given that even 'the fool' understands what he hears when he hears "something than which nothing greater can be thought," an immediate extension of (2c) would be (2d) God exists in the understanding even of 'the fool.' The question, then, is whether that object also exists in reality.

[18] I am following Thomas Williams' translation in Anselm (2007).

[19] My emphasis. Gaunilo's rendition of Anselm's definition as 'greater than everything else' (*maius omnibus*) sounds less problematic in the sense that it does not contain the mediation of thinkable greatness, but Anselm rejects Gaunilo's rendition because it does not have the force of 'that than which a greater cannot be thought' in proving the real existence of the object that is spoken of. See Gaunilo's Reply on Behalf of the Fool 1, and Anselm's Reply to Gaunilo 5.

I take the rest of Anselm's argument to be a *reductio ad absurdum*. He initiates the *reductio* with the assumption that God exists only in the understanding. But there are two further premises that make the *reductio* run as Anselm intends it. The first is a disjunction:

(3) If something exists in the understanding, it exists either only in the understanding or both in the understanding and in reality.

The second is a less obvious principle:

(4) If something exists in the understanding, the same thing can be thought to exist in reality also.

The rest is quite straightforward.

(5) God exists only in the understanding. (Assumption)
(6) God can be thought to exist in reality also. (From 4, 5)

If so, then the substitution of 'God' by the original definition leads up to an absurdity:

(7) "The very thing than which a greater cannot be thought is something than which a greater can be thought." (From 1, 2, 4, 6)

The assumption must be false. It cannot be the case that God exists only in the understanding. The first disjunct in (3) is ruled out. And from (2d), we know that God exists in the understanding. Therefore:

(8) God exists both in the understanding and in reality. (From 2d, 3, 7)

Let us now examine how Anselm's argument fits in the general framework of an ontological argument I proposed earlier. As I said before, (2) is what effects the first step. But since (2) is not explicitly stated by Anselm and its role in the argument is contested by some contemporary scholars, I should first explain why I think it is there.[20]

First of all, Anselm does not state (2) as a distinct premise because he takes it to be an intuitive metaphysical principle. Anselm appends (2) to (4) without even announcing it: "if [God] exists only in the understanding, it can be thought to exist in reality as well, which is greater." That this addition is not at all trivial is clear from the role it plays in reaching the absurdity in (7). For it is precisely because God's existence in reality, which can be thought, is greater than God's existence in the understanding alone, that the latter would be something than which a greater can be thought.[21]

Beyond its role in reducing the assumption that "God exists in the understanding only" to absurdity, (2) is Anselm's way of executing the first essential step of any

[20] Williams (2016).

[21] It is worth noting that Anselm does not reject Gaunilo's reconstruction of the argument, which explicitly employs an equivalent of (2). However, the language of Gaunilo's rendition of (2), "to exist in reality is greater than to exist only in the understanding" (Gaunilo's Reply 1), suggests that he interprets the principle underlying (2) as entailing that *anything* that exists in reality is greater than *anything* that exists only in the understanding. It is not clear that Anselm embraces this strong version of (2). What is clear, however, is that he does not need to do so to accomplish what he aims in his argument.

ontological argument, namely, introducing existence, or in this case, real existence, into the concept of God. Anselm never explains what exactly he means by 'greater'; neither in (1) where he defines God, nor in (2) where he introduces a hierarchy between two different modes of existence of the same object. Now, what makes Anselm's God 'that than which noting greater can be thought' is the superlative nature of divine attributes: with respect to whatever attribute we wish to think Him, God has that attribute to the highest degree. Accordingly, the natural interpretation would be to take the term 'greater' to signify a higher degree of perfection with respect to an attribute that the object in question has. Yet the difficulty here is that Anselm never makes the comparison of greatness between different degrees of attributes, but rather between objects that have those different degrees of attributes. In (1), the comparison is between God and any other thing that can be thought. In (2), the comparison is between two different modes of existence of the same object. An alternative formulation of (2) would be: If two objects are identical in every respect except that one exists in reality (and in the understanding) and the other exists only in the understanding, the former is greater than the latter. Therefore, it is not at all an interpretive stretch to conclude that Anselm treats existence in general as an attribute that admits of various degrees. Yet, even if we abstain from that claim, it is clear that on any possible reading of (2), real existence is something that contributes to the 'greatness' of the thing in question, whatever 'greatness' may be taken to mean. Again, on any interpretation of 'greatness,' (2) introduces real existence into the concept of God defined in (1), through the mediating relation 'greater than.' And once we understand the first two premises of Anselm's argument as defining existence into the concept of God, we recognize that the *reductio* effects the second step, the inference from God's existentially loaded definition to his actual existence.

This argument in chapter 2 of the *Proslogion* is what is usually referred to as Anselm's ontological argument. Less frequently studied is the apparently distinct argument for the *necessary* existence of God presented in chapter 3:

> This [being] exists so truly that it cannot even be thought not to exist. For it is possible to think that something exists that cannot be thought not to exist, and such a being is greater than one that can be thought not to exist. Therefore, if that than which a greater cannot be thought can be thought not to exist, then that than which a greater cannot be thought is not that than which a greater cannot be thought; and this is a contradiction. So that than which a greater cannot be thought exists so truly that it cannot even be thought not to exist.

Analogous to (2) in the first argument, which introduced a hierarchical distinction between two modes of existence in general, 'existence in the understanding alone' and 'existence in reality,' the second argument introduces a hierarchical distinction between two modes of real existence, between 'necessary and contingent existence.' And interestingly enough, the second argument starts with the assertion of the possibility of thinking a necessary being:

(9) "It is possible to think that something exists that cannot be thought not to exist."

(10) Something that cannot be thought not to exist is greater than something that can be thought not to exist.

The comparison Anselm makes here is of a categorical kind, between any one thing that exists necessarily and any one thing that exists contingently, no matter what the two things may be. So Anselm does not need an analogue of (4) to warrant the identity of the conceptual contents of the compared items.

Then, a simpler version of the *reductio* in the first argument follows:

(11) God can be thought not to exist. (Assumption)

(12) Then, something than which a greater cannot be thought is something than which a greater can be thought. (From 1, 9, 10, 11)

(13) Therefore, God cannot be thought not to exist.

With a full-fledged modal terminology, this argument can be translated as the conditional, 'If a necessary existent is possible (thinkable), then God exists necessarily,' or as the equivalent disjunction, 'Either a necessary existent is impossible (unthinkable) or God exists necessarily,' which, upon the assertion of (9), constitutes an inference in *modus ponens* to the necessary existence of God.

There are several interpretive issues here. The argument in chapter 3 has the force of a self-standing argument in that it does not presuppose the conclusion reached in chapter 2, viz., that God exists in reality, in order to reach its own conclusion that God exists necessarily. In fact, the latter conclusion entails the first one and thus the second argument renders the first argument superfluous. Nevertheless, Anselm seems to intend the second argument to be a continuation of the first argument. The first sentence of chapter 3, "This [being] exists so truly that it cannot even be thought not to exist," suggests that Anselm takes the conclusion of chapter 2, "God exists in reality," and produces a new but parallel line of reasoning to further qualify God's mode of real existence. Not only does God exist in reality; He also exists in reality by necessity. Taken as a whole, Anselm's project in chapters 2 and 3 of the *Proslogion* effectively establishes a systematic gradation of different modes of existence. First, there is a gradation between 'existence in the understanding alone' and 'existence both in the understanding and in reality'; second, between 'contingent existence' (in reality as well as in the understanding) and 'necessary existence' (in reality as well as in the understanding). God always qualifies for the comparatively higher mode of existence on any level of comparison and thus exists in the highest mode or degree of existence that is thinkable, i.e., necessary existence (in reality as well as in the understanding). Thus, Anselm starts out with the notion of God as the *ens perfectissimum* and arrives at the conclusion that God is an *ens necessarium*.

The unity of the two arguments is best highlighted by yet another formulation of the ontological argument that Anselm presents in response to Gaunilo's objection to inferring God's existence in the understanding from the mere fact that 'God' can be thought or understood. Anselm proposes the following original reasoning, aiming to prove that the mere fact that God can be thought suffices to show not only that God exists in the understanding, the lowest mode of existence in Anselm's ontology, but also that He exists necessarily, the highest mode of existence:

If [God] can be thought at all, it necessarily exists. For no one who denies or doubts that something than which a greater cannot be thought exists, denies or doubts that if it did exist, it would be unable to fail to exist either in reality or in the understanding, since otherwise it would not be that than which a greater can be thought. But whatever can be thought, but does not in fact exist, could (if it did exist) fail to exist either in reality or in the understanding. So if that than which a greater cannot be thought can be thought at all, it cannot fail to exist.[22]

Here Anselm takes up the three modes of existence together, mere possibility, contingent (actual) existence, and necessary (actual) existence, and attracts our attention to the modal peculiarity of the concept of God as something than which a greater cannot be thought.

(1') If God exists in reality, then God exists in reality necessarily.

(2') If something is possible, yet does not, as a matter of fact, exist in reality, then it is possible for that thing not to exist in reality. So even if that thing existed in reality, it would exist contingently. Thus that thing cannot be God.

(3') God can neither be contingently existent; nor be merely possible.

(4') If God is at all possible, then He exists necessarily, or He is impossible. God's existence or non-existence cannot be qualified as contingent.

I take this passage to be an excellent formulation of a unified Anselmian ontological argument (chapter 2 and 3 combined): once the fact that God can be thought or understood is accepted, then God's definition as 'something than which a greater cannot be thought' warrants the chain of entailment from possibility to necessity.

Some scholars hold that Anselm's second argument fares better than the first one against Kant's objection that 'existence is not a real predicate.'[23] This view seems to rely on two points. First, the second argument does not employ any premise that entails a commitment to the claim that existence is a predicate. Second, even if existence is not a predicate or perfection, necessary existence may well be one. I will revisit this line of defense against Kant's critique in chapter 8. Yet I would like to quickly note here that on the reconstruction I provided above, both of Anselm's arguments operate upon a scheme of gradation of various modes of existence, each corresponding to a certain degree of 'greatness' or perfection of the thing that it applies to. So Anselm understands modes of existence or modalities, necessary (real) existence included, as contributing (different degrees of content) to the essential constitution of things. Moreover, in each instance, Anselm's *reductio* relies on the contention that our respective thoughts of the same thing under distinct modalities do indeed have different conceptual contents. For instance, the first *reductio* assumes that the thought of something existing in reality is the thought of something greater than the same thing thought as existing in the mind alone, and the second assumes that the thought of something existing necessarily is the thought of something greater than the same thing thought as existing contingently, however we may define 'greater than.' Therefore, Anselm understands modes of existence as included in the intensions of the concepts of things. As I will argue in the chapters

[22] Anselm's Reply to Gaunilo 1.

[23] Malcolm (1960), Hartshorne (1965), Goodman (1996).

to come, such inclusion is what Kant takes to be the most fundamental flaw of the ontological argument.[24]

Before concluding this section, let us review Anselm's conceptions of possibility and necessity in light of the present reconstruction. Anselm defines the possibility of a thing only through its thinkability or conceivability. Given that he presents the self-contradictory character of (7) as an impossibility, Anselm ought to understand thinkability as freedom from logical contradiction and consequently employ a 'logical' notion of possibility in the argument.[25] In other words, he takes the possibility of the thought of a thing to be the sufficient criterion of the possibility (of the existence) of a thing. Importantly, however, Anselm does not seem to equate the (logical) possibility of a thing with what he calls 'existence in the understanding' (*esse intellectu*), nor does his extended argument rely on this equation. For while 'existence in the understanding' requires a mental act of thinking or understanding the object, logical possibility expresses the mere possibility of performing that act without logical contradiction. Of course, the former would entail the latter, and Anselm's second argument in chapter 3 and its reformulation in *Responsio* show that Anselm is cognizant of this entailment.

Anselm also distinguishes between necessary or contingent existence in terms of thinkability or logical possibility. Something necessarily exists if it cannot be thought not to exist, that is, if the non-existence of that thing is logically impossible or, what is the same, a logical contradiction. Something contingently exists if it can be thought not to exist without contradiction. Thus, for Anselm the existence of God is a logical necessity, and his extended argument serves as an *a priori* demonstration of the claim that God's existence logically follows from or is entailed by His very essence or concept as the greatest thinkable or most perfect being. As we will be see, this is common to all of the historical variants of the ontological argument.

1.4. Descartes' Argument

One might hold that Descartes' ontological argument in the Fifth Meditation is a reiteration of the first part of Anselm's argument. For, first of all, textual evidence indicates that Descartes knows of the Anselmian argument.[26] Second, in the First Set

[24] In the terminology of the late medieval debate on the distinction between essence and existence that Anselm predates, Anselm conceives the distinction to be between *res* and *res*. Anselm does not base his argument on any kind of distinction between divine existence and non-divine existence, but on the distinction between different modes of existence. But of course, his extended argument intends to demonstrate that necessary existence logically follows from the divine essence.

[25] Both Malcolm (1960), 45 and Hartshorne (1965), 88 assume, without much of a justification, that Anselm's argument employs logical notions of possibility and necessity. There is, however, considerable resistance in the literature to the suggestion that the notion of possibility employed in Anselm's ontological argument is logical possibility. See, for only a few examples, Barnes (1972); Campbell (1976); La Croix (1972). For an insightful discussion of the issue and a strong defense of the view that Anselm's notion of conceivability in the ontological argument does indeed invoke a logical notion possibility, see Smith (2014), esp. ch. 2.

[26] Descartes' reply to the First Set of Objections suggests that he knows of the argument but thinks that the argument belongs to Aquinas (AT 7:106, CSM 2:77), possibly because Aquinas, when citing the argument as a possible objection to his thesis that the existence of God is *not* self-evident, does not give credit to Anselm (*Summa Theologica* Ia.2.1)

of Objections, Caterus quotes Aquinas' refutation of the Anselmian argument as also applicable to Descartes' argument in the Fifth Mediation, for he holds the two arguments to be identical: "But now please tell me if this is not the selfsame argument as that produced by M. Descartes" (AT 7:98, CSM 2:71). However, Descartes himself does not seem to think that his argument is the same as Anselm's. Descartes even thinks that the latter is invalid in that it fails to justify the transition from the conceptual to the objectual level, that is, from the thought existence of God to the actual existence of God, though, as we will see, his own argument is susceptible to the same charge.

Descartes' purpose for presenting an ontological argument in the Fifth Meditation is to illustrate the power of the epistemic principle that 'whatever I perceive clearly and distinctly is true.' Descartes introduces and validates this principle earlier in the Third and Fourth meditations as a rule for discovering the real natures or essences of things by attending to the innate ideas that supposedly represent them.[27] After reiterating a modified version of the principle, "everything which I clearly and distinctly perceive to belong to [a] thing really does belong to it," Descartes emphatically asks, "Is not this a possible basis for another argument to prove the existence of God?" (AT 7:65, CSM 2:45). In order to understand the kind of argument Descartes has in mind here, a few points must be noted concerning the principle in question and his doctrine of essences.

First, Descartes suggests that the application of the principle does not require an ontological commitment to the actual instantiation of essences. For he holds that innate ideas, through the clear and distinct perception of which we discover the true essences of things, can fulfill their epistemic function even if they do not correspond to anything in actual existence:

> I find within me countless ideas of things which even though they may not exist anywhere outside me still cannot be called nothing; for although in a sense they can be thought of at will, they are not my invention but have their own true and immutable natures. When, for example, I imagine a triangle, even if perhaps no such figure exists, or has ever existed, anywhere outside my thought, there is still a determinate nature, or essence, or form of the triangle which is immutable and eternal, and not invented by me or dependent on my mind.
>
> (AT 7:64, CSM 2:44–5).

The innate ideas we have are neither arbitrary conceptual constructions nor abstractions from our empirical observations of existent things. Rather, they represent essences, which have a positive being even if they are not instantiated in reality. To see what this conception of essences ultimately implies for Descartes' understanding of the nature of modal truths requires a more detailed consideration of some of the major claims of his rational theology. I will therefore revisit this issue later on. Furthermore, Descartes' exact position on the ontological status of essences is far from obvious. The Platonic hypothesis that essences are mind-independent truthmakers of eternal truths, though convenient, would conflict with his claim that eternal truths "have no existence outside our thought" (AT 8A:23; CSM 1:208). On

[27] See AT 7:35, CSM 2:24.

the other hand, construing essences as mind-dependent seems problematic as well. For the view that essences are intentional or mental entities in finite minds, as Descartes' above-cited statement strongly suggests, would seem to conflict with their status as eternal, and the more Leibnizian view that essences are located in God's mind would seem to conflict with their status as creatures that depend on God's will, God's own essence excepted. So leaving aside the question of the exact place of essences in Descartes' ontology, we can move forward on the more modest assumption that Descartes takes essences to ground true logical relations among properties, even if these properties and relations are not actually instantiated by existent things.

In the Fifth Meditation, Descartes proposes a special variant of the 'clear and distinct principle' to serve as a basis for a new proof of God's existence. Accordingly, what we discover through our clear and distinct perceptions are the eternal truths about the essences and essential properties of things, truths that hold independently of the actual existence of the particular things in question. Descartes' ontological argument is a unique application of this principle, because it allows us to infer the existence of a certain thing even though the principle itself is formulated to work at an existentially neutral level. Initially, what Descartes suggests looks less like a formal argument than an intuitive act of attending to our innate idea of God through which we thereby recognize that existence belongs to his essence:

> Certainly, the idea of God, or a supremely perfect being, is one which I find within me just as surely as the idea of any shape or number. And my understanding that it belongs to his nature that he always exists is no less clear and distinct than is the case when I prove of any shape or number that some property belongs to its nature. Hence, . . . , I ought . . . to regard the existence of God as having at least the same level of certainty as I have hitherto attributed to the truths of mathematics. (AT 7:65, CSM 2:45)

As a proof of God's existence, Descartes offers here only the clear and distinct perception of the inseparability of God's existence from his essence. No inferential step is needed to reach the desired conclusion, but only the principle that whatever is clearly and distinctly perceived to belong to the essence of a thing can be truly asserted of that thing. Although the passage can be reformulated as a syllogism,[28] Descartes seems to intend it as an intuitive procedure. For the idea of God that Descartes claims we find to be present in our intellect is no other than the Abrahamic idea of God, 'He Who Is.' The demonstrative force of any ontological argument lies in establishing that God is actually 'He Who Is,' the being whose essence is (or contains) existence. However, the argument must not start with this identity as its initial axiom if it is to prove more than a tautology. What Descartes provides above is therefore not an ontological argument, but merely instructions for the intuitive procedure of discovering first how existentially loaded our innate idea of God is, in order then to switch to the discursive level and assert the existence of God as licensed by his epistemic rule of truth.

[28] (1) Whatever is clearly and distinctly perceived to belong to the essence of something is true of that thing. (2) I clearly and distinctly perceive that existence belongs to the essence of God. (3) Therefore, God exists.

In this respect, when Descartes compares this procedure to geometrical or mathematical demonstrations, he should not be understood as suggesting an argumentative analogy. Descartes grounds his comparison on two points: (i) The level of certainty that the proposition 'God exists' shares with geometrical and mathematical truths; (2) the logical inseparability common to God's essence and existence, on the one hand, and to the triangle's essence and any of its essential properties, on the other hand. The non-geometrical example of the idea of a mountain and a valley also suggests that Descartes' comparison concerns the logical connections between the concepts of things and their essential predicates.[29] Again, leaving a broader discussion for later, we can note for now that this comparison suffices to show that Descartes treats existence as a constituent of the intension or content of the concept of God.

Descartes presents what is traditionally identified as his version of the ontological argument only in response to one of the possible objections to the initial, intuitive proof:

> Whenever I do choose to think of the first and supreme being, and bring forth the idea of God from the treasure of my mind as it were, it is necessary that I attribute all perfections to him... And this necessity plainly guarantees that, when I later realize that existence is a perfection, I am correct in inferring that the first and the supreme being exists.
>
> (AT 7:67, CSM 46–7)

Here the emphasis shifts from the simple, direct perception that existence belongs to the essence or concept of God, to the relatively indirect procedure of introducing existence into the concept of God, namely, by defining God as the *ens perfectissimum* and existence as a perfection. The inference that God exists then follows from these two definitions without appeal to the principle of clear and distinct perception. The first and second logical steps, indispensable to any ontological argument, are thus effected. This line of reasoning is usually rendered thus:

(1) God is the supremely perfect being, possessing all perfections.
(2) Existence is a perfection.
(3) Therefore, God exists.

The modality of Descartes' conclusion is not unambiguous. Unlike Anselm, Descartes does not introduce a complementary argument for the necessary existence of God. Yet although he uses existence *simpliciter* when stating his conclusion, there are numerous passages where he claims that there is nothing more self-evident than this conclusion;[30] that the kind of existence that belongs to God's essence is *necessary* (and eternal) existence;[31] and that existence *necessarily* belongs to the essence of God in addition to the other divine attributes.[32] Whatever Descartes exactly means by 'necessity' or 'necessary existence,' he clearly sees the necessary existence of God as

[29] In a letter of 1/19/1642 to Gibieuf (CSM 3:202), Descartes seems to be saying that all he means by a mountain is an uphill slope and all he means by a valley is a downhill slope.

[30] AT 7:167, CSM 2:118.

[31] AT 7:119, CSM 2:85; AT 7:166–7, CSM 2:117; AT 7:383, CSM 2:263.

[32] AT 7:68 (French version).

the implied, stronger conclusion of his argument.[33] Thus, Descartes' version of the ontological argument involves, albeit implicitly, an identification between the *ens perfectissimum* and the *ens necessarium* notions of God.

Although his argument appears to be simple in its form, Descartes' elucidations in response to the objections, taken with his theological commitments, offer a rich and intricate modal backdrop. I suggested earlier that Descartes distinguishes 'clear and distinct ideas' from 'chimerical ideas' that amount to nothing in reality on the basis that the former are not arbitrary conceptual constructs of our minds but determinate conceptual structures representing true and immutable essences. Each essence is constituted by a set of inseparable properties and thus the ideas representing essences express necessary truths about things in the form of subject–predicate relations. I have also noted that Descartes assigns some sort of positive being to true and immutable essences even if they are not instantiated in reality.[34] The uninstantiated essences are not nothing, because, Descartes holds, even if they do not actually exist, they are nevertheless capable of existing. If this is true, then 'clear and distinct ideas' express possibilities as well as necessities about things. It might appear, then, that truths about possibility and necessity depend on immutable essences, or in other words, essences are truth-makers for modal claims.

However, there is another layer to Descartes' doctrine of eternal essences and truths, further complicating his account of modality. Descartes holds that eternal essences and the truths anchored in them depend on God's will. For instance, in response to Gassendi, he writes, "I do not think that the essences of things, and the mathematical truths which we can know concerning them, are independent of God. Nevertheless I do think that they are immutable and eternal, since the will and decree of God willed and decreed that they should be so" (AT 7:380, CSM 2:261). In conversation with Burman, he makes a similar point with respect to possibility, "[God's] will is the cause not only of what is actual and to come, but also of what is possible" (AT 5:160, CSMK 3:343). Accordingly, something is possible or necessary in virtue of God's willing it to be so. Descartes even embraces what might strike many as the embarrassing implications of this view: "I would not dare to say that God cannot make a mountain without a valley, or bring it about that 1 and 2 are not 3" (AT 5:224, CSMK 3:358–9). So God's will and creative power are not bound or informed by prior modal truths about necessities or possibilities. On the contrary, all possibilities and necessities are dependent on God's free will: a modal truth is made true by God's willing it to be so in the first place. Thus, as Newlands suggests, "God's power, prior to creation, is a kind of premodal form of power."[35] This, of course, raises the question of what it really means for a truth to be necessary (and eternal and immutable) if God could have chosen to make it false. After all, this seems to conflict

[33] At AT 7:119, CSM 2:85, he says that a thing exists necessarily if and only if it exists "by its own power."

[34] This is, however, not uncontroversial. Descartes also seems to commit himself to the view that that which does not actually exist is nothing: "I perceive that the objective being of an idea cannot be produced merely by potential being, which strictly speaking is nothing, but only by actual or formal being." (AT 7:47, CSM 2:32).

[35] Newlands (2013), 160.

with our most natural modal intuitions, according to which the necessity of a truth consists precisely in its inability to have been otherwise. Similarly, one cannot help but ask what possibility means if God could in fact bring about what is impossible?

There have been various attempts in the literature to make sense of the counter-intuitive consequences of grounding modality in God's free 'premodal' volitions.[36] I will forgo discussion here. However, it is important to point out that Descartes, no matter how problematic his conclusions might appear, does not go the logicist route of accounting for modal truths only in terms of the laws of logic. The notion of possibility Descartes employs is thicker than that of 'merely logical possibility.' For first of all, it expresses neither what is merely thinkable, nor even an independent standard of what God could bring about, but what God wills to be possible. Second, the epistemic test Descartes offers for cognizing possibility is not based on the absence of logical contradiction, but on the clarity and distinctness of the perception of an idea. Third, while logical possibility applies to any logically consistent chimera or conceptual construct of the mind, and thus does not have any ontological implication, Descartes' notion of possibility applies to clear and distinct ideas of true and immutable essences. Fourth and finally, since Descartes attributes "capability of existence" to the non-instantiated true and immutable essences, the notion of possibility that is at stake here best squares with a 'metaphysical' or 'real notion of possibility.'

One important issue concerning the metaphysical character of Descartes' notion of possibility is his distinction between 'formal' and 'objective reality.' The formal reality of a thing amounts to what that thing is as an actual thing.[37] In accordance with Descartes' tripartite hierarchical ontology, an actual or existent thing can be either a mode, or a finite substance, or an infinite substance. So, while the formal reality of an actual horse is a finite substance, the formal reality of the idea of a horse is a mode. For the idea of a horse, regardless of whether or not it is instantiated by an actual horse (finite substance), is a mode or act of my mind (finite substance). Objective reality, on the other hand, pertains only to ideas or representations, and amounts to the reality an idea has in virtue of what it represents.[38] So while the formal reality of the idea of a horse, insofar as it is an *idea*, is a mode, the objective reality of that idea, insofar it is an idea of a *horse*, is a finite substance. From a modal point of view, what makes Descartes' account of formal/objective reality distinction crucial is a principle that he combines with it. Descartes claims that the causal principle that "there must be at least as much (reality) in the efficient and total cause as in the effect of that cause" holds "not only in the case of effects which possess... actual or formal reality, but also in the case of ideas, where one is considering only (what they call) objective reality" (AT 7:40–1; CSM 2:28–9). Thus, just like an actual horse cannot exist unless it is efficiently caused or produced by another, actual thing, which has at least as much formal reality as a horse, i.e., a finite substance, an idea such as the idea of a horse cannot exist unless it is ultimately caused by an actual horse or some other actual entity with at least as much formal reality as a horse. So the idea of a thing is

[36] For an extensive discussion, see Cunning (2014).
[37] See AT 7:41–2, 102–4; CSM 2:28–9, 74–5.
[38] See AT 7:42; CSM 2:29.

dependent on actuality in two ways. First, insofar as its own formal reality as an idea is concerned, it depends on the actuality of a mental substance, i.e., the mind. Second, insofar as its objective reality is concerned, it depends on the actuality of the cause of its representational content, i.e., something that is either equal to or higher than the thing represented in the order of formal reality.

The second point deserves special attention. Descartes' reasoning here is that the content of an idea cannot have come from nothing, for the essences that our clear and distinct ideas represent are not nothing, even if they are not actually instantiated. In other words, those essences are capable of existing or are really possible. Therefore, in order for an idea to represent a really possible object, it must be causally grounded in an actual entity that can causally produce the object that it represents. Of course, given Descartes' aforementioned commitment to God's 'premodal' volitional freedom to create all immutable essences and the truths about them, we can infer that all real possibilities must ultimately be grounded in God's actual volitions.

Descartes' commitment to the causal dependence of objective reality on formal reality entails that there is an indirect yet rigorous sense in which essences depend on existence. For even though a true and immutable essence, e.g., that of a triangle, need not be instantiated in actuality by a particular triangular object in order to express the logical relations between the predicates of a triangle, its conceptual content must still be caused by something actual. Accordingly, were we to remove all that actually exists, no essence and thus no real possibility would remain. Hence, for Descartes there is no essence or mere real possibility without existence. I suggest that Descartes here offers the germ of a principle that will come to assume an pivotal role in both Leibniz's and Kant's modal theories, and which Kant will employ as the crux of an alternative ontological proof. In the rest of this book, I will call the generic form of this principle the 'Actualist Principle': The facts about real possibility are grounded in the facts about actuality. It is further striking that in the Third Meditation Descartes also constructs his first proof of God's existence on this very principle. He argues that the idea of God, insofar as its representational content or objective reality is concerned, requires the actuality of something that has at least as much formal reality as God, i.e., an infinite substance. Therefore, God himself, as the only infinite substance, must exist in order to for the idea of God to be grounded. Only God himself can cause finite minds to have the clear and distinct idea of God.

Descartes sometimes also uses the term 'possible existence' to refer to the ultimately contingent existence of non-divine beings (including immutable essences and the 'necessary' truths concerning them), i.e., those dependent on God's will as opposed to the absolutely necessary existence of God, whose exceptional essence does not depend on anything else but its own actuality. For instance, in his First Set of Replies, he states, "we must distinguish between possible and necessary existence. It must be noted that possible existence is contained in the concept or idea of everything that we clearly and distinctly understand; but in no case is necessary existence so contained, except in the case of the idea of God" (AT 7:116, CSM 2:83). Again, in his Second Set of Replies he writes, "The ideas of all other natures contain possible existence, whereas the idea of God contains not only possible but wholly necessary existence" (AT 7:163, CSM 2:163).

Descartes also holds that we conceive everything as actually existing. He makes this explicit as the tenth axiom in his Second Set of Replies:

> Existing is contained in the idea or concept of every single thing, since we cannot conceive of anything except as existing. Possible or contingent existence is contained in the concept of a limited thing, whereas necessary existence and perfect existence is contained in the concept of a supremely perfect being. (AT 7:166, CSM 2:117)

However, this statement might be misleading; in particular, it might undermine the fundamental difference between contingent and necessary existence and the ontological distinction between creatures and God, both of which Descartes wants to maintain. The key here is that it is a feature of our faculty of thought that, on the level of abstract thought and in total abstraction from matters of fact, we entertain things as existing without committing ourselves (or 'assenting') to their actual existence. Descartes himself warns that "even though our understanding of other things always involves understanding them as if they were existing things, it does not follow that they do exist, but merely that they are capable of existing" (AT 7:117, CSM 2:83.). It is the privilege only of the necessary being that its actual existence unconditionally follows from our mere thought or 'understanding' of it. Thus, there is still a significant difference between the ways we conceive contingent and necessary things such that the thought of the former does not logically entail their actual existence but that of the latter does.

As some of the quotes above demonstrate, Descartes often conceives necessity and contingency as two predicates contained in the concepts of things in their own right. But sometimes he also formulates the difference between necessity and contingency in terms of the difference in the modality of the connection between actual existence *simpliciter* and the other, non-modal predicates of the thing in question: "actual existence is not necessarily conjoined with other properties in other things, but it is necessarily conjoined with the other attributes of God..." (AT 7:117, CSM 2:83). This latter formulation is more in line with the medieval tradition, defining the individual modalities of things in terms of the essence/existence relationship (distinction or identity): Existence is (really, modally, or conceptually) distinct from the essence of everything, with the exception of God, whose essence is or contains existence. Accordingly, the former are contingent beings whereas the latter is the necessary being. It seems that Descartes either fails to see the difference between the two formulations or takes them to be the same.

In any case, Descartes construes existence (be it necessary or contingent) as a predicate that enters into the conceptual contents of things. We can already see how the Cartesian version of the argument is highly vulnerable to Kant's main objection that existence cannot be treated as such a predicate. In response to Gassendi's criticism of his argument with regard to its treatment of the concept of existence, Descartes himself admits in frustration[39]:

> Here I do not see what sort of thing you want existence to be, nor why it cannot be said to be a property just like omnipotence—provided, of course, that we take the word 'property' to stand

[39] As I will explain in chapter 3, I do not hold that Gassendi's objection to Descartes' argument anticipates Kant's. Despite the appearance of similarity, the two objections are fundamentally different.

for any attribute, or for whatever can be predicated of a thing; and this is exactly how it should be taken in this context. (AT 7:382–3, CSM 2:262–3)

Obviously, Descartes here assumes that a term's predicability is sufficient for it to qualify as an attribute, but this is not the only way he defines attributes. For Descartes, the idea that existence is an attribute is not just a crucial premise of the ontological argument, but also an important part of his metaphysics of attributes. In a letter to an unknown correspondent he writes:

Existence, duration, size, number and all universals are not, it seems to me, modes in the strict sense . . . They are referred to by a broader term and called attributes . . . because we do indeed understand the essence of a thing in one way when we consider it in abstraction from whether it exists or not, and in a different way when we consider it as existing; but the thing itself cannot be outside our thought without its existence. (AT 4:349, CSMK 3:280)

This passage is striking in multiple ways. First, Descartes denies that existence is a mode; thereby, he explicitly differentiates his position from that of the Scotists, who construe existence as a mode of the essence. Second, Descartes denies that existence is a mode because he thinks it is more intimately related to the essences of things. He positively defines existence as an attribute, because he holds that there is a difference between the conception of an existing essence and that of the same essence undetermined with respect to existence. There is an ambiguity in this text as to whether the relevant 'difference' here is a difference between manners of conceiving the same conceptual content or a difference in the conceptual content itself. However, given his commitment to the idea that existence is a predicate contained in the conceptual contents of things, the latter is at least as natural a reading as the former.

For Descartes, then, existence is a special kind of attribute indistinguishable from the actual thing in reality—and this may be one reason why he somewhat surprisingly insists that we conceive things as existing—and yet existence is still an attribute in the sense that it is conceptually or rationally distinct from the essence of the thing itself. The only exception is the case of God whose essence is neither really nor conceptually distinguishable from his existence. But then the sense in which existence is an attribute of God is different from the sense in which it is an attribute of any created thing. Thus Descartes writes, "In the case of God, necessary existence is in fact a property in the strictest sense of the term, since it applies to him alone and forms a part of his essence as it does of no other thing" (AT 7:383, CSM 2:263).

The view that the concept of a thing as existing is different from the concept of the same thing undetermined with respect to existence is of great importance in Kant's analysis and critique of the traditional metaphysics of modality. And the question of whether the proponents of such a view are also committed to the view that the concept of a thing as actually existing 'contains more' than the concept of that thing as merely possible will be particularly crucial in assessing the novelty and fairness of Kant's critique of the tradition. We saw earlier that Anselm's gradation among different modes of existence provides a fair ground for the Kantian charge. In the last block quote, Descartes contrasts the concept of that which exists with the concept of that whose existence is not determined, but does not further qualify the distinction between the two. But in his reply to Gassendi there is a passage suggesting that Descartes, too, might have in mind some kind of gradation among modes of existence:

> I do not, however, deny that possible existence is a perfection in the idea of a triangle, just as necessary existence is a perfection in the idea of God; for this fact makes the idea of a triangle superior to the ideas of chimeras, which cannot possibly be supposed to have existence.
>
> (AT 7:383, CSM 2:263)

The comparison here is not between the concepts of the same objects with different modalities but between the concepts of different objects with different modalities, i.e., between the idea of a (possible) triangle and that of an (impossible) chimera. But it is still the modalities of these different objects that constitute the basis of the relevant comparison. The idea of a triangle is superior to that of a chimera not because of what a triangle and a chimera are but exactly because the former refers to a possible and the latter to an impossible thing. Thus, in this modal sense of superiority, mere possibility is superior to impossibility. By the same token actual existence should be superior to mere possibility, and necessary (actual) existence to contingent (actual) existence. Descartes' language of 'superiority' and 'perfection' is strongly reminiscent of Anselm's talk of 'greatness' and seems to share its motivation of underlining a gradation among modalities. From the minor premise of Descartes' ontological argument, we are already familiar with the definition of existence as a perfection. But here Descartes adds a modal qualification to his definition and suggests that both contingent and necessary existence are perfections in relation to things that they are predicated of. This qualification brings a slight revision (or useful explication) to the original presentation of Descartes' ontological argument. For if what we are seeking to prove is God's existence, then the kind of existence that is to be a perfection must be necessary existence. The ensuing inference is thus not just merely to existence *simpliciter*, but to God's necessary existence. The modal ambiguity of the conclusion vanishes. Thus, by introducing this modal qualification to the thesis that existence is a perfection, Descartes provides a shortcut to the strong conclusion that 'God necessarily exists,' directly identifying the notion of God as the *ens perfestissimum* with that of God as the *ens necessarium*.

* * *

In this chapter I have argued that the ontotheological tradition is the primary historical context for Kant's engagement with modal thought. The rational theological insight underlying every version of the ontological argument is the identity or inseparability between God's essence and existence in contrast to the distinction between essence and existence in created beings, thereby distinguishing God as the necessary being from the radical contingency of everything He created *ex nihilo*. I have offered a general framework for analyzing the argument's logical mechanics. First, the argument consists in a twofold step: (i) introducing existence into the concept of God, and then (ii) inferring the existence of God from this existentially loaded concept of God. Second, the argument moves between two distinct notions of God, *ens perfectissimum* and *ens necessarium*, and aims to establish an identity between the two. I have analyzed the classical version of the ontological argument presented by Anselm and Descartes in accordance with this framework. I have focused on two particular questions: Does the argument treat existence as a predicate? What conception of modality is at work in the argument? I have argued that Anselm's argument treats existence as a predicate that could be contained in the concept of a thing by defining existence as a 'greatness-increasing' property of things and introducing a hierarchy between various modes of existence, according to which real existence is 'greater' than merely thought existence and necessary real existence is 'greater' than contingent real existence. The concept of God, therefore, as 'the greatest conceivable thing,' contains necessary real existence, from which Anselm concludes that such a thing cannot fail to exist in reality. I have further suggested that Anselm employs logical notions of possibility and necessity, both because of his appeal to the criterion of conceivability and the logical contradictions he generates as the motor of his reductios. I have argued that while Descartes' argument treats existence as a predicate by directly defining it as a perfection, Descartes seems to embrace more than a merely logical conception of modality. According to Descartes, truths about possibilities and necessities are grounded on immutable essences, which are in turn grounded on God's free will and power, unconstrained by any law of logic. Thus, modalities are ultimately grounded on God's active, 'premodal' volitions. I have also claimed that Descartes' principle that 'objective reality' (i.e., the representational content) is causally dependent on 'formal reality' (i.e., the actual existence of a being) entails that there could be no mere possibility in the absence of some antecedent existence. This is a striking anticipation of the 'Actualist Principle,' which states that the facts about real possibility are grounded on the facts about actuality. I will demonstrate in the following chapters that this principle comes to play a central role both in Leibniz's and Kant's attempts to revise the ontological argument and in their modal theories in general.

2

Ontotheology and Modality II

The Modal Version of the Ontological Argument

I have two main aims in this chapter. First, I will examine the 'modal' version of the ontological argument. While the modal version can be traced back to the second part of Anselm's extended argument, this version finds its more developed expressions in eighteenth century ontotheology. This period of ontotheology is particularly important, since it is represented by Kant's immediate predecessors such as Leibniz, Wolff, and Baumgarten, who formulate the pillars of the rationalist *Schulmetaphysik*. This is the school of thought that constitutes the major portion of the background of Kant's own intellectual development and philosophical terminology. Therefore, the foremost target of Kant's critique of ontotheology and the accompanying conception of modality is to be sought within this period. While I will retain my general methodology of teasing out the conception of modality in question from the particular construction of the ontological argument, I will go beyond ontotheology and look at other metaphysical contexts in order to draw a more comprehensive picture of the prevalent conception of modality in the German rationalist school.

This brings me to my second aim. In laying out the prevalent conception of modality in the school metaphysics, I will examine two claims that form a narrative about the school metaphysicians: (i) they are committed to a logicist account of modality, according to which modal claims can be exhaustively explained through merely formal-logical principles;[1] (ii) they are committed to the idea that existence is a real predicate.[2] The assessment of this narrative is crucially important for the broader purposes of this book in Parts II and III, for it entails, if not actually coupled with, another narrative: the novelty Kant brings to his predecessors' treatment of modality should be sought in his introduction of a non-logical or metaphysical account of modality and rejection of the conception of existence a real predicate. I will show, however, that both of these two claims are too simplistic and ultimately untenable and what they jointly imply as to where the novelty of Kant's theory of modality lies is misguided.

[1] This claim has been voiced in various forms by Poser (1983, 1989) Motta (2007), and Kannisto (2012). However, Stang (2016) recently provided its boldest and most rigorous defense.

[2] This view is generally rather silently endorsed, based on the assumption that Kant's critique of the ontological argument applies to Leibniz's and Wolff's versions of the argument. Proops (2015) is an exceptional vocal proponent.

2.1. Leibniz: His Argument and Theory of Modality

2.1.1. The modal ontological argument

There is no doubt that Leibniz, like most of his contemporaries, was familiar with Descartes' ontological argument. And contrary to at least one view, it is also clear that Leibniz knew the contents of the *Proslogion* and Anselm's argument therein.[3] In the *New Essays*, Leibniz provides a brief reconstruction of Anselm's argument, where he attributes to Anselm the thesis that 'existence is something more than not existing, i.e., existence adds a degree to the greatness or the perfection,' a rendition of (2) in my reconstruction, and highlights the parallelism between this thesis and the minor premise of Descartes' argument, 'existence is itself a perfection.'[4] However, neither there, nor in other places where he brings up the classical version of the argument does Leibniz level an objection against this premise. This makes it seem as though Leibniz does not have an issue with the first step of the argument, which employs the conception of existence as a perfection.[5] Curiously enough, Leibniz does not raise the Aquinas–Caterus line of objection either.[6] Instead, he claims that the real problem with the argument is not that it is fallacious but rather that it is 'incomplete.'

Leibniz's reason for holding that the classical version of the argument is incomplete is that it presupposes the possibility of God without providing a proof of it. As it stands, the argument merely proves the truth of the conditional, 'if God is possible, He exists necessarily,' and is in need of a complementary proof of the antecedent. Therefore, this conditional should be taken as the first premise of a *modus ponens*, which would yield the necessary existence of God as its conclusion once the possibility of God is proved.[7]

Leibniz is not in fact the originator of the incompleteness charge. Gaunilo criticizes Anselm's presupposition that understanding a term entails that the existence of what is signified by that term (at least) in the understanding, by pointing out that we could understand "any number of false things that have no real existence at all in themselves."[8] Again, in the Second Set of Objections, the objectors state that "it does not follow from [Descartes' argument] that God in fact exists, but merely that he would have to exist if his nature is possible, or non-contradictory" (AT 7:127, CSM 2:91). Sharing Gaunilo's concern, Leibniz underscores the insufficiency of the mere ability to think a term in establishing the possibility of its alleged object, for what we think may be a false idea representing an impossibility: "we often think of impossible chimeras —for example, of the highest degree of speed, of the greatest number, of the intersection of the conchoid with its base or rule" (AG 56).[9] Leibniz also claims that

[3] Hartshorne (1965), 178.

[4] *New Essays*, 437. Hartshorne (1965), 178 is aware of this passage, but seems to think that Leibniz's reconstruction is so misguided that he could not have read the *Proslogion* directly.

[5] Most notably, see *Discourse on Metaphysics*, §23 (AG 55–6). Adams (1994), 135, n. 1 provides a long list of relevant passages from Leibniz's entire corpus.

[6] In the aforementioned passage from the *New Essays*, 437, Leibniz even goes as far as to say that this objection was a "great mistake" on the part of Aquinas and his followers.

[7] See Hartshorne (1965), 177.

[8] Gaunilo's Reply 1.

[9] We saw in chapter 1 that Descartes takes the clarity and distinctness of an idea as a sufficient condition for the possibility of its object and thus deems himself justified in taking the possibility of God for granted.

Anselm's definition of God as something than which nothing greater can be thought is only a nominal definition, the possibility of whose definiendum is dubitable, as opposed to a real definition, which contains an account of how the definiendum is possible.[10]

In any case, Leibniz thinks that the classical version has to be complemented by a proof of the possibility of God. Various reconstructions of Leibniz's 'modal version' of the ontological argument are available in the literature.[11] In the following I will provide yet another reconstruction, placing emphasis on the items on my own agenda: (i) The way in which Leibniz's argument fits in my dual logical framework; (ii) Its solution to the problem of the identification of the two notions of God, a full account of which is missing in other reconstructions; (iii) The metaphysics of modality underlying it. Leibniz offers different formulations of his modal argument in various texts. Here, I will take *Monadology* (§§38–45) to be presenting a complete and official formulation of Leibniz's argument.

Although Leibniz holds that the truth of the conditional 'if God is possible, then he exists necessarily' is successfully demonstrated by the classical version of the ontological argument, in the *Monadology*, he first deduces the notion of God as the *ens necessarium* through a cosmological argument (§36–8), and derives the conditional from the unique modal character of the *ens necessarium*. Hence, the starting point of Leibniz's own argument:

(1) God is the *ens necessarium*. (§38)

Given that Leibniz construes absolute necessity here in terms of something's containing in its essence the sufficient ground of all existence, including its own (§39), the first premise carries out the first logical step of introducing existence into the concept of God. The conditional from possibility to existence appears as the modal privilege of such a being: God's existence follows necessarily from His essence. Proving the possibility of the *ens necessarium* is proving that there is in fact such an essence. A brief comment on the relationship between possibility and essence in Leibniz is in order.

Leaving a discussion of Leibniz's conception of essences for later, we can say that he posits a grounding relationship between essences and possibilities: "The essence of a thing is the specific reason [*ratio*] of its possibility" (A VI, iii, 583).[12] Individual possibilities (and truths about them) are thus ontologically anchored in essences. This entails that the possibility of something is a reliable indication that there really is an essence of that thing grounding its possibility. The modal privilege of God is just that if there really is such an essence, its actual instantiation necessarily follows.

However, as Henrich (1960), 48 notes, Leibniz holds that the clarity and distinctness of an idea has to be first established by an objective test of possibility that does not circularly rely on our assumption of the clarity and distinctness of that idea.

[10] See, for instance, (AG 57). Leibniz gives full credit to Hobbes for this distinction between nominal and real definitions (AG 26).

[11] For a few notable examples, see Henrich (1960), 45–55; Adams (1994), ch. 5; Blumenfeld (1995), 353–81; Hindrichs (2008), 76–102.

[12] Adams (1994), 138 takes this to mean "that essence is a quasi-logical structure that grounds the (unconditional) truth that the existence of that thing is possible."

For the essence in question is that of the *ens necessarium*; it cannot be possible but unactualized. Either there really is the essence of the *ens necessarium*, and it is actually instantiated (God exists), or there is no such essence, and the *ens necessarium* is a chimerical idea (God is impossible). The second step in my framework, the transition from the (existentially loaded) concept of God to the actual existence of God *qua* the real thing, will then be accomplished, once God is proven to be possible.

(2) If God, the *ens necessarium*, is possible at all, then He necessarily exist.

In the *Monadology*, Leibniz follows an indirect strategy to prove the possibility of the *ens necessarium*.[13] He first aims to prove the identity between the *ens necessarium* and the *ens perfectissimum*, and then to prove the possibility of the latter, which would transitively prove the possibility of the former:

(3) The *ens necessarium* is the *ens perfectissimum*.
(4) The *ens perfectissimum* is possible.
(5) The *ens necessarium* is possible. (3, 4)

The first stage of this strategy, the proof of the identity between the two notions of God, constitutes the most crucial part of Leibniz's whole argument. Leibniz makes various attempts at such a proof. The draft he shows to Spinoza in 1676 employs the premise "(necessary) existence is included among perfections" in order to demonstrate that the *ens perfectissimum* would have necessary existence as one of its predicates (A VI, iii, 578).[14] Of course, this premise would make the argument directly susceptible to Kant's objections. In the *Monadology*, Leibniz carefully avoids using this premise and constructs a proof based on the idea that God as the *ens perfectissimum* is the ground of all possibility. However, it is worth noting that Leibniz appears to have worked out some elements of this proof in yet another, earlier version of the argument written in 1676. Following his demonstration of the possibility of the *ens perfectissimum*, Leibniz writes:

Hence now it seems to be proved, further, that a Being of this sort, which is most perfect, is necessary. For it cannot be unless it has a reason for existing from itself or from something else. It cannot have it from something else, because everything that can be understood (*intelligi*) in something else can already be understood in it—that is, because we conceive it through itself (per se), or because it has no requirements outside itself (*nulla extra se habeat requisita*). Therefore either it cannot have any reason for existing, and so is impossible, contrary to what we have shown above, or else it will have it from itself (*a se*), and so will be necessary.[15]

[13] Leibniz seems to have worked on a direct proof of the possibility of the *ens necessarium*, before giving his argument its final shape in the *Monadology*. Both Adams (1994), 142 and Blumenfeld (1995), 363–4 refer to a 1701 passage (G IV, 406), where Leibniz asserts the negative conditional, if the necessary being is not possible, then there is no being (necessary or contingent) that is possible. Any mundane possibility, inferable from existence, would then be sufficient to prove the possibility of the necessary being. However, this a posteriori input that something exists would prevent this sub-proof from serving as a component of a genuine ontological argument.

[14] See Adams (1994), 148. We will see that this problematic strategy resurfaces in Wolff's and Baumgarten's variants of the modal ontological argument.

[15] This extremely important passage is quoted by Adams (1994), 151 and here I adopt his translation from Latin.

Leibniz offers here an alternative interpretation of the principle of sufficient reason (PSR), according to which the principle operates not upon causal but conceptual dependence relations, and thus 'reason' or 'ground' is to provide not the efficient cause of something but the material, i.e., the positive predicates or determinations, required for the construction of its concept.[16]

For Leibniz, perfection is a fundamental, absolute, positive, and maximal quality. He also calls such qualities "natures" or "forms" (AG 35), and sometimes, following the scholastic terminology, "realities" (AG 218). Then, the *ens perfectissimum*, as the subject of all perfections, is also the *ens realissimum*, the subject of all realities, all possible fundamental positive predicates. Such a being is conceptually self-sufficient in that it can be exhaustively conceived through its own predicates without any need for an appeal to a positive predicate that is not had by it. In virtue of his interpretation of the PSR, Leibniz takes such conceptual or logical independence to be a mark of ontological or existential independence. Accordingly, the *ens perfectissimum*, if at all possible, must be an *ens a se*, a being which contains its own sufficient reason to exist. As I will further discuss later, Leibniz repeatedly defines necessary existence of an individual being in terms of its containing its own sufficient reason to exist. Therefore, the *ens perfectissimum*, if possible, is the *ens a se* as well as the *ens necessarium*. Here, the key is that necessary existence figures not as one of the intrinsic predicates (perfections) that make up the concept of the *ens perfectissimum*, but as a consequence following from the logical completeness of the concept of such a being. However, this version, though promising, remains a mere draft, leaving the nature of the dependence relationship between individual conceptual contents and God underdetermined.

In the *Monadology*, Leibniz switches the focus from the ontological independence of the *ens perfectissimum* to the ontological dependence of everything else, possible as well as actual, on the *ens perfectissimum*. After deducing the *ens necessarium* as the ultimate sufficient reason for the series of contingent beings, Leibniz states its identity with the *ens realissimum*:

> We can also judge that this supreme substance which is unique, universal, and necessary must be incapable of limits and must contain as much reality as is possible, insofar as there is nothing outside it which is independent of it. (§40, AG 218)

And subsequently, he extends the identity to the *ens perfectissimum*: "From this it follows that God is absolutely perfect —perfection being nothing but the magnitude of positive reality considered as such" (§41, AG 218). Hence,

(3) The *ens necessarium* is the *ens perfectissiumum*.

Two routes are available to Leibniz for justifying this identity claim, based on two different interpretations of the PSR. In §42, applying the causally interpreted PSR to the existence of imperfect things, Leibniz suggests that these things "get their perfections from the influence of God" (AG 218). Thus, the *ens perfectissimum* is

[16] Leibniz sees the two kinds of dependence as tightly connected. This is strongly suggested by his axiom: "The effect is conceived through its cause." (A V, iii, 514).

necessary for imperfect things to exist. However, this route has to presuppose the actual existence of contingent beings, and thus would amount to a cosmological proof. The second route, which is the one Leibniz actually takes, operates on the application of a specific form of PSR to the possibility, rather than the actuality, of things:

> It is also true that God is not only the source of existences, but also that of essences insofar as they are real, that is the source of which is real in possibility. This is because God's understanding is the realm of eternal truths, or the realm of the ideas in which such truths depend. Without God's understanding there would be nothing real in possibles, and not only would nothing exist, but also nothing would be possible. (§43, AG 218)

First of all, Leibniz here puts forth the claim that there would be no possibility without the existence of God (the *ens perfectissimum*) or that all possibility is ontologically dependent on God. I mentioned earlier that Leibniz holds that individual possibilities (and the truths about them) are grounded in essences. Essences, in turn, are grounded in God, which makes God the ultimate ontological ground of all possibility. Here he adds more substance to this view. An individual possibility is grounded on an essence with respect to its 'real' or material content, as opposed to its form which may be constituted by a formal constraint on this content such as logical consistency. Crudely put, for a thing to be possible, the positive predicates that constitute its concept must themselves be grounded. So essences ground the possibilities of individual things by grounding the positive predicates contained in their concepts.

But what kinds of beings are essences, and in virtue of what do they provide the matter of *possibilia*? In §43, Leibniz maintains that the realm of being for essences and the modal truths grounded by essences is God's mind.[17] In contrast with Descartes, who anchors essences in God's will and actual volitions, Leibniz anchors them in God's mind and actual thoughts. Accordingly, essences are intentional or "internal objects" of Gods mind (§46), constituted by the contents or, in Cartesian terms, objective realities of God's ideas.[18] It is therefore ultimately God's mind which grounds the 'real' or material content of all possibility by his acts of thinking. Without God's mind, no material content would be given in the first place, and thus no individual possibility would be left.

Although Leibniz's starting point in the 1676 draft seems to be the modest claim that the availability of positive content in the logical space is an epistemic condition of our thinking a possibility, here he aims to make a stronger claim regarding an ontological condition of possibility: God's active mind and ideas are not just necessary for a possible thing to be thinkable (by non-divine discursive minds), but for that thing to be possible at all (regardless of the existence and structure of non-divine minds). In §44, Leibniz underscores this point by introducing a crucial metaphysical principle: "If there is reality in essences or possibles, or indeed, in eternal truths, this reality must be grounded in something existent and actual" (AG 218). This is the

[17] Compare *Theodicy*, §184. [18] See Newlands (2013), 165.

'actualist principle,' a version of which we previously observed in Descartes's thesis that objective reality is always dependent on some formal reality. The idea here is:

(3a) Every possibility is (materially) dependent on the actual existence of something.

Leibniz specifies this dependence or grounding relationship in intentional terms: God grounds possibilities by actually thinking or entertaining ideas that constitutes their material contents.[19] So he holds:

(3b) Every possibility is materially dependent on the divine mind's actually thinking its material content.

Leibniz's sub-proof of the identity between the *ens perfectissimum* and the *ens necessarium* (3) is thus based on this dependence of possibility on God's active mind. The *ens perfectissimum*, which, in addition to the divine perfections it instantiates, originally entertains the material content of all possibilities, is the necessary condition for there to be any possibility at all. In other words, the most perfect being is the necessary being, without which nothing would be possible.

There are three issues I wish to note regarding this most important step of Leibniz's argument. First of all, although Leibniz applies the actualist principle (3a) in arriving at the claim that God's active mind is the ultimate ground of possibility (3b), it remains unclear what justifies the actualist principle itself. Why must possibilities be grounded in something actual in the first place? Second, even if the actualist principle is accepted, is it warranted to assert a single ground of all possibility? Why is the ground a single divine mind entertaining the material content of all possibility single-handedly rather than a committee of minds doing the same job of grounding collectively? Third, the move from the proposition that the most perfect being (or the divine mind) is the ground of all possibility to the conclusion that this being necessarily exists seems overly hasty. For the former expresses at best the conditional necessity of the ground of possibility: if anything is possible, then, necessarily, its possibility is grounded. But one might still ask 'why must there be any possibility at all?' or 'why is absolute nothingness itself not a possibility?,' a question, perhaps, more intriguing than Leibniz's question, "why is there something rather than nothing?" (AG 210). These questions threaten the validity of Leibniz's justification for (3) and remain unanswered at least in his explicit statements of the argument. As I will argue in chapters 4 and 8, Kant will grapple with these questions, and while he will address the second and third questions by formulating a new conception of absolute necessity, the shift in his views on the first one will result in his critical revaluation of his own argument.

We can now look at the final step in Leibniz's argument, namely, the proof of the possibility of the *ens perfectissimum*, which, given the identity between the *ens perfectissimum* and the *ens necessarium*, also proves the possibility of the latter. In

[19] As I will discuss from chapter 4 on, Kant will base his own 'only possible argument' on yet another interpretation of the actualist principle. For comparisons of Kant's and Leibniz's different interpretations of the principle, see Adams (2000), Henrich (1960), Nachtomy (2012).

the *Monadology*, this proof is based on three premises. The first and implicit premise is about what it means for the *ens perfectissimum* to be possible:

(4a) For such an entity as the *ens perfectissimum*, possibility is the internal (logical) compatibility of its predicates, i.e., perfections.

Since the *ens perfectissimum* is both conceptually and ontologically self-sufficient, the possible existence of such a being cannot be prevented by anything external to its essence. The impossibility of such a being can thus follow only from an intrinsic reason, which, Leibniz holds, can only be a logical contradiction. Thus, this premise is an extension of Leibniz's logic of divine necessity: God is either impossible in himself or else He exists necessarily. Here Leibniz does not simply commit himself to the view that freedom from logical contradiction is the only condition of possibility in general, but appeals to it as the only condition applicable to the exceptional object in question. However, he still implicitly makes the general assumption:

(4b) Logical contradiction (among the predicates of a thing) is the only kind of internal conflict that can cancel the intrinsic or absolute possibility of a thing.

The third and last premise in this step is expressly stated in §45 through a concise and negative formula: "Nothing can prevent the possibility of what is without limits, without negation, and consequently without contradiction" (AG 218). Since a perfection is a simple, absolute, positive quality, no perfection can logically contradict another perfection.

(4c) All perfections are logically compatible in a single subject.

Since possibility for the *ens perfectissimum* is absolute and logical possibility, the logical compatibility of its predicates warrants its possibility.

(4) The *ens perfectissimum* is possible.

In virtue of the identity between the *ens perfectissimum* and the *ens necessarium* (3), the possibility of the latter is also established:

(5) The *ens necessarium* is possible.

This is what was initially required to satisfy the antecedent of the conditional: 'if God, the necessary being, is at all possible, then he exists necessarily.' With the possibility of this being proven, the *modus ponens* is completed and the conclusion follows:

(6) God exists necessarily.

2.1.2. Leibniz's theory of modality

The conception of modality underlying Leibniz's argument gets further detailed through some of his central metaphysical doctrines. In this section, I will thus shift the focus of my account of Leibniz's theory of modality from its place in his ontological argument to that in his general metaphysics. I will take up the notions of possibility, necessity, and existence respectively in the following three subsections.

POSSIBILITY: INDIVIDUALS, WORLDS, AND GOD

It would be most helpful to examine Leibniz's treatment of possibility at two levels: first, at the level of things considered in themselves; second, at the level of things considered in their relations with other things. These two levels of consideration correspond to two basic types of possibility: 'absolute possibility,' which is constituted by intrinsic grounds, and 'relative possibility,' which is constituted by extrinsic as well as intrinsic grounds. In Leibniz's metaphysics, these two types of possibility can be applied to two kinds of objects: individual things or substances, and whole worlds comprised of compossible individuals (and laws). The respective absolute and relative possibility considerations regarding these two kinds of objects and God, the ultimate ground both of all possibility and actuality, yield a rich and complicated account of possibility.

Absolute possibility For Leibniz, the possibility of something, when considered absolutely or in itself, is to be conceived in terms of an internal compatibility or conflict. Leibniz seems to have in mind only a logical sort of internal conflict. Accordingly, "anything that, in itself, implies no contradiction... [is] possible in its nature" (AG 21).[20] So the absolute possibility of something rests formally on the principle of contradiction, and warranted by the logical compatibility or lack of logical contradiction among the predicates that make up the concept of that thing. Yet Leibniz also offers a material condition of absolute possibility: for something to be possible, the predicates that make up its concept must be grounded in actuality in addition to being logically compatible with one another. The predicates or material contents of concepts must be grounded in essences, which are the intentional objects of God's actual ideas. Thus, the actuality that materially grounds the absolute possibility of individual things is the actuality of God's mind or his act of thinking. As I noted earlier, Leibniz construes the absolute possibility of God to be just his logical possibility, which he explicitly cashes out in terms of the logical compatibility between God's non-intentional predicates or perfections. It is fair to infer that God's intentional predicates, his thoughts or ideas, which altogether ground the contents of all non-divine possibilia, would also be logically compatible with each other.

One can also consider the absolute possibility of sets of 'compossible' individual things, which Leibniz calls 'worlds.' Like individual possibilia, possible worlds are materially grounded in God's ideas. They are, so to speak, different world-plans that God contemplates. Now, Leibniz does not explicitly mention any other sort of incompatibility than logical contradiction that would restrict the contents of God's mind, which seems to leave the principle of non-contradiction as the only constraint on the contents of world-plans. However, he also emphasizes that worlds do not just comprise individuals but also laws of nature that those individuals must conform to.[21] This suggests that the compossibility of the individual members of a world also involves certain non-logical compatibility relations informed by the laws of nature at work in that world. A set of individuals that do not logically contradict one another

[20] See also *Theodicy*, §173. [21] See Look (2013), G II 51.

may well be incompatible for cohabitation in a single world if certain laws are introduced. That world-plan, then, would be absolutely impossible. Therefore, the absolute possibility of a world is constituted, materially, by the contents of God's actual thoughts, and, formally, by logical and non-logical compatibility between its individual members.

We can thus conclude that in its application to individuals (finite individuals and God), absolute possibility amounts to logical possibility. Granted, Leibniz takes it as a requirement of absolute possibility that the material content be grounded in actuality, understood as God's actual thoughts. Nevertheless, he does not offer any other formal constraint on the contents of God's thoughts than logical consistency, which might preclude a logically consistent individual from being absolutely possible. Furthermore, since Leibniz also seems to hold that God, through his infinite mind, actually contemplates (and thus materially grounds) all logically consistent combinations, the two conditions of absolute possibility, material and formal, collapse into one another in the case of individuals, or more precisely, the logical possibility of an individual is the sufficient condition of its absolute possibility. As a result, the domain of absolutely possible individuals overlaps with that of logically possible individuals.

There are two other points here. The coextensiveness of absolute and logical possibility cannot be claimed as easily in the case of the worlds, whereby the laws of nature may require certain non-logical, extrinsic compatibility considerations for absolute possibility. I will revisit this issue. Second, absolute possibility is the result of just the first level of inquiry concerning the possibility of things. Only in the exceptional case of God does absolute possibility exhaust the inquiry of possibility. For God is the only ontologically independent being, whose possibility (as well as actuality) does not depend on its relations with other beings but is an entirely internal matter. In the case of individuals and worlds, on the other hand, absolute possibility is only the first question. There are a good (near infinite!) number of absolutely possible individuals and worlds, which do not and indeed, in a quite robust sense, 'cannot' exist. For the realm of the non-divine, the stronger notion of metaphysical possibility (of existence), as opposed to merely logical possibility, would follow from relative possibility considerations.

Relative possibility Individual things or substances, with the exception of God, can only exist in the context of a world comprising individuals and certain laws of nature. Therefore, for an absolutely or internally possible individual to be really possible, it must be also a member of at least one possible world. This notion of relative possibility, which I will call here 'R1,' consists in an individual's possibility of being a member of some (absolutely possible) world-plan in God's mind, and depends on its compatibility with the content, i.e., other individuals, and laws, of that world.

> Relative Possibility (R1): An individual is relatively possible if and only if it is compatible with at least one absolutely possible world.

The relative possibility of a world, on the other hand, would be its possibility relative to God's creative will, which, as I will discuss, always chooses the 'best' of all absolutely possible worlds. Since only one of the infinitely many worlds is thus chosen and actualized by God, only the actual world is in fact really possible relative to God's will. This yields another, an even stronger notion of relative possibility for individuals, i.e., possibility relative to the actual world. I will call this notion 'R2.'

Relative Possibility (R2): An individual is relatively possible if and only if it is compatible with the content and laws of the actual world.

In both of these preliminary definitions of R1 and R2, I leave 'compatibility' unspecified. The specification of the meaning of compatibility in each case and its implications for the respective domain of possibility requires a closer look at of some of Leibniz's metaphysical commitments.[22]

Leibniz is committed to the principle of complete determination, which will also be adopted, in various forms, by Wolff and Baumgarten: only individuals that are completely determined with respect to all of their predicates can exist. When talking of a possible individual such as Adam, Leibniz says, we must not consider "a vague Adam," an incomplete notion which could represent multiple Adams, but take him as a determinate individual and "attribute to him a notion so complete that everything that can be attributed to him can be deduced from it" (AG 73). So both possible individuals and the entire possible worlds must be understood as completely determinate concepts in God's mind with no vagueness left as to their details. God thinks worlds in their full specifications, as completely and thus uniquely determined configurations of compossible individuals. The result is that each individual is 'worldbound,' i.e., bound with one and only one determinate set of compossibles, corresponding to a unique world-plan in God's mind. Unlike some contemporary accounts of possible worlds such as Plantinga's, Leibniz's account excludes the possibility of 'transworld identity' between individuals inhabiting distinct worlds.[23] This means that each individual can be a member of only one specific world and thus is possible relative to only one specific world.[24]

Leibniz's idea that individuals are worldbound is also tightly connected with his '*in-esse*' doctrine of truth: "In every true affirmative proposition, necessary or contingent, universal or singular, the concept of the predicate is included in some way in the concept of the subject" (L 337).[25] In the *Discourse*, Leibniz draws important metaphysical and theological implications from the '*in-esse*' doctrine. At the center

[22] Against the charges made by Chignell (2009) that Leibniz fails to distinguish between logical and real or metaphysical possibility, Newlands (2013), 181–3 claims that Leibniz has the resources to make a tripartite division in modal domains: logical, metaphysical, and "divinely actualizable," corresponding respectively to the principle of non-contradiction, the principle of sufficient reason, and the principle of the best. However, Newlands does not explain how exactly the PSR (or other metaphysical principles that, he claims in passing, Leibniz uses to demarcate a domain of possibility) demarcates a domain of metaphysical possibility that is distinct from that of actualizability. Newlands also suggests that the aforementioned three domains may turn out be coextensive, but without explaining how. Although my account here also offers a tripartite division (absolute possibility, R1 and R2), corresponding roughly to Newland's division, I will provide an explanation of both how these domains are constructed by Leibniz's metaphysical doctrines and principles, and whether and why they turn out to be coextensive.

[23] Plantinga (1974). See, for instance, Adams (1994), 53–7; and Mates (1972). 90–2.

[24] A 'Judas who does not sin,' for instance, is not a logically contradictory concept, but given the constitution of this world, with the determinate combination of possibilities that qualifies it as the best among all possible worlds, the existence of an innocent Judas is not possible. Once again, in the way Leibniz understands possible worlds, it does not make sense to posit an innocent Judas in another possible world.

[25] See also AG 11, 30, 45–6.

lies his idea of 'complete notion': "we can say that the nature of an individual substance or of a complete being is to have a notion so complete that it is sufficient to contain and to allow us to deduce from it all the predicates of the subject to which this notion is attributed" (AG 40–1). Accordingly, crossing the Rubicon and sinning are contained in the respective complete notions of Julius Caesar and Judas.[26] A complete notion individuates its object uniquely by determining it with respect to every possible predicate, and in relation to every other individual thing, event, and law in the world that object is a member of. Thus, each individual substance in its complete notion "is like a complete world and like a mirror of God or of the whole universe, which each one expresses in its own way" (AG 42).

So the complete notion of each individual reflects the world to which it is relatively possible and/or bound in complete detail.[27] Since the complete notion of an individual contains all of its (extrinsic) relations in a world, each individual is bound to each and every other individual cohabitating in that world as well as the laws of nature regulating the interrelations of the cohabitants of that world.[28] For different laws would generate different properties and consequently different individuals.[29] Therefore, a world is an either all or nothing deal. Each individual and its respective world are in fact mutually bound. Picking one would mean picking the other.

It is crucial to add here that although the world-boundness of individuals does not necessarily entail it, Leibniz seems to hold that each absolutely possible individual inhabits one world. This primarily follows from the infinity of God's active mind and the infinite space of logical possibility generated by it. Now it is logically possible that any absolutely possible individual can be accommodated by at least one world (e.g., the world consisting only of that individual) among the infinite number of worlds grounded in God's mind. All logical possibility must be grounded in God's mind, and if there is anything logically possible, God would not fail to actually think and ground it. Then it is in fact necessary that for any individual that is possible in itself there is at least one world that accommodates it, or that every absolutely possible individual is possible relative (R1) to some world. It is as though God's infinite mind creates a distinct world-scenario built around each and every absolutely possible individual. Thus, possible worlds are not only uniquely bound with every individual in their respective exclusive sets of individuals, but they also jointly occupy the entire domain of logical possibility. They are, as Mates aptly puts, both "mutually exclusive and jointly exhaustive."[30]

This mutual fate of individuals and worlds has somewhat disappointing implications for relative (R1) possibility. I have established earlier that in the absence of a non-logical account of internal conflict, absolute possibility amounts to logical possibility. I have also defined the relative possibility (R1) of an individual as its

[26] AG 43, 61.

[27] Leibniz does not restrict the '*in-esse*' doctrine to truths concerning the actual world alone, nor does he imply, like Baumgarten does, that the notion only of that which exists is complete. Moreover, his account of possible worlds in the *Theodicy* (§§414–17) makes it clear that Leibniz takes the '*in-esse*' and consequently the conception of complete notion as applying to any possible world and possible individuals in it.

[28] This is in line with the views defended by Newlands (2013) and Yong (2014) that on Leibniz's account God's intellect also grounds relations.

[29] See Look (2013).

[30] Mates (1972), 92.

(extrinsic) compatibility with the content and laws of some world. The latter, at least *prima facie*, seemed to offer a domain of possibility distinct from that of logical possibility, as one can putatively think of non-logical, e.g., causal, sorts of compatibility relations in the context of a world. For instance, an individual *i*, which is logically possible in itself, cannot be part of the world *w* if it is causally incompatible either with the individual inhabitants of *w* or with its natural laws. This suggests that each specific world excludes a large number of individuals that are both logically possible in themselves and logically compatible with the content of that world. However, although it certainly calls for non-logical compatibility considerations, R1 too ultimately fails to delineate a domain of possibility distinct from that of logical possibility in the case of individuals. That is, the notions of R1 and logical possibility pick out the same individuals and thus are coextensive. For each and every logically possible individual necessarily inhabits a specific world, with which it is mutually bound. Consequently, each and every logically possible individual is (necessarily) possible (R1) relative to some (in fact, only one) world. The promising question of a non-logical extrinsic compatibility (or conflict for that matter) remains a moot one in the case of R1.

This account of R1 tells us that for the non-logical compatibility (or conflict) considerations to become effective constraints on possibility in general such that we can demarcate a domain of metaphysical possibility narrower than that of mere logical possibility, relative possibility must be defined against the background of a specific world with specific content and laws of nature. I suggest that this fixed background of relative possibility should be the specific world that the actual world is. For what motivates our efforts to formulate a metaphysical notion of possibility is our modal intuition that some logical possibilities cannot obtain in reality. So a metaphysical notion of possibility should successfully accommodate the sense in which some logical possibilities are not really possible. Relative possibility can satisfy this expectation only if it is defined with respect to the actual world, the one specific world the incompatibility with which demarcates a domain of logically possible individuals that can never exist in reality. This is indeed the primary advantage of R2.

Leibniz's actualism about possibility appears to apply to R2 as well. While absolute (or logical) possibility is grounded in God's actual thoughts of infinitely many worlds, relative (or real) possibility (construed as R2) is grounded in the actual world, which, in turn, is grounded in God's active choice of one world among infinitely many: "This regard or relation of [God] to simple possibilities can be nothing other than the *understanding* which has the ideas of them, while to fix upon one of them can be nothing other than the act of the *will* which chooses" (*Theodicy*, §7). We can thus redefine R2 as possibility relative to God's will.

> Relative Possibility (R2*): Something is relatively possible if and only if it is compatible with God's actual will.

This is applicable to individuals and whole worlds alike. For God's decision to actualize an individual would amount to the decision to actualize the world that individual is bound with.

In response to the further question of what grounds God's decision to actualize one among infinitely many possible worlds, Leibniz offers the 'principle of the best':

God, in virtue of his supreme goodness, always acts in the best possible way and thus chooses to actualize the best among all possibilities. He often formulates the 'bestness' of a possible world in terms of the degree of economic rationality for God's choice to actualize it. In 'On the Ultimate Origination of Things,' he appeals to the rule of seeking "maximum effect at the minimum cost" (AG 150), or similarly, in the *Discourse*, to the combination of "the simplest in hypotheses and the richest in phenomena" (AG 39), in the *Monadology*, to the obtaining of "as much variety as possible, but with the greatest order possible" (AG 220).[31] The individual evils are then the necessary cost to pay to get the best series of compossibles. God's choice concerns not just the absolute qualities of individuals but also the comparative qualities of the world-plans these individuals are bound with, and this is how, for instance, Judas can be part of the best of all possible worlds.

However, Leibniz also expresses commitment to another doctrine regarding the question of why this world rather than any other is the actual world, some elements of which might appear incompatible with his doctrine of divine choice. This is the so-called doctrine of '*Daseinstreben*' or the striving of possibles toward existence, according to which an unactual possibility represents more than a mere neutral capacity for existence, a mere idea in God's mind waiting to be actualized, but a positive 'urge' for existence. Leibniz develops this doctrine in several of his late texts such as the *Theodicy* and *Monadology*, but his most commonly cited statement appears in his 1697 paper, 'On the Ultimate Origination of Things':

> [...] we must first acknowledge that since something rather than nothing exists, there is a certain urge for existence or (so to speak) a straining toward existence in possible things or in possibility or essence itself; in a word, essence in and of itself strives for existence. Furthermore, it follows from this that all possibles, that is, everything that expresses essence or possible reality, strive with equal right for existence in proportion to the amount of essence or reality or the degree of perfection they contain, for perfection is nothing but the amount of essence.
>
> From this it is obvious that of the infinite combinations of possibilities and possible series, the one that exists is the one through which the most essence or possibility is brought into existence. (AG 150)

A couple of points are in order. First, Leibniz's motivation in developing this doctrine seems to be to offer a more metaphysical (as opposed to theological) explanation of why this world exists. Setting aside God's creative will, Leibniz seems to want to ground existence in the essences of things themselves. Second, Leibniz formulates this latter kind of grounding in terms of a claim that each possible lays on existence in proportion to the amount of its material content (positive reality, essence, or perfection). Actual existence, then, results from a fair contest of possibles based on an ontological principle of desert. Hence, he writes, "as the possibility is the principle of essence, so perfection or degree of essence is the principle of existence" (L 487–8). Therefore, the *Daseinstreben* doctrine, by explaining the existence of things in terms

[31] In the *Theodicy*, §110, Leibniz also defines the degree of the perfection of a world in terms of the extent to which it would please God, the architect of all possible world projects.

of their intrinsic qualities, at least appears to constitute an alternative to God's external creative activity.[32]

This, however, is only an appearance. The doctrines of *Daseinstreben* and divine creation need not be mutually exclusive. The key here is that Leibniz uses the *Daseinstreben* as a metaphorical device not to refute but to further explicate the principle of the best. This gets particularly clear in the *Theodicy*. For instance, he writes:

> One may say that as soon as God has decreed to create something there is a struggle between all the possibles, all of them laying claim to existence, and that those which, being united, produce most reality, most perfection, most significance carry the day. It is true that all this struggle can only be ideal, that is to say, it can only be a conflict of reasons in the most perfect understanding, which cannot fail to act in the most perfect way, and consequently to choose the best. (§201)[33]

God's creative activity itself is not at all contested here, but it is rationalized by an appeal to a reason or ground intrinsic to those possibilities (i.e., their perfection) that God actualizes. Leibniz never says that this intrinsic ground is a sufficient one for a (contingent) possibility to come into existence. God's role as the ultimate sufficient reason of all existence remains undisturbed. Nevertheless, Leibniz does say, for instance, in the *Monadology*, §53, that the perfection or the 'best-ness' of the actual world is the "sufficient reason for God's choice," a reason which morally determines but does not metaphysically necessitate God's preference to actualize it among infinite number of possible worlds.[34] The *Daseinstreben* makes it easier to see that the principle of the best is in fact an extension of the PSR in the case of God's choice: God is a perfectly rational agent, and his decision to create this particular world rather than another must have a sufficient reason pertaining to this world. The latter, Leibniz argues, provides us with an objective reason to praise God's work.[35] Thus, the doctrines of divine choice and striving possibles both employ the PSR and produce not incompatible but complementary answers to the question of why this world rather than another exists.[36]

Leibniz's account of the actual world thus implies that possibility relative to the actual world (R2) ultimately involves compatibility with metaphysical principles such as the PSR and the principle of the best. At the end, something is relatively possible (R2) if it can be part of the world which God decides to actualize for the reason that it is the best of all (absolutely) possible worlds. Furthermore, the relative possibility

[32] This appearance leads some scholars like Lovejoy (1936), ch. 5 to the view that the *Daseinstreben* is incompatible with the doctrine of creation, which Leibniz indubitably subscribes to, and that the former must be an inconsistency on Leibniz's part. On this discussion, see also Shields (1986).

[33] See also §§8, 52, 110, 119, 149, 168, 201, 335, 414, 416.

[34] For Leibniz's distinction between moral determination and metaphysical necessitation, see *Theodicy*, §§168, 174, 201.

[35] *Discourse on Metaphysics*, § 2 (AG 36). For a similar view on Leibniz's motivation concerning the doctrine of *Daseinstreben*, see Blumenfeld (1973), 176–7.

[36] Shields (1986), 354–5 also defends the compatibility of the doctrines of *Daseinstreben* and creation without following Blumenfeld's suggestion to read the former metaphorically, and by taking even the urge or exigency of essences for existence as something endowed upon them by God. On the account I present here, such an additional move is not necessary to establish the compatibility of the two doctrines.

(R2) of an individual would also require its compatibility with the content and laws of the actual world. Now, this compatibility involves logical compatibility between the individual and all other inhabitants of the actual world, as the absolute possibility of the actual world requires the logical consistency of its content in the first place. But given that the actual world also consists of laws of nature, it is tempting to cash this compatibility in terms of the causal compatibility of the individual with the actual world.

However, Leibniz's account of causation poses some complications. Leibniz rejects both causal interaction between finite individual substances and occasionalism, which denies causal powers to finite substances altogether. Instead, he argues, there is causation within substances (intrasubstantial causation): each substance is a world apart and follows its own causal laws in a closed system.[37] He explains the apparent causal coherence between substances in terms of a pre-established harmony. Accordingly, God creates the individual inhabitants of the world in a predetermined regulation with respect to one another such that even though there is no real interaction among individual substances and each one acts in its own spontaneity, the states of all individual substances at any given time are in perfect harmony, giving the appearance that they are in thoroughgoing causal interaction.

Despite the complications, the compatibility of an individual with the actual world must still involve causal considerations in two ways, directly and indirectly. First, intrasubstantial causation entails that the changes in the states of individual substances form a temporally ordered causal series. The state of a substance S at time t2 is causally determined by its state at time t1 and its internal causal power in accordance with its own causal law. Second, pre-established harmony entails that at any given time, the respective states of individual substances (in the same world), which are all causally determined from within, must conform to each other. If two substances, S1 and S2, are to be compossible in the same world, at any given time t, the state of S1 with respect to S2 must conform to the state of S2 with respect to S1. This, in turn, entails a dual causal conformity: each substance must conform to its own causal series directly, and also to any other fellow substance's internal causal series indirectly, even though it is not at all influenced by the latter.[38] We can thus conclude that intrasubstantial causation and pre-established harmony place non-logical constraints on the content of the actual world. Accordingly, an absolutely (logically) possible individual may well fail to be compatible with the actual world and relatively possible (R2). Therefore, R2 successfully demarcates a domain of metaphysical possibility that is narrower than that of logical possibility. As I will argue, both Wolff and Kant will adopt different formulations of R2 in their attempts to define metaphysical or real possibility.

[37] See, for instance, *Discourse on Metaphysics* §14 (AG 47), *New System* (AG 144–55) *Monadology*, §7 (AG 213–14).

[38] The same kind of dual conformity applies also to the relation between an individual substance or 'monad' and the 'organized mass' or material body it grounds: the substance directly conforms to its own causality which determines the changes in its states or perceptions, but since the perceptions of the substance corresponds to the states of the material body, where the substance's unique perspective lies, the substance also conforms to the mechanical causal laws that govern the particular material body it grounds as well as other material bodies in the phenomenal realm. (G IV 484–5/AG 143–4).

However, one might wonder whether this domain of metaphysical possibility is too narrow. We have seen that the metaphysical compatibility of an individual substance with the PSR (or the principle of the best) is established through its causal (as well as logical) compatibility or harmony with the actual world. But for an individual to be so harmonious with the actual world means that the distinct temporally ordered internal causal series that constitutes the individual substance is in perfect harmony with all the distinct temporally ordered internal causal series constituting all other individual substances in the world. That is to say, the individual substance is in perfect conformity with all the events that make up the entire history of the actual world. But considering his commitments to the principles of complete concept, the world-boundness of individuals, and the exclusivity of worlds, Leibniz must hold that being so harmonious with the world's history amounts to being an actual part of it. Accordingly, the complete concept of an individual substance contains the entire history of the world that it is bound with, from its own point of view, just as much as the complete concept of that world contains all distinct histories of individual substances in it. This mutual boundness is exclusive and completely determining in that the concept of the individual and that of the world expresses one and the same history. This is why, for instance, Leibniz claims that "all human events can be deduced . . . by assuming the creation of an Adam determined with respect to all these circumstances" (AG 72). In the Leibnizian picture, then, compatibility with the actual world and being part of it boil down to one and the same kind of relation.

The worry here is that this account would leave no room for merely relatively possible (R2) individuals. All individual substances and events that are relatively possible (R2) are determined to become actual, since to be relatively possible (R2) is to be part of the history of the actual world. Leibniz would happily embrace determinism. Yet the real danger for Leibniz is that the exclusion of unactualized relative possibilities (R2) suggests that his account might be getting too close to a form of necessitarianism. Leibniz seems aware of this danger, and tries to address the worry in his account of contingency and necessity, which is where I will turn to next.

But before then, it is important to recap the results of this analysis from the viewpoint of the charge that Leibniz's account of possibility is merely logicistic. On my reading, Leibniz's account accommodates three distinct notions of possibility: absolute, Relative 1, and Relative 2. Absolute possibility and R1 are coextensive with logical possibility in the case of individuals, but for different reasons. Absolute possibility turns out to pick out the same set of individuals as logical possibility because even though Leibniz explicitly endorses the actualist principle that the material content of possibility must be grounded in actuality, he does not offer any other sort of internal incompatibility than logical contradiction. R1 is coextensive with logical possibility because even though its definition involves non-logical or causal compatibility with a world, the infinity of possible worlds in God's infinite mind warrants that each logically possible individual is (logically or otherwise) compatible with, and thus is possible relative (R1) to a world. In other words, although they end up being coextensive with logical possibility, both absolute possibility and R1 are defined with reference to metaphysical principles that are not reducible to the principle of contradiction (i.e., actualism and causation). Furthermore, R2, based on, again, metaphysical principles such as the PSR, the principle of the best and causation, demarcates a domain narrower than that of logical possibility, though one that is coextensive with that of actuality. So even if one might say that Leibniz fails to offer a notion of metaphysical possibility that is not coextensive with actuality, it would be too dismissive to categorize his account of possibility as logicistic.

NECESSITY AND CONTINGENCY: OF THINGS AND PROPOSITIONS

I will examine Leibniz's account of contingency and necessity first in their *de re* forms, as applied to individual things, and in their *de dicto* forms, as applied to propositions.

There are places where Leibniz appears to follow the scholastic formulation of the *de re* necessity/contingency of a being in terms of the containment of existence in the essence of that being.[39] But I do not take these passages to reflect Leibniz's considered view on *de re* modalities, for as I will argue later on, Leibniz comes to reject the conception of existence as a predicate that can be contained in the essence or concept of a thing. The formulation of *de re* modality that I take to be more in line with Leibniz's most mature views on existence is based on the PSR.

> Necessity (*de re*): Something is a necessary being if and only if its essence contains its own sufficient reason for existence.
>
> Contingency (*de re*): Something is a contingent being if and only if its sufficient reason lies outside its essence.

For something to have its sufficient reason in its own essence means that such a being is ontologically independent and its existence follows from its essence alone. On the other hand, for something to have its sufficient reason outside of its essence means that its existence is dependent on an extrinsic source. However, Leibniz's previously discussed doctrines about actuality might lead to the worry that there is still a strong sense in which contingent existence is necessary: If the fact that this world is the best among all possible worlds is the sufficient reason for God to actualize it, how can this world be conceived as contingently actual? Besides, if God, as Leibniz sometimes puts it, cannot help but choose the best among all possible worlds, where 'best-ness' is not an arbitrary attribution but an objective quality of a world, how can God's choice be conceived as free?[40] The tension between divine freedom and Leibniz's various explanations of the actuality of this world has been a subject of significant scholarly interest.[41] I will be content here with stating only some of the pertinent points following from my analyses of possibility and *de re* modality in Leibniz.

i) All essences other than God's are devoid of sufficient reason for their existence. Existence, for all non-divine essences, is an endowment by God. This endowment does not proceed directly at the individual level, but individuals are endowed with existence through their inclusion in the best world-plan.

[39] See, for instance, AG 28, 218. [40] See, for instance, *Theodicy*, §8, 168.

[41] In addition to the more radical position held by Lovejoy (1936), 281 that the *Daseinstreben* and the doctrine of creation present two irreconcilable explanations of the existence of the world, Blumenfeld (1973), 176–7 and Rescher (1967), 29–30 hold that the *Daseinstreben*, literally understood, is incompatible with divine freedom, and suggest a figurative reading of the former.

Therefore, the difference between the (*de re*) necessity of God and the (*de re*) contingency of any other individual can be understood as the difference between an immediate and mediate relationship between existence and essence.[42]

ii) The 'best-ness' of the actual world is not an absolute property of it, but a property relative to other possible worlds. Even if 'the best of all possible worlds exists' is a *de dicto* necessity (provided that God necessarily actualizes the best possible world), whether this particular world is the best one is contingent upon other worlds' degrees of perfections.

iii) The existence of this world does not follow from its 'best-ness' in the same way as the existence of God follows from his essence. Leibniz accounts for this difference by means of a distinction between two kinds of necessity (or 'following'), namely *moral necessity* which rests on the principle of Wisdom and Goodness, and *metaphysical necessity* which rests on the principle of contradiction. In the *New Essays*, he claims that "metaphysical 'followings' necessitate, but physical and moral ones incline without necessitating" (178–9). In the *Theodicy* he emphasizes that that God chooses the best of all possible worlds is only a "moral" and not a "metaphysical" ("brute," "blind") necessity (§§168, 169, 174, 201). The actualization of a less perfect world does not imply a logical contradiction in itself but only raises a doubt about the wisdom and goodness of God (AG 36–7). Therefore, the moral necessity of the actualization of the best possible world does not diminish divine freedom; on the contrary, it shows that God's choice is not arbitrary but based on a moral-rational law to which God subscribes freely.

There are two ways in which Leibniz explicitly formulates the *de dicto* modality as applied to propositions or truths. The first one is rather straightforward.

Necessity (*de dicto*): A proposition is necessary or a 'proposition (or truth) of reason' if and only if its negation is logically impossible.

Contingency (*de dicto*): A proposition is contingent or a 'proposition (truth) of fact,' if and only if its negation is logically possible.[43]

Propositions of the former sort concern essences without any existential import and rest on the principle of contradiction, whereas propositions of the latter sort concern existences (except that of God) and rest on the principles of the best and sufficient reason.[44] In other words, propositions of reason express the absolute or inner logical structures of essences and their interrelations. Since all essences, as anchored in God's mind, have an eternal being, the propositions that express truths about essences hold by (*de dicto*) necessity even if those essences are not actually

[42] Shields (1986), 343–57 defends a similar interpretation and suggests that Leibniz is in fact trying to account for the contingency of the actual world by means of the *Daseinstreben* doctrine.

[43] See AG 217.

[44] See AG 19; *New Essays*, 433; G VI, 612. For a more extensive list of places where Leibniz correlates necessary with essential, and contingent with existential propositions, see Adams (1994), 45. Similarly, Curley (1972), 73 states: "it seems clear that necessary truths carry no commitment to the existence of anything, whereas contingent propositions are related in some specific way to existence and time."

instantiated. Propositions of fact, on the other hand, express the actual properties of things and actual relations between them, and thus, depend, for their truths, upon the actuality of whatever particular relations they express, which in turn, ultimately, rest on whether or not the latter are included in the best world-plan in God's mind.

However, Leibniz points out one sense in which propositions of fact can be construed as necessary, aside from the moral necessity of God's actualization of the best world. Namely, propositions of fact are necessary *ex hypothesi*.[45] This is necessity relative to the actual world and correlates with relative possibility (R2).

> Hypothetical or Relative Necessity: A proposition of fact is relatively necessary if and only if it is true in the actual world.

For instance, 'Judas sins' is absolutely contingent in itself, since its contrary, 'Judas does not sin,' does not imply a logical contradiction, but it is necessarily true given that Judas the sinner is included in the best possible world and that God actualizes it. This is how Judas is still free in his sinning, even though that he sins is certain. Similarly, all physical facts of this world are hypothetically necessary and thus are "certain" without being absolutely or logically necessary in themselves. It should be noted that given that all facts are relatively necessary, and that relative possibility (R2) is coextensive with actuality, it can be argued that in a specific sense necessity, actuality, and possibility are all coextensive. When defined as relative to the actual world, modal distinctions collapse. We will see this important feature of relative modality pointed out more explicitly by Wolff and Kant.

Leibniz's second formulation of *de dicto* modality rests on his '*in-esse*' theory of truth and defines an epistemic distinction between necessary and contingent propositions in terms of the *a priori* demonstrability of their truths:

> And with this secret the distinction between necessary and contingent truths is revealed... For in necessary propositions, when the analysis is continued indefinitely, it arrives at an equation that is an identity; this is what it is to demonstrate a truth with geometrical rigor. But in contingent propositions one continues the analysis to infinity through reasons for reasons, so that one never has a complete demonstration, though there is always, underneath, a reason for the truth, but the reason is understood completely only by God, who alone traverses the infinite series in one stroke of mind. (AG 28)

Therefore, even if in every true affirmative proposition the predicate is contained in the subject, this containment is demonstrable in a finite number of steps of conceptual analysis only in the case of necessary propositions.

> Necessity (*de dicto*)*: A proposition is necessary if and only if its truth is demonstrable *a priori* or reducible to an identity in a finite number of steps.

Here lies the connection between the two formulations of the *de dicto* conception of necessity. The reason the first formulation states that necessary propositions are true in virtue of the principle of contradiction is indeed the fact that they are reducible to identities whose negations are logical contradictions. Following Descartes, Leibniz

[45] AG 20–2, 45, 150.

has in mind here geometrical and arithmetical propositions that express existentially neutral truths about essences.

On the other hand, contingent propositions express existence-dependent facts and are true in virtue of the principles of the best and sufficient reason. The demonstration of the truth of 'Judas sins' requires not only an analysis of the complete notion of Judas, but also a comparative survey of all of the infinitely many possible worlds so as to determine whether the particular possible world that uniquely matches up with the possible individual, Judas, is the best of all possible worlds.[46] For the proposition 'Judas sins,' which implies 'Judas the sinner exists,' is true only on the condition that Judas is involved in the best possible world that God is morally inclined to actualize, and the 'best-ness' is not an absolute quality that can be contained in the complete notion of any possible world, but a quality a world can have only relative to all other possible worlds. Hence,

> Contingency (*de dicto*)*: A proposition is contingent if and only if its truth is not demonstrable *a priori* or not reducible to an identity in a finite number of steps.

Therefore, the second formulation of the distinction between *de dicto* necessity and contingency turns on the epistemic limits of human mind. For the impossibility of the demonstration of contingent truths by means of conceptual analysis applies to our finite minds that cannot maintain the required survey of all possible worlds.[47] Then, for God, who can run such an infinite survey 'in one stroke,' there will not be any distinction between necessary and contingent truths.

However, although the impossibility of demonstration does not apply to God, there is still a difference even for him between a necessary proposition such as 'A triangle has three sides' and a contingent proposition such as 'Judas sins' in virtue of the first formulation: the negation of the former is a logical contradiction, whereas that of the latter is not. Therefore, Leibniz's first formulation implies that he takes the conceptual containment in necessary propositions to be of a different nature than the one in contingent propositions. I will now turn to this difference in the framework of another serious interpretive question: whether existential propositions are exceptions to the '*in-esse* principle.' This is a question of crucial importance in determining whether Leibniz treats existence as a real predicate in his metaphysics.

EXISTENCE

The question of whether existential propositions are exceptions to the *in-esse*, and the correlated question of whether such an exception would be justified have led to an interpretive divide in Leibniz scholarship.[48] One major view, defended primarily by Russell and Curley, is that Leibniz is required to hold this exception in order to

[46] Curley (1972), 92–3 rightly points out that truths about unactualized possible individuals require also infinite analysis for demonstration, and such requirement cannot be a sufficient condition for contingency. But as I suggested above, truths about possibles that do not involve any existential commitment require the analysis of only one world, but existentially loaded propositions require the survey of all of the infinitely many possible worlds. Leibniz might just be thinking that the analysis of a single world could be finite, as some worlds could have meager content.

[47] For this worry, see also Curley (1972), 77.

[48] For a more detailed discussion of these questions see my (2013).

account for the contingency of the actual world. The opposing view, defended by Couturat and Adams, is that neither does Leibniz's text verify the claim of exception, nor does he need such an exception in order to solve the problem of contingency.[49]

I hold that while the Couturat–Adams view is the more tenable one, it requires a reinterpretation of the *in-esse* that neither Couturat nor Adams articulates. For the *in-esse* to apply to all propositions (including the existential ones), a strict concept–containment interpretation should be given up in favor of a metaphysically more moderate interpretation which states only that all true predication has some basis in the nature of things.

Leibniz's text offers different and even conflicting statements on what existence is. However, these statements can still be read as suggesting a developing account of existence, gradually moving away from Descartes' conception of existence as a perfection or simple quality and approximating Kant's rejection of it on the basis that existence is not a real predicate. Leibniz's developing views on existence can be classified under three distinct conceptions of existence: i) simple quality, ii) degree of reality, iii) extrinsic denomination.

As I mentioned in 2.1., some of Leibniz's early discussions of the ontological argument indicate his endorsement of Descartes' conception of existence, either explicitly, as in, for instance, the 1676 draft he shares with Spinoza, or implicitly, as in the *Discourse on Metaphysics*, §23. However, this cannot be Leibniz's considered view, since there are other places where he expresses discontent with construing existence as a simple, first-order predicate *on a par* with other such predicates that make up the essence or concept of a thing, and he avoids using this conception of existence as a premise in the final version of his ontological argument in the *Monadology*. Most notably, in a 1677 exchange with Eckhard regarding Descartes' ontological argument, Leibniz levels an explicit objection to the idea that existence is a perfection understood as a simple positive predicate.[50] Yet after he is convinced by Eckhard that perfection should instead be conceived as "a degree or quantity of reality or essence," Leibniz concludes, "It is plain that Existence is a perfection, or increases reality; that is, when *existent A* is conceived, more reality is conceived than when *possible A* is conceived" (AG 150).[51]

Leibniz thus replaces the conception of existence as an ordinary, first-order quality with that as a special quality increasing the overall degree of positive reality contained in an essence or concept as a whole, echoing Anselm's premise that what exists is greater than what is merely possible.[52] The conception of existence as a reality-increasing quality of essences implies that existence adds something novel to an essence that it does not have prior to its actualization. Soon after the conversation

[49] See, respectively, Russell (1937), Curley (1972), Couturat (1972), Adams (1994).

[50] A II, I, 313.

[51] See also Adams, (1994), 157.

[52] Henrich (1960), 51 is right in that while Leibniz does not count existence among perfections, he construes it as an indication of a perfect essence. I think, however, Henrich is wrong in suggesting that Leibniz's motivation here is to show that an unactual essence cannot be perfect. As I will show below, the ultimate implication of Leibniz's doctrine of the striving possibles is not that unactual essences are incomplete or imperfect, but that the world they belong to is less perfect than the one that actualized essences belong to. The conception of existence as a complement of essences or possibles, which Henrich seems to have in mind here, will emerge only with Wolff.

with Eckhard, Leibniz comes to feel uneasy with this implication: "If existence were something other than an essence's demand [exigentia], it would follow that it has some essence or superadds something new to things, about which it could be asked again whether this essence exists, and why this one rather than another" (G VII, 195). Leibniz's concern parallels the one that Suárez raises against the Thomistic and Scotistic interpretations of the 'real' distinction between essence and existence. If existence is conceived as adding to the very content of the essence, then a kind of essentiality should also be attributed to existence itself, which raises the question of whether this newly added bit of essence (existence) exists!

One point is worth noting here. Although Leibniz attributes a pre-existential being to essences as intentional entities located in God's eternal mind, his commitment to the doctrine of creation compels him to admit that existence is some kind of an addition *onto* essences through God's creative act. The talk of existence as a demand of possible essences, an *exigentia essentiae*, makes sense only if existence is something that essences do not already have prior to creation.[53] So what Leibniz rejects is the conception of existence as an addition *into* essences in the form of a first-order quality operating at the level of the real contents of essences. Leibniz strongly anticipates here Kant's thesis that 'existence is not a real predicate that could add to the concept of a thing' (A599/B627; *OPA*, Ak. 2:72).

Leibniz's anticipation of Kant on existence does not consist only in his rejection of the idea that existence adds content to essences, but also in his positive formulations of a new conception of existence. In an undated memorandum, he writes:

> [I]f we consider more accurately, [we shall see] that we conceive something more when we think that a thing A exists, than when we think that it is possible. Therefore it seems to be true that existence is a certain degree of reality; or certainly that it is some relation to degrees of reality. Existence is not a degree of reality, however; for of every degree of reality it is possible to understand the existence as well as the possibility. Existence will therefore be the superiority of the degrees of reality of one thing over the degrees of reality of an opposed thing. That is, that which is more perfect than all things mutually incompatible exists, and conversely what exists is more perfect than the non-existent, but it is not true that existence itself is a perfection, since it is only a certain comparative relation [comparatio] of perfections among themselves.[54]

Here Leibniz is in search of a conception of existence that would resolve a dilemma: existence is neither a perfection in its own right nor a reality-increasing quality and yet the degree of reality of the actual is still superior to that of the merely possible. It appears that Leibniz's *Daseinstreben* doctrine motivates this new conception of existence. For this doctrine implies a reversal of the scholastic conception of the relation between the existence and the perfection of an essence by construing existence as following from the degree of reality of an essence instead of adding to the reality contained in an essence.

[53] This seems to be the sense of addition Leibniz has in mind in his later statements such as those in his 1686 essay "General Inquiries": "an existent is an entity, that is, a possible, and something besides"; "[actual existence is] something superadded to the possibility or Essence." See Adams (1994), 167.

[54] Translated and cited by Adams (1994), 165.

This, however, is a very indirect kind of 'following' (except in the case of God). The possibles or essences lay claim on existence in proportion to the degree of reality or perfection they contain. Then, the degree of reality, which functions as an objective standard in the contest for actualization, is something the essences already have at the level of mere possibility in God's mind. The contest for actualization is not run between individual possibles but between possible worlds or distinct combinations of compossibles. Thus, the pre-existential perfection that offers an objective standard for God's choice is a quality not of individuals, but of whole worlds, namely, the degree of a world's success in comprising maximum reality with the simplest structure.

Furthermore, God's act of creation consists in a modal switch, making the best possible world actual, and does not increase the degree of perfection that the best world already has prior to actualization. Leibniz writes, for instance:

> God's decree consists solely in the resolution he forms, after having compared all possible worlds, to choose that one which is the best, and bring it into existence together with all that this world contains... it is plain to see that this decree changes nothing in the constitution of things: God leaves them just as they were in the state of mere possibility, that is, changing nothing either in their essence or nature, or even in their accidents, which are represented perfectly already in the idea of this possible world.[55]

So, not only is the actualization of the best possible world a reward by God for its pre-existential perfection, but also this reward does not make it a better world. So, strictly speaking, the actual A (a world or an individual) is not more perfect than the merely possible A. Since the individual possibilia are world-bound, capable of being involved in one and only one possible world, the decision to actualize a single essence is the same as that to actualize the entire world it belongs to. Thus, it is in fact meaningless and even impossible to compare an actual A with a merely possible A. For an A either belongs to the best world and consequently is actual, or not and is merely possible. Yet the actual (something) is more perfect than the merely possible (something else), in the sense that the former is included in the best possible world while the latter is not.[56]

The dilemma I posed earlier is thus resolved: The superiority of the 'perfection' of the actual over that of the merely possible means just that the former is included in the world that comprises maximum compossible reality with the simplest structure, but the latter is not. Accordingly, existence is neither a simple nor a reality-increasing quality of individuals, but it is an indication of superior degree of reality in the indirect sense of being included in the most perfect set of compossibles. Hence from the early 1680s onwards, we find in Leibniz such definitions of the "existent" as "the compossible with the most perfect," "that series of possibles which involves more of reality, and whatever enters into," "[that which] is compatible with more things than

[55] *Theodicy*, §52, See also, §183, 416.

[56] As Curley (1972), 87 says, "the difference between what exists and what does not exist is not like the difference between a singed painting and an unsigned painting, but like the difference between a good painting and a bad one. Things which differ in respect of existence cannot differ only in that respect, there must be some further difference on which that difference is based."

any other which is incompatible with it."[57] On this account, existence amounts to a higher-order quality, or better, as Curley dubs, a "consequential" or "supervenient" quality: the existence of an individual supervenes on the first-order qualities that make that individual belong to the most perfect of all possible worlds.[58] Existence is then not intrinsic even to the complete concept of an individual, which only reflects the world it is bound with, but, as requiring a comparative consideration with other possible worlds, is radically extrinsic to it.

One might ask whether Leibniz takes this account of existence to apply to God as well. Obviously, God's existence cannot be explained through his inclusion in the best possible world. Yet, the general principle that existence is an extrinsic denomination supervening on the intrinsic qualities of an individual still applies to God. The final version of his ontological argument in the *Monadology* (§§38–45) does not at all rely on the traditional contention that existence is contained in or an intrinsic quality of God's essence. As I showed earlier, Leibniz's argument is instead based on the idea that God's existence *follows* from his essence, i.e., from his intrinsic qualities or perfections, immediately and necessarily, and thus can be proven without any extrinsic and relational considerations regarding possible worlds.

We can finally turn to the question of whether Leibniz takes existential propositions to form an exception to the *in-esse* principle. First of all, as Couturat and Adams point out, there is no definitive textual evidence showing that Leibniz maintains such an exception to the *in-esse*.[59] However, Leibniz's mature conception of existence suggests that he ought to be thinking so. If existence is indeed a higher-order and extrinsic denomination that follows, indirectly, in the case of any creature, or directly, in the case of God, from other, first-order, intrinsic qualities contained in the essence, then the relationship between the essence and existence of a thing cannot be one of containment or inherence. On this account, essences are pre-existentially complete entities, and existence does not complement essences or add anything to them that they lack at the level of mere possibility. Therefore, the *in-esse*, understood strictly as the containment of predicate in the subject, cannot apply to existential propositions of the form 'S exists/is.' The special character of existence as just not the kind of predicate that could be contained in the concept of any subject seems to make it necessary for existential propositions to be an exception to the *in-esse*.

Yet this need not be the end of story. The conflict between the peculiarity of existence as a predicate and the containment of the predicate in the subject that is dictated by the *in-esse* principle can be alleviated by adopting a less literal interpretation of the *in-esse*. Leibniz's motivation in subscribing to the *in-esse* is to unite all truths, necessary and contingent, under a single principle. Moreover, some of his statements on truth in general such as "all true predication has some basis in the

[57] Respectively cited in Adams (1994), 165 (Gr 325) and in Curley (1972), 88 (C 360).

[58] Curley (1972), 89.

[59] There is one passage Curley (1972), 83–4 refers to as "indicating" such an exception: "all truths about contingents, i.e., about the existence of things, depend on the principle of perfection. All existences, the existence of God excepted alone, are contingent. But the reason why a contingent thing exists, rather than others, is not sought from its definition alone, but from comparison with other things" (Gr 288).

nature of things" (AG 45), and "it is common to every truth that one can give a reason for every nonidentical proposition" (AG 28–9) suggest that Leibniz sees the *in-esse* as a version of the PSR. That is, there is always a sufficient reason or truth-maker in reality that explains why a proposition is true rather than not. When Leibniz talks about the issue in this manner, he embraces a broader understanding of the *in-esse* in terms of "some connection between the subject and the predicate" (AG 28) rather than the containment of the latter in the former.

The *in-esse*, interpreted in these broader terms, applies to existential as well as any other true affirmative proposition. Existence, though not a predicate containable in the concept of anything, is an extrinsic predicate that follows, directly or indirectly, from the ordinary, first-order predicates contained in the complete concept of a thing. The existence of a thing, then, has always a ground in the nature or concept of that thing. In this sense, Couturat and Curley are right to say that even existence is not "purely" or "completely" extrinsic.[60] In the case of God, this internal ground is sufficient; in the case of non-divine beings, the internal ground is never sufficient but requires both the comparative perfection of the entire world the individual thing belongs to and God's free commitment to the principle of the best. There is at least one passage in the *New Essays* where Leibniz clearly construes this ground-consequence relationship between the concept and the existence of a thing as the 'connection' stated by the *in-esse* principle, interpreted broadly: "When it is said that something exists or possesses real existence, this existence itself is the predicate, i.e., the notion of existence is linked with the idea in question, and there is a connection between these two notions" (358). Thus, in this qualified sense, there is no exception for Leibniz to the *in-esse* principle of truth.[61]

2.2. Wolff: His Argument and Theory Of Modality

Christian Wolff's position on the ontological argument displays significant similarities to Leibniz's. Wolff shares Leibniz's criticism of the Cartesian argument that it is incomplete and offers another variant of the 'modal argument.'[62] However, while Leibniz develops an original argument that does not rely on Descartes' conception of existence as an intrinsic predicate of God, Wolff's completion of the Cartesian argument seems to still embrace this conception of existence, and consequently represent a backward move from the viewpoint of Kant's thesis that existence is not a real predicate of anything. However, Wolff's views on existence and modality in general cannot be judged solely on the basis of his ontological argument or rational

[60] Couturat (1972), 23; Curley (1972), 86.

[61] In my (2013), I argued that Leibniz's mature conception of existence as an extrinsic denomination that could not be contained in the concept of anything and yet is somehow connected to the subject-concept in a true affirmative proposition would commit him to Kant's view that existential propositions are synthetic (A 598/B626). For as I will discuss in chapter 7, the ground of Kant's claim is precisely the special status of existence as a predicate: it is 'only a logical predicate' that could occupy the predicate position in a proposition but not 'a real predicate' that could be contained in the subject-concept.

[62] It is evident from the letter exchange between Wolff and Leibniz that the former was inspired by the latter on the modal criticism of the classical Anselmian-Cartesian version of the ontological argument. See Bissinger (1975), 244.

theology. Wolff offers sophisticated accounts of modal notions both in his ontology or "general metaphysics" and rational cosmology. My aim in this section is to examine Wolff's theory of modality by paying attention to these different contexts of its presentation. Such an examination will prove crucial in fully understanding the nature of Kant's critique of the rationalist school metaphysics with regard to modality. In laying out Wolff's theory of modality, I will also try to establish Wolff's continuities and discontinuities with Leibniz, and consider whether the charge of modal logicism fairly applies to Wolff. I will bring up Baumgarten's accounts of existence and modality to the extent that they diverge from or misrepresent Wolff's views.

2.2.1. Wolff's ontological argument

Although Wolff defines natural theology as "the science of that which is possible through God" (NT I, §1) and thus understands its proper subject matter to be God's attributes and whatever can occur through them, he follows the scholastic tradition in taking the proof of God's existence as the first task of natural theology.[63] His *Theologiae Naturalis* offers, respectively, an a posteriori proof and an *a priori* proof of God's existence in its two volumes. While these two proofs work independently, they form a conceptual unity that is reminiscent of the way Leibniz sets up his ontological argument in the *Monadology*. The first one proves the existence of an *ens necessarium* through a cosmological argument (NT I, §24).[64] The second one proves, through an ontological argument, that the *ens perfectissimum* exists necessarily and thus is an *ens necessarium* (NT II, §§6–21). The same subject, God, is proved to exist both cosmologically, through its relation to the actual world, and ontologically, through its mere conceptual structure, and the two traditional conceptions of God, the necessary being and the most perfect being, are identified.

Here, I will focus on Wolff's second, ontological argument, but one point is worth mentioning regarding its connection with the first, cosmological argument. The conclusion of the latter is that a necessary being actually exists (NT I, §24). In demonstrating that God is in fact the *ens necessarium*, Wolff emphasizes that whatever deserves the title of a necessary being must be existentially independent (§25) or self-sufficient (§26) or contain the sufficient ground of its existence in its own essence or possibility (§§29–31). Such a being must therefore be an *ens a se*, a being through itself (§§32–3), which "exists, because it is possible" (§34). Wolff has therefore every conceptual tool to follow the Leibnizian route to prove God's existence through an equivalent of the modal conditional, 'God exists, if he is at all possible,' and yet, as I will show below, he does not. The statement '*ens a se* exists, because it is possible' plays no demonstrative role in either of Wolff's ontological and cosmological arguments.

Wolff's ontological argument in the second volume of *Theologiae Naturalis* perfectly fits my framework. It displays both the first logical step of introducing existence into the concept of God and the second step of inferring the existence of God from that existentially loaded concept, as well as the identification of the two

[63] See also Corr (1973), 108.

[64] The same proof appears also in DM, §928.

conceptions of God, that of the *ens perfectissimum* and that of the *ens necessarium*.[65] Wolff starts the argument with a nominal definition of the most perfect being:

(1) "The most perfect being (*ens perfectissimum*) is that in which all compossible realities are involved *(insunt)* in the absolutely highest degree." (NT II, §6)

By 'reality' (*realitas*), Wolff means the kind of predicate that goes into the true objective content of a subject, "all that which is understood to truly inhere in a thing (*inesse*), rather than that which seems to do so due to our confused perceptions" (§5). And 'compossibles' are those realities "which can inhere in (*inesse*) the same subject together" (§1). This is Wolff's first step toward establishing the absolute or inner possibility of the *ens perfectissimum*, since he holds that

(2a) Something is possible, considered absolutely or in itself, iff the realities that inhere in it (i.e., internal predicates) are compossible.

Wolff initially states that a set of realities are compossible when they are possible in themselves and are not in conflict or "repugnance" with one another (§2), then cashes out "repugnance" in terms only of a logical sort of incompatibility, namely logical contradiction (§12). Therefore, like Leibniz, Wolff does not seem to offer a formal constraint on absolute or internal possibility beyond logical consistency.

(2b) Something is (absolutely or internally) possible iff its predicates are logically consistent with one another.

The question now is whether the predicates of the *ens perfectissimum* are logically consistent. (1) suggests that Wolff construes realities as coming in degrees. Accordingly, the realities contained in the *ens perfectissimum* are in their absolutely highest degrees (*in gradu absolute summo*), that is, they are perfections, excluding any defect, limitation or negation (§11). Since logical contradiction occurs only when "a predicate is affirmed and negated of the same subject simultaneously" (§12), the lack of negation in these realities entails that there cannot be any logical contradiction between them:

(2c) All perfections (realities in their maximal degrees) are logically consistent or compossible.

It follows that a subject that contains only such logically consistent predicates is absolutely or internally possible.

(2) The *ens perfectissimum* is (absolutely) possible. (§13)

Subsequently, Wolff states the identity between the notions of God and the most perfect being:

[65] Wolff's ontological argument has not attracted much attention in the literature. Henrich (1960), 55–62 gives an account of the argument from the perspective of his reading of the history of ontotheology; Corr (1973) also presents a brief account of it. The most thorough reconstruction of the argument is offered by Bissinger (1970), 259–67 and (1975), 243–7.

(3) God is the *ens perfectissimum* (§14).

Wolff does not offer a separate justification for this identification, but only notes that "this is a nominal definition of God just like the one used in (§67) of the first volume" (§14). It is thus clear that he takes the various natural theological conceptions such as "*ens necessarium*," "*ens a se*," "*ens perfectissimum*" to be all defining different aspects of God. This identification enables him to apply whatever can be said of the *ens perfectissimum* to God: God contains (*continet*) all compossible realities in the highest degree (§15), and God is (absolutely) possible (§19).

The rest of Wolff's argument, attempting to prove the actual existence of what has been proven to be absolutely possible, drastically diverges from Leibniz's modal argument. For Wolff adopts the classical version's straightforward first logical step of introducing existence into the concept of God as a predicate. In (§20), he first puts forth the Cartesian premise:

(4) Existence (necessary or contingent) is a reality.

Wolff defines reality as whatever truly inheres in a thing (§5). While this definition implies that realities are those kinds of predicates that can be contained in the concept of a thing, in reaching (4) Wolff moves from a broader conception of reality as anything that is truly (as opposed to mistakenly) predicable of a real subject. Accordingly, (4) means only that there really exists something as opposed to a mere illusion. In order to justify that both necessary and contingent existence are realities, Wolff reminds us of his cosmological proof in the first volume (§24), which starts from the existence of contingent beings such as the human soul and the world, and concludes that a necessary being exists.

Now assuming that existence, as a reality, is subject to gradation, Wolff suggests that "one cannot conceive of a higher degree of existence than necessary existence" (§20). For existence is either contingent or necessary, and since contingent existence is only hypothetically necessary, it must be inferior in degree to necessary existence. Necessary existence is the highest degree of existence, and thus, a reality of the highest degree.

(5) Necessary existence is a reality of the absolutely highest degree.

Wolff also holds that existence is compossible with every other reality that inheres in a thing. For existence contradicts only non-existence, which is not a reality as such. It then follows that God, the *ens perfectissimum* that contains all compossible realities to the highest degree, contains existence in the highest degree, or, what is, in virtue of (5), the same thing:

(6) God contains necessary existence.

Wolff takes the second logical step of inferring the actual existence of God from the existentially loaded concept of God somewhat silently. For although (6) is the only obvious logical conclusion of Wolff's argument as I reconstructed, he apparently takes (6) to entail:

(7) God necessarily exists (§21).[66]

Wolff sums up his entire proof in the following way:

God contains (*continet*) all compossible realities in the absolutely highest degree (§15). He is, however, possible (§19). On account of the fact that the possible can exist (*Ontologia*, §133), existence can inhere (*inesse*) in it; consequently, since existence is a reality (§20), and since realities are compossibles, which could inhere in one and the same thing (§1), it [existence] is counted among the compossible realities. Furthermore, necessary existence is the absolutely highest degree (§20). God, therefore, contains (*competit*) necessary existence, or, what is the same, God necessarily exists. (§21)

That God is an *ens necessarium* (§22) is, of course, also the conclusion of the cosmological argument. Thus, by connecting the two conceptions of God, *ens perfectissimum* and *ens necessarium*, Wolff's ontological argument connects itself to the cosmological argument and confirms the latter's conclusion through merely conceptual means.

This is now a good place to open up a brief parenthesis for Baumgarten's version of the ontological argument, which, some claim, is the version that directly bases Kant's critique of the argument.[67] The only notable deviation of Baumgarten's version from Wolff's is that it does not employ an equivalent of (5), which operates on the assumption that existence admits degrees, and concludes only that God "has existence" (*Metaphysica*, §810) or is actual (§811).[68] It seems that Baumgarten wants to put forth a strictly self-contained ontological argument that works solely through the analysis of the *ens perfectissimum* without the need to appeal to the cosmological argument and the existence of an *ens necessarium*.[69] Baumgarten too states that God exists necessarily, but only as resulting from a separate reflection upon the

[66] In order to make the transition from the *conceptual* (6) to the *objectual* level (7), Wolff must have in mind something like the following principle of predication: 'if existence is contained in (the concept of) a thing as a predicate, then that thing actually exists.'

[67] Henrich (1960), 63; Hartshorne (1965), 208. I will argue later on that there is no reason to believe that Kant only has in mind Baumgarten's version when he criticizes ontotheology.

[68] Stang (2016), 56–65 offers an original reconstruction of Baumgarten's ontological argument. On Stang's reconstruction, Baumgarten's construal of existence as 'complete determinacy' leads him to the conclusion that *ens perfectissimum*, whose essence warrants its complete determinacy, cannot be merely possible must be also actual. There are, however, two issues here. First, Baumgarten's official ontological argument in the *Metaphysica* (§§803–11) makes no use of the conception of existence as complete determinacy, but, just like Wolff's version, utilizes the conception of existence as reality (§810). As I will explain below, Baumgarten refers to complete determinacy in his follow-up proof that God is also a necessary being (§§823–5). However, this latter proof is based on the idea that an incompletely determined being cannot be perfect, and for this proof to work, existence must be conceived as a perfection or reality with respect to which God must be determined. Obviously, this would not bother Stang, who explicitly dismisses this official version as an "uninteresting version of the Leibnizian argument" (2016, 57) in favor of the reconstruction he builds around the doctrine of complete determinacy Baumgarten conveys in the earlier sections of the *Metaphysica*. The second and the more substantive issue with Stang's reconstruction is that the doctrine of complete determinacy originally belongs to Wolff, and as I will argue later on in this section, Baumgarten's theory of existence, which Stang takes to be "more developed than Wolff" (2016, 57), involves a major misunderstanding of Wolff's original doctrines of existence, predication and complete determination.

[69] See Henrich (1960), 64–5: "Baumgarten will den Sinn des Begriffs 'notwendige Existenz' durch das ontologische Argument selbst bestimmen."

implications of the ontological argument proper. He suggests that a non-actual God would be a thing enjoying all realities but one, or a thing determined with respect to all inner perfections but one (§823). But then a non-actual God would not be perfect and thus would not be God, or in other words, 'a non-actual God' is a contradiction in terms. "Therefore, the opposite of divine existence is impossible in itself. God's existence is absolutely necessary. God is the necessary being." (§823) Baumgarten thus takes the absolute necessity of God's existence to be logically following from the assumption that existence *simpliciter* is already contained in the concept of God as an essential predicate. He even goes on to say that "If God were not actual, then the principle of contradiction... would be false" (§824). This approach to divine necessity differs from Wolff's direct inclusion of necessary existence into the concept of God as a distinct essential predicate. While Wolff directly concludes the *de re* necessity of God, Baumgarten either unwarrantedly infers the *de re* necessity of God from the *de dicto* necessity of the proposition 'God exists' or conflates the two.

Closing the parenthesis on Baumgarten, I would like to underline once again that the conditional 'God exists, if He is at all possible,' which constitutes the crux of Leibniz's version, does not play any demonstrative role in Wolff's and Baumgarten's ontological arguments. A logical equivalent of this conditional appears in Wolff only after his ontological argument proper. Paralleling the reflections following his cosmological argument, Wolff moves from the conclusion that God is an *ens necessarium*, and further characterizes him as existing through his own power (NT II, §23), as an *ens a se* (§24), as an *ens independens* (§25), and as having the ground of his existence in his essence (§26), or existing through his essence (§27). The conditional comes up as a modal formulation of the ontological self-sufficiency of God: "God exits, because he is possible" (§28).[70] This statement shows that Wolff takes God's possibility as the sufficient ground of his existence and is logically translatable to Leibniz's conditional 'God exists necessarily, if he is at all possible.' However, Wolff's consistent use of 'because' (*ideo... quia*) as the logical operator in both instances of this statement in his cosmological and ontological arguments (NTI, §34 and NT II, §28), and the fact that in both cases he puts it forth after he has proven God's existence indicates that he intends this statement to express a modal feature of God, but not a ground of the proof that such a being exists.[71] The same is also true of Baumgarten. He makes the statement, "From God's possibility, it is valid to draw the conclusion that he exists" (*Metaphysica*, §820), only after he purports to have proven God's actuality, and again only in the context of explaining God's essential characteristics. Therefore, Wolff and Baumgarten do not in fact follow the Leibnizian strategy of turning the modal conditional into the first premise of a *modus ponens*, but they still offer 'modal' arguments in that they provide a sub-proof of the possibility of God with the intention of completing the classical version.

[70] "Deus ideo existit, quia possibilis. Est enim ens a se (§24). Se dens a se ideo existit, quia possibile (NT I, §34). Ergo Deus ideo existit, quia possibilis."

[71] Henrich (1960), 59 gives the impression that he understands Wolff's ontological argument to be based on this conditional, when he sums up Wolff's strategic reasoning as "Dem vollkommensten Wesen, von dem festeht, daß es möglich ist, kommt die notwendige Existenz zu."

Now, our examination of Wolff's ontological argument shows that he follows Leibniz in conceiving the absolute possibility of a thing in terms of the logical consistency of its (internal) predicates. Moreover, the crucial premise (4), 'Existence is a reality,' appears to employ the Cartesian conception of existence as a reality. It is, however, not obvious whether this premise really commits Wolff to a clear position on the nature of existence as a predicate. For, as I noted earlier, the argument in fact works with a broad epistemological conception of reality, emphasizing the truth in reality as opposed to the falsity in illusion, according to which (4) just means that there really exists something (rather than mere illusion or nothing). Nonetheless, Wolff's narrower definition of reality as what inheres (*inesse*) in a thing (NT II, §5) has strong metaphysical implications, as illustrated by his further statements that a reality is what is involved within (*insunt*) the most perfect being (§6), or what God contains (*continet*) (§15, §21) to the highest degree. The technical language here (*inesse, insunt, continet*) suggests that realities are *intrinsic* to the essences of things, and thus, that Wolff also embraces the scholastic conception of reality as '*res*' or '*quidditas*,' as the real content of the essence or concept of a thing.[72] This metaphysical interpretation of reality would, of course, make Wolff's premise (4) a fair target of Kant's critique that existence is not a predicate that can be contained in the concept of anything.

2.2.2. *Existence as* complementum possibilitatis

While Kant's critique may still apply to Wolff's treatment of existence in his ontological argument, the latter is too restricted a context to reveal Wolff's theory of existence in general.[73] In the specific context of natural theology, Wolff's statements about existence mostly concern the exceptional kind of existence that only God enjoys. Wolff's ontological argument alone does not license us to conclude more than that he treats God's existence as a predicate contained in the essence or concept of God. A more comprehensive understanding of Wolff's conception of existence and modality in general requires us to look at other parts of his metaphysics.

To that end, it is imperative to discuss an oft-cited but not so well understood dictum that Wolff presents in his ontology: "existence is the complement of possibility (*complementum possibilitatis, Erfüllung der Möglichkeit*)" (*DM* §14; *Ontologia* §174).[74] The prevalent interpretation in the contemporary literature holds this dictum to suggest that existence is a predicate or determination that adds a final determination to the merely possible, making it completely determinate, and, thus actual.[75] In fact, Kant seems to adopt this misinterpretation as well, and formulates both of his negative theses about existence, 'existence is not a (real) predicate' and 'the actual contains nothing more than the merely possible,' in response to what he takes this dictum to mean.

[72] See also *Ontologia*, §243, where Wolff defines reality as whatever can be conceived as being, "quicquid est vel esse posse concipitur."

[73] Kant's charge obviously applies to Baumgarten's argument too, and indeed even more so, since Baumgarten simply defines realities as "positive determinations" (*Metaphysica*, §807), and states that "existence is a reality" (§810).

[74] For a detailed discussion of Wolff's dictum, see my (2018).

[75] See note 99.

I have two aims in this final section of the chapter. First, I will demonstrate that the common interpretation of the dictum that is also shared by Kant is both textually and philosophically untenable. Second, I will offer an alternative interpretation of the dictum from the viewpoint of Wolff's cosmology. On this interpretation, the dictum indicates that existence does not add a particular determination to the merely (absolutely or logically) possible, but it does add to the latter an entire world by locating it in the causal-spatio-temporal nexus of all things in the world of human experience. This interpretation, aside from being textually more accurate, has two important virtues. First, it connects existence to Wolff's account of real possibility, which, I will demonstrate, turns around a notion of relative or extrinsic possibility akin to Leibniz's R2. Second, it reveals that Wolff's dictum is not only compatible with but anticipates Kant's conception of existence in significant ways.

The notion of possibility has primacy in Wolff's general metaphysics or ontology. Not only does Wolff define philosophy in general as "*scientia possibilium*," but he also understands ontology to be an existentially neutral study of being, running at the level of mere possibility.[76] For a thing ("*ens*" or "*ding*") must be first considered in itself, and in isolation from the question of its existence and relation to other things. Hence, Wolff defines a thing through its mere possibility: "Everything that can exist, may it be actual or not, we call a thing (ding)" (*DM*, §16).[77] This definition does not express a simple synonymy between being possible and being a thing.[78] Instead it implies that possibility is at once the necessary and sufficient condition for being a thing, and this, being a thing, is the most general sense of being. The impossible, then, amounts to nothing or a non-being (*nihilum*).[79]

The notion of possibility in question here is internal or absolute possibility, which Wolff defines through the principle of contradiction: "possibility is that which contains no contradiction" (*Ontologia*, §79).[80] Yet it does not follow that Wolff reduces absolute possibility to mere logical consistency. For as opposed to nothing "to which no concept corresponds" (*Ontologia*, §57), Wolff characterizes something (or absolutely possible) as that "to which some concept corresponds" (*Ontologia*, §59). So absolute possibility requires a material content before any formal constraint can even be applied. Baumgarten also emphasizes the necessity of material content for absolute possibility. Identifying the impossible or nothing with what cannot be represented, he states: "That which is not nothing (*nonnihil*) is SOMETHING: the representible, what does not involve a contradiction, whatever is not both A and not-A, is possible" (*Metaphysica*, §8). So both Wolff and Baumgarten are aware that logical consistency is only a formal condition of absolute possibility, which presupposes a material condition in the first place, though neither goes as far as committing to Leibniz's actualist principle that the material content of absolute possibility must be grounded in actuality.

Now, based on this conception of absolute possibility, Wolff runs the following reasoning about existence:

[76] See *Logica*, §29. Both Gilson (1963), 175 and Pichler (1910), 10 point out this distinctive character of Wolffian ontology, and describe it, respectively, as "essentialist" and "daseinsfreie."

[77] See also *Ontologia*, §§134, 135.

[78] See Bissinger (1970), 148.

[79] See *Ontologia*, §§101, 136.

[80] See also *Ontologia*, §§ 85, 100, *DM* §12.

(1) While everything actual is also (absolutely) possible, not everything that is (absolutely) possible is actual.[81]
(2) Then, (absolute) possibility is not the sufficient ground of actual existence. The actual existence of a thing requires something more than (absolute) possibility.[82]
(3) The (absolutely) possible must be complemented by something more from outside in order to exist.[83]
(4) What we call existence is this complement of (absolute) possibility (*complementum possibilitatis, Erfüllung der Möglichkeit*).[84]

However, this, Wolff admits, is only a nominal definition of existence.[85] One needs to know what exactly existence adds to mere absolute possibility. Only then can one ascertain whether Wolff's dictum entails that existence is a predicate and that the actual contains more than the possible.

Wolff conceives of a being in general as existentially neutral. Yet in order to understand how he conceives of existence in relation to the existentially neutral nature of being, it is key to look at two important doctrines of his ontology: first, the doctrine of the tripartite inner structure of a being (*ens, Ding*), and, second, the doctrine of complete determination. For these two doctrines appear to be the basis of the popular misinterpretation that Wolff's dictum entails that existence is a predicate.[86]

According to Wolff, in the metaphysical inner structure of a being one finds three layers of predicates or internal determinations. The predicates in the innermost layer, 'essentials' (*essentialia*), constitute the essence (*essentia, Wesen*) of a thing. The essentials are fundamental or primitive predicates that cannot be derived from one another or any other predicates but are logically consistent with one another.[87] Since "a thing is possible through its essence" (*Ontologia*, §153), and nothing can be without an essence, the essentials are the necessary determinations of a thing.[88] 'Essence' has a logical as well as a metaphysical significance here: "Essence is what is first conceived in a thing" (*Ontologia*, §144).[89] The essentials are, then, those predicates through which we define a thing and compose its concept.[90] For instance, the essentials of a triangle are 'being a closed figure' and 'having three straight sides.'

The predicates in the second layer, 'attributes' (*attributa, Eigenschaften*), also belong to a thing by necessity, but are not as primitive as the essentials, for they

[81] See *Ontologia*, §§170–1, *DM* §15. [82] See *Ontologia*, §§171, 173; *DM* §14.
[83] See *DM*, §14. [84] *DM*, §14, *Ontologia* §174. [85] See *Ontologia*, §174.
[86] The contemporary literature on Wolff's conception of existence is fairly limited. I find three studies particularly illuminating, even if my own interpretation diverges from them. Michaelis (1937) is, as far as I know, the only work that is entirely devoted to the subject. Bissinger (1970) examines Wolff's theory of modality in the context of his arguments for the existence of God, and as such, presents the most ontotheologically oriented study. Honnefelder (1990) discusses Wolff's ontology in a broader historical context.
[87] See *Ontologia*, §143. [88] See also *Ontologia*, §144. [89] See also *DM*, §34.
[90] As Bissinger (1970), 155 puts it, being and concept are not just externally analogous here, but there is an internal connection (*inneren Zusammenhang*) between the two, so the model Wolff offers for the systematic analysis of the inner structure of being well applies to the analysis of concepts too.

are determined by the latter.[91] Thus, the attributes cannot contradict one another or any of the essentials. If the essentials are those predicates that sufficiently define the essence of a thing, the attributes are necessary extensions of that essence. The attributes of a triangle are its features that necessarily follow from being a closed figure with three straight sides, such as having a sum of 180 degrees.

Thus, the essence comes with two necessary layers of predicates, essentials and attributes, the first of which determines the second. Wolff defines yet a third layer of internal determinations: 'modes' (*modi*). The modes cannot contradict the essentials, but are not determined by them either.[92] So the modes are rather contingent properties (*accidens praedicabile*), which can but do not necessarily belong to a thing.[93] It follows that only contingent beings that can be determined in diverse ways admit modes. The modes of a triangle would be all of its contingent qualities, such as the particular lengths of its sides.

What ties this account of the internal predicates of a thing to the question of existence is Wolff's other doctrine: "whatever exists or actual is completely determined" (*omnimode determinatum*). (*Ontologia*, §226) 'Determination' refers to the affirming or denying a predicate of a subject.[94] Here Wolff has in mind internal predicates, and claims, accordingly, that whatever exists is determined with respect to all of its possible internal predicates in all three layers. So Wolff presents this doctrine as designating a principle of individuation, and seems to be motivated by the insight that only individual things exist.

The essentials and attributes do not exhaustively determine a thing, and without the determination of the modes, the thing in question will remain incompletely determined. Such an indeterminate thing is an '*ens universale*,' which represents a mere (absolute) possibility and, Wolff insists, cannot exist as such.[95] What actually exists, for Wolff, is always an '*ens singulare*,' a singular thing that is completely determined with respect to all the possible predicates (including modes) it can have. It is important to note here that unlike Leibniz, who explicitly embraces the idea that all possibilia are completely determined in God's infinite intellect, Wolff avoids this talk of completely determined individual possibilities.[96] Yet Wolff does not present complete determinacy as a definitive criterion of existence. The complete determinacy of a thing is, of course, a necessary condition of its existence, since only thus determined singular beings can actually exist, but it does not follow from the latter that complete determinacy is a sufficient condition of existence. Wolff's doctrine of complete determination, by itself, does not exclude Leibnizian mere possibilia. There is therefore a unidirectional necessary connection between the complete determination of the modes of a thing and its actual existence, and if existence is the 'complement' of the absolute possibility or essence of a thing, it must be so in virtue of this connection. Yet even with these two doctrines at hand, it remains unclear how exactly, according to Wolff, existence is related to the complete determination of the modes.[97]

[91] See *Ontologia*, §§145, 146; *DM*, §44. [92] See *Ontologia*, §147. [93] See *Ontologia*, §148.

[94] See *Metaphysica*, §29–31. [95] See *Ontologia*, §235.

[96] On this very point, see Van Peursen (1987), 79–80.

[97] Bissinger (1970), 158 raises the right specific question here: "Ist nun Existenz selbst 'Teil' dieser durchgängig Bestimmung, oder ist sie durch diese *eo ipso* gegeben?"

The prevalent interpretation in the literature is that Wolff construes existence as a mode in its own right, and specifically, as the 'final' or 'ultimate' mode in the determination of a thing through the addition of which the merely possible and incompletely determinate turns into an actual and completely determinate being. To give a few examples, Michealis states that existence is "the final determination" which brings about "the evolution (*Entwicklung*) of the possible toward the actual"; Gilson writes "existence can only be one of [a being's] modes... [existence] add itself to [the possible being] as an ultimate determination"; Blackwell writes, "existence is understood by Wolff to be the final complement in the order of possibility"; Hettche writes "Existence is the completing determination or predicate of a possible thing that brings it into actual reality."[98]

This reading seems to be shared by Kant as well. Kant does not explicitly criticize Wolff on existence, except for charging his dictum of being "very indeterminate" (*OPA*, Ak. 2:76). However, as I will argue in chapters 3 and 8, when he formulates his negative theses, "Existence is not a predicate or determination of a thing" (Ak. 2:72) and "The actual contains nothing more than the merely possible" (A599/B627), Kant intends to undermine the conception of existence as the complement of possibility, construed as adding a specific determination to the mere possibility of a thing. Again, as I will elaborate in chapter 7, Kant's critique of the contention that the extension of possibility is greater than that of actuality, which, he claims, is based on the assumption that "something must be added to the former to constitute the latter" (A231/B284), targets exactly Wolff's aforementioned reasoning leading up to the dictum.

However, the popular interpretation of Wolff's dictum is misguided. First of all, there is no compelling textual ground for this interpretation. In the one place Wolff explicitly states that "the existence of a contingent is nothing but its mode" (*Ontologia*, §316), he means that existence is a mode in the case of a contingent being in the rather broad sense that it does not belong to the necessary composition (the first two layers) of its essence.[99] However, it does not follow from this statement that existence is a particular mode that further determines this being. One might insist that Wolff treats existence as an essential determination of God in his ontological argument. However, as I suggested earlier, even if it were accepted that the argument commits Wolff to the view that (necessary) existence is a determination of God, this would still not tell us whether he commits to a general view of existence as a predicate of actual things (divine or non-divine). Moreover, given that the punchline of the ontological argument is God's exceptional ontological perfection, which makes any 'complement' to his essence superfluous for his existence, it is obvious that the dictum does not apply to God. Therefore, Wolff's ontological argument cannot be regarded as a textual ground for the common interpretation of the dictum.

One likely historical source of this misinterpretation is Baumgarten's *Metaphysics*. Baumgarten formulates the doctrine of complete determination in the following way:

[98] See, respectively, Michaelis (1937), 50, 54, 55; Gilson (1963), 175; Blackwell (1963), ix; Hettche (2014). For other examples of the broader view that for Wolff existence is a mode, see Corr (1975), 255; Bissinger (1970), 179; Burns (1966), 28.

[99] For this general sense of a mode as "accidens praedicabile," *Ontologia*, §148.

"Aside from essence, something possible is either determined with regard to its affections that are also compossible in it, or not. The former is an actual being, while the latter is called a privative (merely possible) non-being" (*Metaphysica* §54). By "affections" Baumgarten means the non-essential internal determinations of a thing, namely its attributes and modes. This, however, suggests a significant deviation from Wolff. For as opposed to Wolff's statement of this doctrine, which, at most, suggests that complete determinacy is a necessary condition of existence and does not exclude completely determined mere possibilia, Baumgarten's statement clearly offers complete determinacy as a sufficient condition of existence and strictly identifies mere possibility with incomplete determinacy.[100]

Right after his interpretation of complete determination, Baumgarten presents his version of Wolff's dictum: "Existence (actuality) is the collection (*complexus*) of affections that are compossible in something; i.e. the complement of essence or of internal possibility" (§55).[101] Since Baumgarten, like Wolff, holds that the attributes are determined by the essence but modes are not, one would expect him to also hold that those determinations that complement the essence or internal possibility of something must be only its modes as opposed to the collection of its modes and attributes.[102] However, later on, in his discussion of contingency, Baumgarten drops the talk of "collection," and identifies existence as a mode: "the existence of a contingent being... is thus neither an essential quality nor attribute, although it is still an internal determination (§55) and therefore a mode (§52)" (§134). This is evidently not just a slip, because Baumgarten reiterates the same conclusion in his account of finitude: "a finite being is internally alterable (§248, 126) and hence it is not a necessary being (§132); its existence is a mode (§134) and the being itself is contingent (§109)" (§257).[103] In both contexts, Baumgarten runs the same argument from elimination. In the case of a necessary being (i.e., God), there are no modes or accidental predicates that can be altered and existence (as well as any other internal determination) must therefore be an essential, necessary determination.[104] On the other hand, in the case of a contingent being, existence, as alterable and accidental, can be neither an essential determination, nor an attribute,

[100] See also *Metaphysica*, §152: "Singular beings are internally entirely determined, and hence are actual."

[101] See also *Metaphysica*, §54.

[102] See *Metaphysica*, §§64, §65.

[103] See also *Metaphysica*, §66, where Baumgarten defines existence as a "reality," which makes it "a positive determination."

[104] See, respectively §§108, 110, 111, and §§259, 260. See also §825. The alternative ontological argument that Stang builds upon Baumgarten's interpretation of the doctrine of complete determination turns around the idea that God's essence entails its complete determinacy, which, since complete determinacy is existence, means that God's essence entails God's existence. Stang (2016), 62 suggests that this argument may avoid being the target of Kant's first negative thesis, since it takes existence to be a "higher-order" property of objects (having a fully determinate set of accidents), and thus does not have to take "existence to be one perfection among others that can partly constitute an essence" like, for instance, Descartes' version does. However, Baumgarten's actual analysis of a necessary being in these passages suggests that he takes God's essence to be not just entailing his existence via complete determinacy as a higher-order property but to be directly containing existence as a distinct essential determination, that is, as a first-order property. Accordingly, God's essence entails complete determinacy, because God, as a necessary being, does not have any modes to be determined and has all of its predicates as its essential determinations.

and thus, must be a mode.[105] It should be noted, however, that Baumgarten goes beyond the modest position that I earlier inferred from a similar statement of Wolff's, namely that the existence of a contingent being is a 'mode' in the sense of not being a necessary determination of the essence of such a thing. For Baumgarten's reasoning relies on the further assumption that existence is to be conceived within the order of a thing's internal structure, and thus as one of its internal determinations. He states that "Every determination of a possible thing is either an essential determination (§39), or an attribute, or a mode (§42), or a relation (§37)" (*Metaphysica*, §52), and makes clear that a relation is an external determination of a thing, considered in a nexus or connection with other things.[106] Therefore, on Baumgarten's account, that existence is a mode means that existence is an internal determination in its own right. This is exactly where the common misunderstanding of Wolff's theory of existence lies. Such misunderstanding may also be why Baumgarten identifies complete determinacy as a sufficient condition of existence. For if existence itself is an internal determination, what is completely determined (with respect to all its internal determinations) must also be determined with respect to its existence.[107]

I suggest that an accurate interpretation of Wolff's dictum that existence is the complement of (absolute) possibility and his construal of the relationship between existence and complete determination requires us to go beyond his ontology, which only aims at an existentially neutral analysis of 'being' considered in itself, and look at his cosmology, which examines contingent beings in the context of their spatio-temporal-causal relations. For it is specifically this relational context where the (absolute) possibility or essence of a being can get extrinsically complemented and thus where Wolff's dictum applies *in concreto*.

Now, as opposed to God, the necessary being, whose essence is completely determined in itself without the addition of modes or any contingent determinations and sufficiently grounds his own existence, a contingent being needs the complete determination of its modes, which are left undetermined by its essence. The actual existence of a contingent being, therefore, cannot be accounted for through an absolute consideration, with respect only to its internal determinations and in isolation from other things. The essence of such a being provides only the 'intrinsic logical ground' (*ratio intrinseca*) of the range of modes it can have, (e.g., a triangle can be large or small, right or equilateral, but cannot be circular); but its actual modes require 'extrinsic grounds' (*ratio extrinseca*) for their determination. In the *Cosmologia*, Wolff underscores this existential dependence of contingent beings upon extrinsic grounds in modal terms, by introducing a distinction between "intrinsic possibility" (*posibilitatis intrinseca*) and "extrinsic possibility" (*possibilitatis extrinseca*) (§111).[108] The notion of possibility we have talked about so far as one that is to be

[105] Honnefelder (1990), 369–70 sees this idea that existence applies to contingent beings only as a contingent predicate as the fallacious ground of the claim that existence should count as a mode in Wolff's system.

[106] See §37. [107] See Kannisto (2016) on this point.

[108] For a helpful explanation of the distinction between these two kinds of possibility in Wolff, see Camposampiero (2012).

complemented (with the determination of modes) is that of intrinsic possibility. In other terms, this is the absolute possibility of a thing, 'considered only in itself,' and is constituted by the logical consistency of its internal determinations, which is sufficiently grounded in the essence alone, as no attribute or mode can contradict the essentials that are compossible with one another.[109] The extrinsic possibility of a thing is its possibility considered in relation to other things, and depends on the availability or actuality of a determinate cause, through which the modes are determined and what is intrinsically possible becomes also actual.[110] The context of such externally determining relations between things is the actual world, which Wolff defines as a nexus of causal as well as spatial (i.e., being next to) and temporal (i.e., simultaneity and succession) relations.[111]

Extrinsic possibility is, therefore, the Wolffian notion of real possibility, and is a version of Leibniz's relative possibility (R2), fortified by an account of direct causal interaction between substances.[112] Wolff refers to the actual world also as the "visible world" (*mundo aspectabili*), the world of our perceptual experience, and describes extrinsic possibilities as "the possibilities of this world" (*possibilia huius mundi*) as opposed to the possibilities in the unrestricted logical space of all absolutely possible worlds (§111). Therefore, existence should be conceived as the groundedness or connectedness of a thing in the thoroughgoing causal, spatio-temporal nexus of all things in the world of our experience. Hence, "What becomes actual (*würklich*)," Wolff writes, "is what is grounded in the nexus (*Zusammenhang*) of things which constitutes the content of the present (*gegenwaertige*) world" (*DM* §572).[113]

So on Wolff's account, existence does indeed complement the intrinsic or absolute possibility of the thing. However, this addition is not simply made in the order of intrinsic possibility, in terms of a specific first-order, internal predicate or mode, in abstraction from the relations with other things. So, what exactly does existence add to intrinsic possibility? This addition can be described in various ways.

(i) Existence adds to the intrinsic possibility of the thing its extrinsic, relative, or real possibility, which is grounded in an actual external cause.
(ii) The essence is complemented by the modes. For the modes of a thing are only determined by extrinsic grounds. We can now pick out the triangle as an individual triangular object with a determinate magnitude, location in space and time, causal connections etc.
(iii) Since this world is 'the visible world,' a connection with our perception is added to the essence or intrinsic possibility of the thing.
(iv) Most importantly, actual existence adds an entire world to the essence of the thing! For the complete determination of the modes of a thing comes with its (extrinsic) determination in relation to the entire content of the actual world.

[109] See *Cosmologia*, §111. [110] *Cosmologia*, §111.
[111] See esp. *Cosmologia*, §§52–8; *DM*, §§546–7.
[112] For a detailed discussion of how Wolff diverges from Leibniz on 'physical influx,' see Watkins (1998).
[113] A similar cosmological understanding of existence can be found in Crusius, who defines existence as "that in virtue of which a thing can be encountered somewhere and somewhen outside the thought" (*Entwurf*, §46).

Therefore, despite the necessary connection between the complete internal determination of a thing and its actual existence, existence itself is neither a mode, nor does it represent a particular internal determination, let alone 'the final determination' which transforms mere possibility into actuality. Instead, existence primarily expresses a relational and extrinsic fact about a thing: that it is sufficiently grounded in, and thus, determined with respect to the spatio-temporal, causal nexus of the actual world. The complete internal determination (of a contingent thing) is only possible through this extrinsic determination. Wolff's conception of existence, insofar as it applies to non-divine beings, is best understood as this extrinsic and higher-order property, and is thus compatible with Kant's negative thesis that existence is not a real predicate.

This reading also shows that Wolff's dictum is compatible with Kant's other negative thesis that the actual does not contain more than the possible, and that it does not entail that the extension of possibility is greater than that of actuality, a contention that Kant emphatically rejects. Intrinsic or absolute possibility does indeed have a greater extension than actuality as the former includes all (universal and singular) beings, while the latter includes only singular beings. It should be clear by now that this extensional difference is what Wolff has in mind when he infers that (intrinsic or absolute) possibility must be complemented by something 'more from outside' in order to become actual. Yet it is also clear that Wolff holds extrinsic or relative possibility to be coextensive with actuality. For whatever has an actual determinate cause is necessarily part of the causal nexus of the world at some time: "What is possible in this world has either already become or is or will in the future become actual" (*DM* §572).[114]

Like we saw in Leibniz, such coextensiveness between this worldly possibility and actuality motivates Wolff to introduce a distinction between two notions of contingency and necessity, absolute and relative. Accordingly, that whose negation involves a logical contradiction is absolutely necessary, and that whose negation does not involve a logical contradiction is absolutely contingent.[115] Only God enjoys existence with absolute necessity, because only he has existence among his essential predicates, which makes his non-existence a logical contradiction.[116] Any other being is absolutely contingent, since its existence is extrinsically grounded and thus can be denied without a logical contradiction.[117] On the other hand, necessity relative to the actual world or 'hypothetical (*hypothetice*) necessity' requires a cosmological consideration, involving extrinsic sufficient grounds. Something is hypothetically necessary if and only if the sufficient ground of its existence (or, what is the same, the sufficient ground of the determination of its modes) is available in causal and spatio-temporal nexus of things in the actual world.[118] Given the thoroughgoing causal and spatio-temporal connection of all things in the actual world, all absolutely contingent things in the actual world are in fact hypothetically necessary: "what is hypothetically

[114] Similarly, Wolff writes, "in mundo extrinsece possibile esse nequit, nisi quod vel extitit, vel actu preasens est, vel olim existet" (*Cosmologia*, §112).

[115] *Ontologia*, §279, *DM* §36.

[116] *Ontologia*, §§308, 309.

[117] *Ontologia*, §310. Compare *Metaphysica*, §102.

[118] *Ontologia*, §§306, 307.

necessary is in itself contingent" (*Ontologia*, §318).[119] Therefore, when conceived in the cosmological context, (relative) possibility, actuality, and (relative) necessity are all coextensive, extending over the very content of the actual world. As I will argue in chapter 7, this coextensiveness is the ultimate implication of Kant's claim that the actual does not contain more than the merely possible.

What is even more striking is that Wolff's conception of existence anticipates Kant's positive claim about what existence is to a significant degree. While Kant emphatically denies that existence adds any particular predicate to the merely possible, he maintains that "Through the actuality of a thing I certainly posit more than possibility, but not in the thing" (A234–5/B287). So, Kant holds, the addition is not to be conceived as a particular, further determination that would enrich the conceptual content of the thing, however determinate that concept may be. The addition is something extra-conceptual: a connection with perception. As I will explain in chapters 7 and 8, this is indeed a vast addition. Through the connection with perception, the concept is posited as corresponding to actual object of experience, "as contained in the context of the entirety of experience" (A600–1/B628–9). In other words, existence adds an entire world to the mere absolute possibility of the thing by locating it within the thoroughgoing spatio-temporal-causal nexus of experience, where the thing is completely determined with respect to all possible empirical predicates. This is essentially what Wolff's (iv) above suggests from a non-epistemological and purely cosmological point of view. Thus, although Kant regards himself as criticizing a conception of existence that is commonly but mistakenly thought to be implied by Wolff's dictum, he echoes Wolff to such an extent that one finds quite an illuminating explanation of Wolff's dictum in Kant's positive account of existence.

* * *

In this chapter, I presented a detailed analysis of Leibniz's and Wolff's views on modality with a view to assessing two claims regarding school metaphysicians: (i) they are committed to a merely logicist account of modality; (ii) they are committed to the idea that existence is a real predicate. Both claims, I argued, are overly crude and in fact do not hold up against the intricacies of Leibniz's and Wolff's respective discussions of modality and existence. If the reading I offered in this chapter is correct, the historical novelty, if any, of Kant's treatment of modality should not simply be sought in his introduction of an account of real or metaphysical modality in response to the inadequacy of logicism, and his rejection of the conception of existence as a real predicate.

In 2.1., I demonstrated that Leibniz anticipates Kant's theory of modality on a number of significant points. First, Leibniz endorses the actualist principle that the material content of possibility must be grounded on something actual, a principle that will form the basis of Kant's conception of real possibility. Second, Leibniz's account accommodates a strong notion of real possibility, R2 or possibility relative to the actual world, which rests not only on the principle of contradiction but also on

[119] Compare *Metaphysica*, §105.

metaphysical principles such as the PSR, the principle of the best, and causality. The apparent downside of this notion of possibility is that it is coextensive with actuality. Yet as I will argue in chapter 7, Kant's own critical notion of real (empirical) possibility will also turn out coextensive with actuality. Third, Leibniz explicitly defines a notion of relative necessity as a correlate of R2, applying to propositions that can be negated without logical contradiction but cannot fail to be true if they express a fact about the actual world. Fourth, in his more mature reflections, Leibniz adopts a conception of existence as an extrinsic or extra-conceptual denomination and thus, at least partly, anticipates Kant's theses that existence is not a real predicate.

In 2.2., I provided an account of Wollf's dictum that existence is the complement of possibility, diverging significantly from the common interpretation that it entails a notion of existence as a real predicate. I argued that when situated in the right context, i.e., cosmology, this dictum proves not only compatible with Kant's thesis that existence is not a real predicate, but also strikingly anticipatory of his conception of existence as the positing of the concept of a thing instantiated in the context of the causal nexus of the actual world, 'adding' to it an entire world, without adding any particular predicate. Wolff's theory of modality cannot be reduced to a mere logicistic account of modality. For while Wolff adopts a logical notion of possibility when he considers things absolutely in his ontology, his discussion of things in relation to other things in his cosmology offers a relative or extrinsic notion of possibility (a version of Leibniz's R2), i.e., the possibility of a thing in terms of its extrinsic (causal) grounds in the actual world. Wolff also offers a notion of relative necessity correlating with this notion of relative possibility, and goes as far as to explicitly recognize the coextensiveness of (relative) possibility, actuality, and (relative) necessity in the context of the actual world, a major conclusion of Kant's critical theory of modality, as will be seen.

PART II

Kantian Modality: Precritical and Revisionist

The overarching aim of Part II is to flesh out Kant's views on modality in his precritical period, by which, following the somewhat standard periodization in Kant scholarship, I mean his body of works roughly prior to 1770. With the exception of a few side discussions that I will underline when necessary, Kant's precritical views on modal notions manifest themselves in the context of his engagement with ontotheology. In this period, Kant develops a rather sympathetic and revisionist critique of the ontotheological tradition from within. He brings forceful and original objections to the previous versions of the ontological argument, but never denies the possibility of an ontological proof and even provides his alternative, 'only possible' proof. The two chapters of this part are respectively organized around the negative and positive aspects of Kant's precritical outlook on the argument. In chapter 3, I will provide an account of Kant's well-known attack on the argument and its place in the history of objections to the argument. In chapter 4, I will offer a reconstruction of Kant's alternative ontological argument.

The guiding question here will be what exactly is novel about Kant's precritical views on modality and what is not. Part of my broader argument in this book is that the novelty of Kant's overall theory of modality has usually been misplaced in his celebrated negative thesis that existence is not a real predicate, upon which he constructs his main objection to the classical ontological argument, and in his formulations of real possibility and necessity, upon which he constructs his own, 'only possible' ontological argument. As I showed in Part I, however, Leibniz explicitly anticipates Kant's negative thesis and Wolff's considered view of existence in general is compatible with it. Moreover, neither Leibniz nor Wolff can be reduced to logicists, since they both offer conceptions of real or metaphysical possibility and necessity. In Part II, I will argue that Kant's precritical conception of modality displays a character that runs parallel to his precritical engagement with the ontological argument: revisionist but not revolutionary. While Kant brings important correctives to the previous formulations of modal notions, in the final analysis he still understands modality from within the prevailing ontological paradigm, as expressing the modes of being of things. This, in fact, is the ground of his continuing commitment to the possibility of an ontological argument. However, I will also argue, the positive theses that Kant introduces in this period regarding existence point to his first discovery of the core idea that will develop into a revolutionary theory of modality in the critical period: modal notions are features of our conceptual representations of things and express the modes in which concepts of things are related to our cognition, rather than the modes in which things themselves are. As I will show in Part III, it is this paradigm shift which is genuinely novel in Kantian modality.

3

Kant and Ontotheology

From Anselm's first presentation of the ontological argument on, the ontotheological tradition faced forceful objections. The contribution that these objections made to the development of modal thought is comparable to that made by ontotheology itself. A fair evaluation of the originality of Kant's views on modality partly depends on accurately locating Kant's objections to the argument in the context of this counter-tradition.

In chapter 1, I offered a framework analyzing this counter-tradition as advancing two main lines of objection, each of which targets one of the two logical steps at work in the ontological argument. The first line aims to block the second step in the argument, namely the inference of the actual existence of God, the real entity, from the concept of God that is existentially loaded by the first step. According to this line of objection, the inference from existence thought in the concept of something to the actual existence of that thing is illegitimate, unless the latter is already presupposed. The second line of objection aims to block the first step of introducing existence, under whatever disguise, into the concept of God, by an appeal to the claim that existence is just not the kind of predicate that can be contained in the concept of anything.

One might want to modify this framework to accommodate finer distinctions in historical objections. For instance, Henrich, who calls the first line of objection the "logical" objection and the second line the "empiricist" objection, suggests that there is a third "critical" line of objection.[1] According to the 'critical' objection, the very concept of an absolutely necessary being is impossible to conceive with clarity, and a certainty of existence can never be concluded from this concept. Henrich argues that Kant's overall critique of the ontological argument falls under the second and third lines of objection. I approach Henrich's reading with some caution. First, as I will argue in chapter 8, Kant ultimately endorses a version of the first line of objection in the *CPR* as well. Second, it is not clear that the 'critical' objection constitutes an independent line of objection to the ontological argument. First of all, Hume, who, Henrich thinks, is the originator of this objection, raises the worry that we cannot distinctly conceive of a necessarily existent being as an objection not to the ontological but to the cosmological argument.[2] As I will discuss extensively in chapters 4 and 8, Kant's relation to this particular worry is rather complicated. The precritical Kant is openly critical of the way the concept of a necessary being is constructed in the cosmological argument as a self-grounding being and states that such a being is

[1] Henrich (1960), 75–6.

[2] See *Dialogues Concerning Natural Religion*, 54–7.

inconceivable. He even uses this criticism to motivate his own 'only possible' ontological argument, but never formulates it as a distinct objection to the ontological argument. Admittedly, however, there is a way in which the inconceivability of an absolutely necessary being could be damning for the ontological argument. For one major implication of the ontological argument is that the non-existence of God is a logical contradiction. But if the very idea of absolutely necessary existence itself is inconceivable, then there is nothing whose non-existence is logically contradictory. In fact, Kant's first salvo at the ontological argument in the *CPR* is precisely the cancellability of all and any particular existence without contradiction. Nevertheless, Kant thinks that the problem of misconstruing absolutely necessary existence in terms of logical necessity is rooted in misconstruing existence as a first-order predicate contained in the concept of a thing. In this sense, for Kant, what Henrich sees as a third line of objection is in fact an extension of the second line of objection.

3.1. The First Line of Objection: Gaunilo, Aquinas, Caterus, Crusius

The first line of objection does not question the contention that existence is a predicate that can be contained in the concept of a thing, but offers the insight that the fact that the concept of a thing contains a set of predicates does not warrant that there is a real object actually instantiating that concept and the set of predicates contained in it. The transition from the ideal or conceptual to the real or objectual level of predication requires the presupposition of the actual existence of the real object. The objection targets the second step in the ontological argument, which consists in the inference from the mere concept of God, supposedly containing existence as well as other divine predicates, to the actual existence of God. The condition of the validity of such an inference is the very conclusion of the inference. Therefore, the ontological argument presupposes what it is supposed to prove.

Some scholars attribute the origin of this line of objection to Aquinas.[3] This is why I will also call this objection 'Thomistic.' In fact, however, Gaunilo's reply to Anselm conveys the main insight I described above some two centuries before Aquinas. Gaunilo primarily finds it doubtful that the mere understanding of the term 'something than which nothing greater can be thought' is sufficient evidence for the 'mental existence' (*esse intellectu*) of such a being in the understanding. Yet he insists that even if the mental existence of such a being in the understanding is conceded, Anselm's argument cannot yield the desired conclusion. For "there is no way to derive from this the conclusion that this thing also exists in reality."[4] An *a priori* transition from the conceived existence of a thing to its actual existence is not possible.

The justification Gaunilo offers for this objection is quite instructive. As I pointed out in chapter 1, the motor of the transition from the 'mental existence' to the 'real existence' (*esse reale*) of God is the highest conceivable degree of greatness that is predicated of God *qua* a mental or intentional object: God must exist also in reality,

[3] See, for instance, Sala (1990), 51–2; Henrich (1960), 74. [4] Gaunilo's Reply 1.

because otherwise he would not be the greatest conceivable being. Gaunilo refuses to accept the validity of predicating a real quality of an intentional object whose real existence has not been proven yet. For, he suggests, God cannot be said to be greater than anything real unless he is proven to be a real thing: "I do not yet admit—indeed I actually deny, or at least doubt—that this being is greater than any real being."[5] Ganuilo's rejection anticipates an important discovery about the relationship between real predication and actual existence: definitional predications are inherently hypothetical, and the actual existence of the subject is a prerequisite for the validity or obtaining of real predication.

Aquinas does not go far beyond Gaunilo. Aquinas's objection is usually extracted from his discussion of the question of 'Whether the existence of God is self-evident' in the *Summa Theologiae*. There he criticizes an argument that seems identical to Anselm's argument in the *Proslogion* 2, without acknowledging the latter. Fortunately, a number of other texts, where Aquinas discusses the same argument with explicit reference to Anselm, help us verify that he specifically objects to Anselm's argument.[6] It seems reasonable to assume that Aquinas is also familiar with Gaunilo's Reply. Aquinas formulates his objection in support of his general opposition to the arguments for the self-evidence or logical necessity of the existence of God. He writes:

> Granted that everyone understands that by this word "God" is signified something than which nothing greater can be thought, nevertheless, it does not therefore follow that he understands that what the word signifies exists actually, but only that it exists mentally. Nor can it be argued that it actually exists, unless it be admitted that there actually exists something than which nothing greater can be thought; and this precisely is not admitted by those who hold that God does not exist. (*Summa Theologiae*. Ia q2)

Aquinas grants both the definition of God and the dual treatment of existence under *esse intellectu* and *esse reale* for the sake of the argument. His point is that even then the best the argument can prove is the *esse intellectu* of God, but not the *esse reale* of God, unless it is already presupposed.

In the First Set of Replies to Descartes' *Meditations*, Caterus states that Aquinas's objection to a certain argument (with no acknowledgment of Anselm) is applicable to Descartes' argument in the Fifth Meditation since the two arguments are identical. Caterus writes:

> Even if it is granted that a supremely perfect being carries the implication of existence in virtue of its very title, it still does not follow that the existence in question is anything actual in the real world; all that follows is that the concept of existence is inseparably linked to the concept of a supreme being. So you cannot infer that the existence of God is anything actual unless you suppose that the supreme being actually exists; for then it will actually contain all perfections, including the perfection of real existence. (AT 7:99, CSM 2:72)

[5] Gaunilo's Reply 1.

[6] Wippel (2000), ch. 10 names four other texts in which Aquinas talks about Anselm's ontological argument. These are Aquinas's Commentary on Sentences, d.3, q.1, a.2; commentary on the *De Trinitate*, q. 1, a.3; *De veritate*, q. 10. A12; and *Summa contra Gentiles* I, c. 10. Wippel (2000), 398 also makes a good case for Aquinas's acquaintance with Anselm's second argument in the *Proslogion* 3.

Like Gaunilo and Aquinas, Caterus finds the inference from the existentially loaded concept of God to the actual existence of God fallacious. The reasoning behind this transition in Descartes' argument is an epistemological principle: "That which we clearly and distinctly understand to belong to the true and immutable nature, or essence, or form of something, can truly be asserted of that thing" (AT 7:115–16, CSM 2:83).[7] Descartes takes this principle to be the license to move from existence as an attribute in the essence of God to the real existence of God. Yet such ideal or definitional predication does not warrant the actual existence of its subject. The real predication of a quality (whether existence or any other attribute) to a subject requires the actual existence of that subject, which, however, is precisely what the argument is set out prove.

One finds another statement of this line of objection in Crusius, one of Kant's immediate predecessors. In criticizing Wolff, Crusius repeatedly states that the principle of contradiction is never a sufficient means to arrive at conclusions about real existence. Crusius takes the Cartesian version of the ontological argument, which, he seems to think, is endorsed by Wolff, to be displaying an example of such faulty use of the principle of contradiction. He writes:

> In the major premise [i.e., The *ens perfectissimum* has also the perfection of existence] the talk is about the existence in the understanding, because a concept in the understanding contains in itself existence in such a way that if it [the concept] is thought or posited, existence as a part of it must also be thought or posited with it. In the conclusion, however, the talk is about the real existence outside the thought. The premises are all ideal-propositions but the conclusion has to be a real-proposition. Therefore, there is indisputably more in the conclusion than there is in the premises. (*Entwurf*, §235)

All historical statements of the first line of objection suggest that this is essentially an epistemological objection, claiming that a concept (or an intentional object in the mind) alone cannot yield the cognition of whether that concept is instantiated by an actual object outside the mind, and that such a transition from ideal to real existence requires something additional.[8]

3.2. The Second Line of Objection: Gassendi?

The second line of objection to the ontological argument aims to block the first step of introducing existence into the concept of God and consists in the rejection of the minor premise that existence is a perfection (or reality). I hold that this objection finds its true expression in Kant's first negative thesis that existence is not a real predicate. However, it has been suggested that Gassendi's objection to Descartes'

[7] Descartes sometimes formulates the 'clear and distinct' principle as the broader principle that whatever we clearly and distinctly understand to be true is true, both in thought and in reality. Then it would seem that he does not need to presuppose the actual existence of anything to make the transition from the ideal to the real, and thus he is immune to the first line of objection. However, the broad interpretation of the principle does not in fact solve the problem of the illegitimacy of the transition but only conceals it. Worse yet, it makes the ontological argument too dependent on a boldly idealist epistemology.

[8] For an alternative formulation of the same claim, see Forgie (2007), 517 and (2008), 119.

ontological argument anticipates Kant's thesis and consequently, his objection.[9] I will argue here that there is a significant difference between Gassendi's and Kant's theses as to what existence is not, which means that there should also be a difference between their respective objections to the ontological argument. I will show that Gassendi's objection cannot represent the second line, but in fact falls back on the first line.

Descartes' argument purports to establish the inseparability of existence from the essence or concept of God through a comparison with the inseparability of the property of having three angles equal to two right angles from the concept of a triangle. In the Fifth Set of Objections, Gassendi complains that this comparison is problematic because it compares incommensurate items, namely (God's) existence and a property (of triangle): "Existence is not a perfection either in God or in anything else; it is that without which no perfections can be present" (AT 7:323, CSM 2:224). There are two theses about existence here, negative and positive. The first one, existence is not a property (or perfection) of any object (God or another), is what is thought to anticipate Kant's negative thesis that existence is not a real predicate.[10]

I will present my detailed account of Kant's negative thesis in 3.4. But in order to readily expose the difference between Kant's and Gassendi's apparently similar negative theses, I will just state what I take to be the upshot of the former here. Kant's thesis is that existence is not a first-order predicate of objects, and thus could not be contained in the intension of the concept of any object. Accordingly, the concept of God, like the concept of anything else, cannot be thought of as containing existence as a predicate. This rules out not only the minor premise of the ontological argument, which introduces existence into the concept of God, but also what underlies this logical step, namely the dual treatment of existence as mental and actual. Since existence cannot be contained in the concept of anything, there is no such thing as the merely conceived existence of a thing as opposed to the actual existence of that thing. The concept of a thing does not amount to any form or degree of existence, but only an existentially neutral mental representation.

In Gassendi, on the other hand, we see neither such an unambiguous exclusion of existence from the concepts of things, nor an abandonment of the dual treatment of existence. Gassendi does not seem to think that his negative thesis that existence is not a property commits him to the extra-conceptuality of existence. The key here is to recognize that Gassendi presents his positive thesis as both a justification and an explanation of his negative thesis.

For surely what does not exist has no perfections or imperfections, and what does exist and has several perfections does not have existence as one of its individual perfections; rather, its *existence is that in virtue of which both the thing itself and its perfections are existent*, and *that without which we cannot say that the thing possesses the perfections or that the perfections are*

[9] Malcolm (1960), 44–45; Stang (2016), 39n108. See also Forgie (2007) for an insightful criticism of this view.

[10] Although Gaunilo's example of the perfect "Lost Island" seems to be designed to illustrate the absurd consequences of treating existence as an ordinary, first-order perfection, Gaunilo does not state that such is in fact a misconception of existence.

possessed by it. Hence we do not say that existence 'exists in a thing' in the way perfections do; and if a thing lacks existence, we do not say it is imperfect, or deprived of a perfection, but say instead that it is nothing at all. (AT 7:323, CSM 2:224–5)

So, according to Gassendi, existence is not a property of any object, because it is a prerequisite for an object's having any property and thus, for being an object at all. Gassendi's motives here seem to go deeper than refuting a particular sort of argument for God's existence. By defining existence as the necessary condition of being a subject of predication, he rejects a fundamental pillar of the Cartesian ontology of 'true and immutable essences.' As I argued in chapter 1, even though Descartes ultimately subscribes to the position that the distinction between essence and existence is only an abstract one and the two cannot be really separable in actual beings, he still assigns a pre-existential ontological status to true and immutable essences and treats them as subjects of predication even if they are not actually instantiated. Gassendi differs from Descartes in that he interprets the same position as entailing that essences that are not instantiated cannot be subjects of predication: what does not exist is 'nothing at all' and thus cannot have any real properties, positive or negative, perfections or imperfections.

Throughout his critique of Descartes' ontological argument Gassendi reiterates this Suárezian idea of the essence/existence distinction as a merely conceptual (and not real) one:

...you should have said that the existence of God can no more be separated from his essence than the existence of a triangle can be separated from its essence...Real separation is impossible no matter how much the mind may separate them or think of them apart from each other. (AT 7:323, CSM 2:224)

You say that existence is distinct from essence in the case of all other things, but not in the case of God. But how, may I ask, are we to distinguish the essence of Plato from his existence, except merely in our thought? Suppose that Plato no longer exists: where now is his essence? Surely in the case of God the distinction between essence and existence is just of this kind: the distinction occurs in our thought. (AT 7:323, CSM 2:225)

Gassendi's central emphasis is that the real inseparability of essence and existence is not an exclusive privilege of God but holds for anything. The problem, however, is that this inseparability does not justify Gassendi's negative thesis that existence is not a property of any object. On the contrary, it entails that existence, though perhaps not an ordinary property commensurable to having three sides, is still a property of objects. More specifically, Gassendi should hold that existence, as a necessary condition of being anything and having any properties at all, is a first-order or object-level but universal predicate of all objects, like, for instance, the predicate of being self-identical or that of having no contradictory predicates. This significantly differs from Kant's idea that existence is not a predicate of objects at all. Furthermore, the claim that existence is a universal predicate of objects implies that existence is in fact contained in all concepts of objects. Gassendi seems to embrace this implication. For instance, in response to Descartes' claim that it is impossible to think of "a wise and powerful God without thinking of him as existing," Gassendi challenges him to explain "how it is possible for us to think of a sloping mountain and a winged horse without thinking of them as existing" (AT 7:324, CSM 2:225). The idea that all concepts of objects contain existence is flatly opposite of what I take Kant's negative thesis to amount to: no concept of an object contains existence.[11]

Gassendi does not present his objection to Descartes' ontological argument in the form of a systematic refutation. The view that Kant's and Gassendi's objections to the ontological argument are the same operates on the assumption that Gassendi's thesis on what existence is not is the same as or anticipates Kant's. However, if, as I argued above, this assumption is false, then one should be cautious about categorizing Gassendi's objection together with Kant's. First of all, if Gassendi holds that existence is contained in the concept not only of God but of all objects, he cannot simply take the second line of objection, which intends to block the introduction of existence into the concept of God. Second, even if we stick to Gassendi's own wording and accept his view of existence as a necessary condition of having any properties, this view alone would not upset the ontological argument. In fact, as Forgie rightly points out, it would make the argument flow even better.[12] For if existence is a necessary condition of having any properties or perfections at all, the first premise that God has all perfections alone leads to the conclusion that God exists, even without the second premise that existence is a perfection on its own.

With the second line of objection ruled out by his own view of existence, Gassendi has to fall back on the first line of objection. As we saw above, this objection turns on the idea that even if existence is contained in the concept of something, it does not necessarily follow that this concept is instantiated by an actual object. Whether the object actually exists and instantiates the concept with all the predicates thought in it (including, one might add, the predicate of existence) is a further question that cannot be answered at the mere conceptual level. The proponents of the first line insist that allowing such an inference from the ideal or conceptual to the real (without an independent proof of existence) would unleash the absurdity of applying the ontological argument to any object conceived as existing in virtue of its other conceived predicates or perfections. Gassendi explicitly resorts to this line of reasoning and goes even as far as employing the example of a 'perfect Pegasus,' an analogue of Gaunilo's 'Lost Island':

> [A]lthough you say that both existence and all the other perfections are included in the idea of a supremely perfect being, here you simply assert what should be proved, and assume the conclusion as a premiss. Otherwise I could say the idea of a perfect Pegasus contains not just the perfection of having wings but also the perfection of existence... It seems that there is no point that you can raise in this connection which, if we preserve the analogy, will not apply to Pegasus if it applies to God and *vice versa*. (AT 7:325, CSM 2:226)

Therefore, Gassendi's objection to the ontological argument should rather be categorized under the first line. Despite his strong statement that existence is not a property, Gassendi fails to anticipate the essence of Kant's negative thesis and the

[11] Forgie (2007) argues for a similar view of the difference between Kant's and Gassendi's conceptions of existence. Accordingly, Kant conceives of existence as a second-order property, that is, a property of concepts, while Gassendi still conceives of existence as a pseudo-property operating at the first level.

[12] Forgie (2008), 121.

second line of objection based on it. The differences between Gassendi and Kant that I underlined in this section are crucial in understanding Kant correctly, for, as I will point out later on, the conflation of Kant and Gassendi on existence has led some in the literature to adopt a particular kind of misinterpretation of Kant's first negative thesis and his objection based on it.

3.3. Kant's Objections

There is no doubt that Kant's better known objection to the ontological argument embodies the second line of objection. From the early 1760s on Kant explicitly criticizes the first step of the argument on the basis of his thesis that existence is not a real predicate, and retains this particular criticism throughout. However, categorizing Kant's reaction to ontotheology is not so straightforward a task. Kant appears bafflingly sympathetic also to the first, Thomistic line of objection, not only in some of his precritical texts, but also, as I will show in chapter 8, in his refutation of the ontological argument in the *CPR*.

Since Part II is limited to Kant's precritical period, I will only examine here Kant's critique of ontotheology in his precritical works such as the *New Elucidation* (*NE*) of 1755 and *The Only Possible Argument* (*OPA*) of 1763. I will first demonstrate that the precritical Kant's flirt with the first line of objection displays serious ambivalence, as he expresses both clear endorsement and equally clear rejection of this sort of objection in different places. Second, I will turn to Kant's precritical statement of the second line of objection and start laying the ground for my proceeding analysis of the theory of existence on which this objection relies.

3.3.1. New Elucidation

In the *NE*, Kant levels an objection against the Cartesian version of the ontological argument:

> ... the claim is made that the existence of God is determined by that concept. It can, however, easily be seen that this happens ideally, not really. Form for yourself the concept of some being or other in which there is a totality of reality. It must be conceded that, given this concept, existence also has to be attributed to this being. And accordingly, the argument proceeds as follows: if all realities, without distinction of degree, are united in a certain being, then that being exists. But if all those realities are only conceived as united together, then the existence of that being is also an existence in ideas. (Ak. 1:394)

Now, this objection echoes the insight voiced earlier by Aquinas, Caterus, and Crusius: the conceived existence of something does not warrant the actual existence of that thing. Given that in this section of the essay Kant develops a critique of the Wolffian principle of sufficient reason, a critique for which he professes to be inspired by Crusius, and engages in the talk of ideal/real distinction, it is likely that Kant borrows this objection directly from Crusius.[13] It is therefore clear that Kant is at least

[13] On this point, see Sala (1990), 51, 52.

sympathetic toward the first line of objection in the *NE*.[14] This also suggests that as of 1755 Kant accepts the dual conception of existence, 'mental' (*esse intellectu*) and 'real' (*esse reale*). He does not yet seem to offer anything along the lines that existence is not a predicate.

However, some claim otherwise. Schönfeld, for instance, reads the passage I quoted above as containing a "reject[ion] of existence as a property."[15] Schmucker points out that although Kant does not employ the thesis that existence is not a real predicate in order to refute the Cartesian argument, he does not assert anywhere in the *NE* that existence is a predicate either. Schmucker further argues that Kant's account of possibility in Proposition VII implies that existence does not belong to the complete determination of a thing.[16] I hold that neither in Proposition VI nor in Proposition VII can one find such an implication. What both Schönfeld and Schmucker seem to overlook is that Kant's explicit commitment to the dual conception of existence precludes any commitment to the thesis that existence is not a predicate. For the former entails that existence can be contained in the concept of a thing, even if Kant thinks, along with the proponents of the first line of objection, that this containment alone does not warrant the inference to the actual existence of that thing.

There is also compelling textual evidence that Kant himself understands the dual conception of existence and the thesis that existence is not a predicate as mutually exclusive. It is worth noting that Kant does not raise the first line of objection at all in the *OPA*, which begins with a clear statement of the thesis that existence is not a predicate. More decisively, in a series of reflections from the early 1760s, he openly dismisses the first line of objection as vain, and claims that the proponents of this objection presuppose that existence is a predicate. Once existence is considered among those predicates that could belong to a thing, Kant argues, "then certainly no proof that would be more conclusive and at the same time more intelligible than the Cartesian one could be demanded for demonstrating the existence of God" (*R* 3706, Ak. 17:240). That is to say, once you allow the first logical step of including existence into the definition or concept of a possible being, then there is no way to block the second step of inferring from that concept to the actual existence of that being. Kant adds: "Against [the Cartesian argument] one objects in vain that such a possible thing includes existence within itself only in the understanding, i.e., only as soon as the thing is posited in thought, but not outside of thought." (Ak. 17:240)

[14] Sala (1990), 51, 57 and Henrich (1960), 182 are with me on this point. Schmucker (1980) diverges from this reading by claiming that Kant's objection is not to the transition from the ideal to the real existence of God, but to the arbitrary construction of the concept of God. Schmucker's reading, however, is not tenable. First, there is no explicit complaint in Kant's text about the way the concept of God is constructed. Second, the arbitrariness of the construction of the concept of God would ground a self-standing objection of the Cartesian argument, only if the transition from the concept to the actual object were to be allowed. Third, and more importantly, such an objection to the Cartesian argument would be ineffective. For Descartes moves from the conviction that God is not an arbitrary concept but one representing a true and immutable nature, and holds that only in the case of such non-arbitrary concepts can the clear and distinct principle be used to carry out the transition from conceived to actual existence.

[15] Schönfeld (2000), 199.

[16] Schmucker (1980), 65.

The reason why Kant thinks that this objection remains vain is interesting. He argues that the transition from ideal to real predication is illegitimate "only when one arbitrarily combines something with a concept that is not necessarily posited thereby" (Ak. 17:240). On the other hand, he claims, if there is not only a subjective but also an objectively necessary connection between the subject-concept and the predicate-concept, there is no reason for not accepting that this connection holds not only in thought but also in reality. After appealing to the example of a triangle and its geometrical properties for the objectively necessary connection between the concepts of the subject and the predicate, Kant concludes:

> This is how matters also stand with existence, if it could be regarded as a predicate of things. For it would necessarily belong, among all that is possible, to that in which all reality exists, i.e., most real being will necessarily exist. (Ak. 17:241)

Kant here seems just dismissive of the main insight in the first line of objection that the inference from ideal to real predication is not possible without the presupposition of the actual existence of the subject of predication. However, given his familiarity, at least, with Crusius, and his aforementioned endorsement of the first line of objection in the *NE*, it is not a viable option to read Kant as simply dismissing this insight. I suggest that his idea here is that if existence is ever conceived as a predicate of things and thus part of the intension of concepts of objects, this insight loses its force and consequently the first line of objection remains indecisive. A decisive refutation of the ontological argument would involve interpreting the illegitimacy of the *a priori* transition from ideal to real predication as stemming from the fact that existence is not a predicate. As I will argue in chapter 8, the recognition of this dependence of the first upon the second line of objection is what Kant's ultimate position in the *CPR* will come down to.

3.3.2. The Only Possible Argument

The *OPA* is the most important precritical text for understanding Kant's critical reception of traditional ontotheology. However, one does not find there a systematic refutation of the ontological argument. Kant's professed intentions in this text are more constructive than destructive. He starts with the premise that a demonstration of the existence of God, which would satisfy "the subtle scholar's" demand for "exactitude of precisely determined concepts and regularly connected syllogisms," "has not yet been discovered" (*OPA*, Ak. 2:65). But as the tentativeness of the title of the essay suggests, Kant's positive intentions are rather modest. He takes himself to be offering only a general basis for the missing demonstration, "a main draft" or a "ground of proof," which can be utilized in constructing the specific proof(s) of God's existence (Ak. 2:66). In section 1, Kant provides this 'ground of proof' proper. In section 2, he claims that this 'mode of proof' can be applied to revise the method of traditional physicotheology and to produce a new teleological argument, which moves from the necessary unity and systematicity of things in nature in their mere possibility to a supreme principle. In section 3, he presents his argument as to why this and no other basis of proof can succeed.

Only in this final and third section do we see some discussion of the previous arguments for the existence of God. However, this discussion does not contain an explicit refutation of any of the versions of the ontological argument. Kant does not seem so much concerned about the specific flaws of the previous arguments. Instead, he points out the general reason why any argument which "proceeds from the concept of the merely possible as a ground to existence as a consequence" is bound to fail: "existence is not a predicate at all" (Ak. 2:156). Kant holds that this mode of argumentation works only if existence is contained in the concept of the possible as a predicate, for only then is existence derivable as a consequence from the concept. This may be why Kant states the negative thesis that existence is not a predicate in the beginning of the first section, though this thesis will not take part in the construction of the alternative ontological argument he develops later in that section. One might also take the location of Kant's negative thesis about existence in the *OPA* as an indication that his critique of traditional ontotheology mainly targets its modal background. The first section of the *OPA* comprises four reflections. The first three reflections criticize the traditional conceptions of existence, possibility, and necessity, respectively, and propose alternative conceptions or revisions. What Kant actually takes to be his alternative ontological argument, the '*Beweisgrund*,' comes in the fourth reflection, which suggests that this new argument does not employ any of the previously criticized misconceptions of modal notions.

In the following, I will examine Kant's thesis that existence is not a predicate in its original statement in the *OPA* and discuss it against the background of the history of ontotheology. I will focus on Kant's precritical views on possibility and necessity as laid out in the second and third reflections and how these views inform his 'only possible argument' in the next chapter.

3.4. Kant's Theses on Existence in *The Only Possible Argument*

Kant has a lot to say about the notion of existence in the *OPA*. Although it has attracted the most attention in the literature, the thesis that existence is not a predicate is in fact part of a series of theses that Kant formulates in a broader effort to put forth a new account of existence, and cannot be fully and correctly understood in isolation from this account. I suggest that there are, at least, five such distinct theses explicitly stated in the *OPA*. Two of these are negative and concern what existence is not (T1 and T4), and the other three are positive and concern what existence is (T2, T3, T5):

T1: "Existence (*Dasein*) is not a predicate or a determination (*Prädicat oder Determination*) of a thing" (Ak. 2:72).

T2: "Existence is a predicate not so much of the thing itself as of the thought which one has of the thing" (Ak. 2:72).

T3: "Existence is the absolute positing (*absolute Position*) of a thing" (Ak. 2:73).

T4: "Nothing more is posited in an existent thing than is posited in a merely possible thing" (2:75).

T5: "More is posited through an existent thing than is posited through a merely possible thing" (2:75).

In this section, I will examine each of these theses and their interconnections, and explain how the account of existence emerging from these theses support Kant's objection to the ontological argument, though I reserve my more comprehensive discussion of Kant's refutation for later. As I will demonstrate in the rest of this book, these theses form the basis of Kant's overall treatment of modality. As I will also argue extensively, while Kant's precritical views, despite offering certain novel elements, are still within the general traditional paradigm of understanding modal notions as modes in which things themselves are, Kant's critical views on modality display a historically revolutionary character, which consists in reframing modal notions as expressing the various modes in which the concepts of things are related to the cognitive subject. However, in this section, I will show that some of the theses on existence Kant presents in the *OPA* anticipate the main insight underlying the revolutionary aspect of Kant's critical theory of modality, even if they remain rather revisionist without the critical epistemology.

3.4.1. T1: Existence is not a predicate or determination

Kant offers the following argument for T1:

> Take any subject you please, for example Julius Caesar. Draw up a list of all the predicates which may be thought to belong to him, not excepting even those of space and time. You will quickly see that he can either exist with all these determinations or not exist at all. The Being who gave existence to the world and to our hero within that world could know every single one of these predicates without exception, and yet still be able to regard him as a merely possible thing, which in the absence of that Being's decision to create him, would not exist. (Ak. 2:72)

It is clear that Kant is using a Leibnizian conceptual framework here. In fact, if the reader did not already know Kant's *OPA*, they could easily mistake this passage for one from Leibniz's *Theodicy*.[17] As I discussed in chapter 2, on Leibniz's account each and every possible individual is grounded in God's mind in its completely determinacy. For God represents these possibilia in their complete concepts, as completely determined with respect to the entire world they are bound with, and thus completely determined with respect to all predicates they would have if they were to be actualized by him. Existence itself, however, is not part of the complete determination and thus the complete concept of any possible individual in God's mind. For whether a possible individual or the world it inhabits gets actualized depends on God's will. God's act of creating a possible world (with all its determinate content) does not add any new predicate or determination to the constitution of this world. The best of all possible worlds does not become even a better (or more perfect) world through creation. God creates exactly what he represents in his mind as a possibility in its complete determinacy. Likewise, Kant writes:

> ... who can deny that in the representation which the Supreme Being has of them [millions of merely possible things] is not a single determination missing, although existence is not among them, for the Supreme Being cognizes them only as possible things. It cannot happen,

[17] Compare, for instance, *Theodicy* §52, §183, 416.

therefore, that if they were to exist they would contain an extra predicate, for; in the case of the possibility of a thing in its complete determination, no predicate at all can be missing.

(Ak. 2:72)

So, is Kant merely repeating Leibniz here? I argued earlier that Leibniz's conception of existence as something extrinsic to the essential constitution of things anticipates Kant's T1, and I will revisit this point later in this section. However, Kant's allusion to Leibniz's doctrine of creation here seems to be motivated by that it provides strong imagery that is both familiar to his contemporary audience and useful in explaining the meaning of T1, which otherwise does not stand or fall with this doctrine. In order to accept T1, one does not have to commit to Leibniz's notion of a complete concept of a thing either. Kant's idea is that if it were possible for an omniscient mind to form a complete concept of an individual with all its predicates including even spatio-temporal ones, existence would still be not contained in that concept. The reference to an omniscient mind and its capacity to represent things in their complete concepts is to underscore that existence is not contained in the concept of anything as a matter of principle and not because of a limitation in our faculty of conceptual representation. Existence is absolutely extra-conceptual, regardless of how determinate the concept in question may be and what kind of object it may designate. Whether something exists is always a further question and cannot be settled through looking at the intension of a concept. Kant's argument here can be summed up in the following way:

1. If existence were a predicate of a thing, then a complete concept of a thing would contain existence (or non-existence).
2. No (logically consistent) concept, regardless of its level of determinacy, can contain existence (or non-existence).
3. Therefore, existence is not a predicate of a thing.

Kant intends his justification of the thesis to work against the background of the principle of complete determination. 'Determination' is the general term referring to any (internal or external) predicate that could be affirmed or denied of a subject.[18] As I argued in chapter 2, Wolff interprets the principle as stating that whatever exists is completely determined with respect to its internal predicates, i.e., essentials, attributes, and modes, but does not take it to entail that complete determinacy warrants existence. Wolff's interpretation of the principle is compatible with Leibniz's completely determined mere possibilia or uninstantiated complete concepts. On Baumgarten's interpretation, on the other hand, the principle deems the complete determinacy of a thing as a sufficient condition of its existence, and thus excludes completely determined mere possibilia. What motivates Baumgarten's interpretation may be his view that existence itself is a mode or internal predicate. For if existence is itself a predicate, then both whatever is completely determined must also be determined with respect to existence and whatever is not

[18] Kant will later specify 'determination' as a synthetic predicate that could add to the subject-concept what is not already contained in it (A598/B626). I will discuss Kant's more specific conception of determination in chapter 8.

determined with respect to existence (i.e., merely possible) must remain incompletely determinate. The way Kant justifies his thesis that existence is not a predicate above suggests that he takes it to follow from an antithesis of Baumgarten's interpretation of the principle of complete determination. So Kant's argument can also be formulated in the following way:

1. If existence were a predicate of a thing, then anything that is completely determined would be actual (or impossible).
2. However, complete determinacy does not warrant actual existence (or non-existence), and/or there are completely determined mere possibilia.
3. Therefore, existence is not a predicate of a thing.

This reconstruction also relies heavily on the existential neutrality of concepts. If concepts cannot contain any mark of existence, the level of their determinacy is immaterial to the question of existence.

Earlier in this chapter, I suggested that Kant's T1 is sometimes conflated with Gassendi's claim that existence is a precondition of having any predicates at all and thus is a universal predicate of all actual objects. In fact, however, Kant's idea of the existential neutrality of concepts entails not just that existence is not contained in the concept of a merely possible object and thus is not a predicate of a merely possible being, but the much stronger position that existence is not a predicate of an actual thing either! For if existence cannot be contained in the concept of an object, no matter how determinate the concept may be, then existence would not be a predicate that an object would instantiate if it actually existed. Julius Caesar, for instance, though an inhabitant of the actual world, does not instantiate a predicate of existence, since the complete concept of him did not contain such a predicate in the first place. Thus, in contradistinction with Gassendi's claim that existence is a universal predicate of all objects, Kant's T1 states that existence is not a first-order predicate of (actual) objects in the first place.

Overlooking this point leads to the following fairly common misinterpretation of Kant's T1 among prominent scholars: any object that instantiates a concept necessarily also instantiates the predicate 'exists' (presumably because existence is a precondition of having or instantiating any predicate and thus being an object at all). For instance, Malcolm interprets T1 to mean that one and the same person could satisfy two separate descriptions of "the most perfect chancellor" that are identical except that one includes existence among the attributes and the other does not.[19] Barnes takes the "substance" of T1 to be the claim that "For any property F and any object x: x is F if and only if x is an existent F."[20] Plantinga suggests that T1 amounts to the claim that if C1 is the whole concept of Taj Mahal, and C3 is the same as C1 only except existence, "[T]here are no possible circumstances in which C3 but not C1 has application; it is a necessary truth that if C3 is exemplified, so is C1."[21] Again, Van Cleve argues that T1 comes to mean that "there is no concept C such that $\Diamond\exists x\ (Cx\ \&\ \text{-}Ex)$," that is, it is not possible that a concept is instantiated without thereby existence

[19] Malcolm (1960), 44. [20] Barnes (1972), 46. [21] Plantinga (1967), 35.

is also instantiated.[22] Stang offers the most recent version of this reading of T1, claiming that it means that 'existence' and 'non-existence' do not divide the extension of any concept such that any object that instantiates (or falls under the extension of) a concept C necessarily also instantiates (or falls under the extension of) 'exists.'[23] Most notably, Stang argues that this is because "there are *no* non-existing objects"; existence is a first-order predicate that universally or unrestrictedly applies to all (and not a subset of) objects.[24]

The problem with this interpretation is that it infers a false conclusion from an accurate observation. It is true that Kant's T1 entails that 'exists' does not divide the extension of any concept, as Stang puts it, or that Malcolm's two descriptions of the most perfect chancellor can be satisfied by the same person, or that Plantinga's C1 and C3 have exactly the same extension. However, this is not because T1 defines existence as a universal predicate, but because it strictly holds that existence is not and cannot ever be part of the intension of the concept of anything. What cannot be part of the intension of any concept cannot divide or narrow the extension of any concept. Such existential neutrality of concepts of objects does not entail that existence is a first-order, albeit a special, i.e., universal, predicate applying to objects. In fact, Kant's T1, 'existence is not a predicate or determination of a thing,' makes the opposite claim that existence is just not a first-order or an object-level predicate.

It has to be noted, however, in both of the reconstructions I presented above, Kant's argument for T1 relies on the claim that concepts are existentially neutral no matter how determinate they may be. In fact, I identified the ultimate meaning of T1 with this claim. Therefore, unless Kant offers an independent argument for the existential neutrality of concepts, his argument for T1 remains to be question begging.

3.4.2. *T2: Existence is a predicate of concepts*

Despite his claim that existence is not a first-order predicate of objects, Kant observes the linguistic phenomenon that we nevertheless use the term 'existence' (or 'exists') as a predicate in our daily discourse. Kant's second thesis provides a response to this linguistic observation: "When existence occurs as a predicate in common speech, it is a predicate not so much of the thing itself as of the thought which one has of the thing" (Ak. 2:72).

So, existence is not a predicate of objects, but a predicate of our conceptual representations of objects, a second-order or concept-level predicate. But what exactly does this tell us about existential statements? If existence is not a predicate of objects, then an existential statement does not assert or deny a predicate of the object designated by the subject-concept. What an existential statement asserts or denies instead is that the subject-concept is instantiated by an actual object.

[22] Van Cleve (1999), 188. [23] Stang (2016), 39–40.

[24] To be fair, Stang's position is in fact more nuanced than others. He does admit that on Kant's account the 'fundamental' role of the predicate 'exists' is to apply to concepts as second-order predicates, but insists that if Kant's objection to the ontological argument is to succeed, Kant must also hold à la Gassendi that 'exists' is a predicate that applies to all objects. See Stang (2016), 38, n. 108, 39.

To use Kant's own example, 'Sea-unicorns (narwhals) exist' does not tell us anything about the sea-unicorns themselves, but that the concept 'sea-unicorn' is instantiated by an actual object out there.[25] Kant thus suggests that the existential statements of the form 'S exists' are a defect of our ordinary language. Instead, we ought to say "the predicates, which I think collectively when I think of a sea-unicorn (or narwhal), attach to a certain existent sea-animal" (Ak. 2:73).

I will come back to this extremely important and yet underappreciated thesis at the end of this chapter, but a few implications are worth noting at this stage. First and most importantly, this thesis makes a reference to the cognitive subject and the faculty of conceptual representation in explicating the meaning of existence. Second, this thesis suggests that existence has a locative reference. That is, existential assertions connect the conceptual representation of an object with the perception of that object *outside* (or distinct from) that representation. Third, the locative reference to outside implies that the truth of existential assertions must be bound by epistemic conditions regarding the subject's cognitive relation or access to the external realm. Even at this precritical stage, Kant states that the truth of "Sea-unicorns exist" means that "the representation of sea-unicorn is an empirical concept" and understands the cognitive relation to the external realm of objects in terms of experience: "If one wishes to demonstrate the correctness of such [existential] a proposition, one examines the source of one's cognition of the object. One says 'I have seen it' or 'I have heard about it from those who have seen it'" (Ak. 2:72–3). Fourth, and finally, this thesis, defining existence as a property of concepts, i.e., the property of being instantiated, characterizes existence as extrinsic to the intension of the concept and yet applying to the totality of the predicates contained in the intension of the concept.

3.4.3. T3: Existence is absolute positing

Kant may be the first to make a philosophical use of the German term 'Setzung.'[26] However, his contemporary audience would already be familiar with the Latin equivalent, '*ponere, positio*,' as it is quite often used in Baumgarten's *Metaphysica* in the sense of asserting a thing, a ground, a proposition or a predicate as real or true, as opposed to, for instance, cancelling or negating (*tollere, negatio*) something. In his early Latin works Kant follows Baumgarten's use of the term. For instance, in the *NE*, he writes "To determine is to posit (*ponere*) a predicate while excluding its opposite" (*NE*, Ak. 1:391). According to this traditional use, 'positing' is primarily a cognitive act of assertion by a cognitive subject. In introducing T3 in the *OPA*, Kant expounds upon the notion of positing:

[25] Bennett (1974), Forgie (2000), Rosefeldt (2011) have defended the view that Kant anticipates Frege here. Frege (1979), 75 indeed expresses a very similar thesis: "The existence expressed by 'there is' cannot be a characteristic mark of a concept whose property it is, just because it is a property of it. In the sentence 'There are men' we seem to be speaking of individuals that fall under the concept 'man,' whereas it is only the concept 'man' we are talking about. The content of the word 'exist' cannot well be taken as the characteristic mark of a concept, because 'exists,' as it is used in the sentence 'Men exist,' has no content." Rosenkoetter (2010) and Kannisto (2018) have criticized different aspect of this 'Kant–Frege anticipation view.'

[26] See 'Setzung' in *Historisches Wörterbuch der Philosophie* (1971–2007), vol. 9, 697–8.

The concept of positing or setting (*Position oder Setzung*) is perfectly simple: it is identical with the concept of being in general (*Sein überhaupt*). Now, something can be thought as posited merely relatively, or, to express the matter better, it can be thought merely as the relation (*respectus logicus*) of something as a characteristic mark (*Merkmal*) of a thing. In this case, being, that is to say, the positing of this relation, is nothing other than the copula in a judgment. If what is considered is not merely the relation but the thing posited in and for itself, then this being is the same as existence. (*OPA*, Ak. 2:73)

The initial statement confirms that Kant is still engaged in the traditional use of the term 'positing.' For by 'being in general,' he means the veridical and the most fundamental use of the verb 'to be' or 'is' in the sense of 'is true' or 'is really the case.' So, positing is the act of asserting that something is really the case.

What is more instructive here is the distinction Kant introduces between two ways of positing that something is the case: relative and absolute, corresponding, respectively, to predicative or copulative and existential senses of being. Relative positing is the assertion of a merely logical relation between two concepts, a subject-concept and a predicate-concept, or more precisely, it is the assertion of the latter as a 'characteristic mark' (*Merkmal*) of the former, which is carried out in a judgment through the copulative 'is.' Kant understands 'mark' as "that in a thing which constitutes a part of the cognition of it, or—what is the same—a partial representation, insofar as it is considered as ground of cognition of the whole representation" (*JL* (Ak. 9:58). So, he has in mind here only the predicates that are contained in the definition of a thing, for only those predicates amount to partial representations that constitute our cognition of the whole representation of a thing. Then relative positing is the positing of the predicate-concept (a partial representation of the thing) as contained in the intension of the subject-concept (the whole representation of the thing).

Two points already suggest that existential statements cannot be instances of relative positing. First, Kant defines relative positing at an existentially neutral level, in terms of the assertion of a logical relation of containment of one concept in another concept with no ontological commitment to the instantiation of these concepts by actual objects. For instance, he notes that in the judgment "God is omnipotent," all that is posited is the logical relation between the concepts of God and omnipotent, but whether God exists "is not contained in the original assertion at all" (*OPA*, Ak. 2:74). For he holds that "the relations of predicates to their subjects never designate anything existent" and "the proposition 'God is omnipotent' must remain true even for someone who does not acknowledge the existence of God" (Ak. 2:74). So following the proponents of the first line of objection to the ontological argument, Kant both allows the truth of definitional or ideal predication without the presupposition of the actual instantiation of the subject-concept and denies that the former entails the latter. Second, as T1 puts forth, existence is not a predicate that can be contained in the concept of anything. Relative positing, however, is just the assertion of a predicate-concept as contained in a subject-concept. The assertion of existence, therefore, requires a different kind of positing.

This other kind of positing, Kant calls 'absolute positing.' As opposed to the relative positing where the logical relation between the subject-concept and predicate-concept is asserted, in absolute positing the subject-concept is asserted 'in and for itself' (*an und für sich*) as instantiated by an actual object. This kind of positing is what 'exists' or 'the is of existence' in a judgment carries out. The third thesis that existence is the absolute positing of a thing should be understood in connection with the first two theses. First of all, since existence is not a predicate itself, the absolute positing of the concept of a thing does not add any further predicate to the content represented by that concept. To bolster this point, Kant resorts to the Leibnizian imagery again and draws a parallel between the cognitive act of absolute positing and God's creative act:

> If I imagine God uttering His almighty '*Let there be*' over a possible world, He does not grant any new determinations to the whole which is represented in His understanding. He adds no new predicate to it. Rather, He posits the series of things absolutely and unconditionally, and posits it with all its predicates; everything else within the series of things is posited only relatively to this whole. (Ak. 2:74)

So as much as the actualization of the best possible world does not make it an even better world, but only makes it an actual object instantiating its complete concept in God's mind, the absolute positing of the concept of a thing by a cognitive subject does not change the intension of that concept in the subject's mind but asserts it as having an actual object out there, instantiating that concept as a whole, with all its intension or predicates that are already contained in it. Thus, like T2, T3 implies that existence comes with a locative reference to the outside and the subject's epistemic relation to it. This also means that absolute positing does not tell anything about what the object is but only that there is an object corresponding to the cognitive subject's conceptual representation of it. Hence, T3 reinforces T2's insight that 'existence,' when used as a predicate, should be construed as a predicate not of the thing itself, but the concept or "the thought which one has of the thing" (Ak. 2:72), i.e., the predicate of being instantiated.

Kant's distinction between relative and absolute positing has interesting implications for his critical theory of judgment. In the *OPA* Kant defines relative positing as the positing of a conceptual relation that holds only in virtue of the one concept (predicate) being contained in another concept (subject). This implies that, in Kant's critical terminology, all instances of (merely) relative positing constitute analytic judgments. Absolute positing, on the other hand, posits a concept in relation to something that is not analytically contained in it, i.e., an actual object. So Kant seems to anticipate here his critical thesis that all existential judgments, i.e., instances of absolute positing, are synthetic. I will revisit this issue in chapter 7. Yet it is important to note that despite the impression Kant gives in the *OPA*, absolute positing and relative positing should not be understood as mutually exclusive. First, every absolute positing presupposes some relative positing. For when a subject-concept (e.g., 'God') is absolutely posited as instantiated by an actual object, each and every predicate that is contained in and thus posited in relation to that concept (e.g. 'omnipotence') is also posited as instantiated.[27] Second, even though Kant's account in the *OPA* dismisses the case of synthetic (non-existential) judgments, such judgments involve both the

[27] See Stang (2016), 78.

absolute positing of the subject-concept and the positing of the relation between the predicate-concept and the subject-concept.

3.4.4. T4 and T5: Is there more in existence than in mere possibility?

Kant's fourth and fifth theses on existence are his two answers to this peculiar question. The former provides a negative answer and the latter a positive answer with qualification. Kant grounds these answers, first, on the basis of his T3, 'existence is absolute positing,' and second, on the basis of a distinction between 'what is posited' and 'how it is posited' or between the content and manner of positing. Now, Kant's T4 turns on the idea that as far as the content is concerned, there cannot be any difference between what is posited in actuality and what is posited in mere possibility, or as he will put it in his critical period, "the actual contains nothing more than the merely possible" (A599/B627).

I noted earlier that Kant's argument for T1 depends on the claim that the concept of a thing is existentially neutral, no matter how determinate its intension may be. Even though Kant does not explicitly state it in the *OPA*, we can still tease out the implicit connection between the existential neutrality of concepts and T4, "Nothing more is posited *in* an existing thing (*in einem Existirenden*) than in a merely possible thing" (*OPA*, Ak. 2:75). The difference between existence and mere possibility does not lie in the difference between the conceptual contents posited in each case. For, if we were able to form complete concepts, we could recognize that the actual object instantiates the exact same conceptual content through which we would represent it as otherwise completely determined but merely possible. The (complete) concept of a merely possible something and the (complete) concept of that same thing as existing have exactly the same content, i.e., the same predicates, or better, they are one and the same concept. Otherwise, the complete concept of an actual thing would not be completely representing it, which is absurd. Thus, concepts must be thought of as existentially neutral, which means that existence is not contained in the concepts even of actual objects, and thus, is not a predicate of objects (T1). This might present a promising line of reasoning for Kant's T1.

But even if we accept that nothing more is posited in the actual than in the merely possible at the level of intensions of concepts, there is literally a world of difference between the actual and the merely possible! In considering something as merely possible, all the predicates that are contained in that concept are posited relatively to that concept, but neither the concept itself nor any predicate contained in it is absolutely posited as actually instantiated. So through the mere possibility of a thing, only the logical relations between the predicates and the concept of the thing are posited, regardless of whether or not they actually obtain in reality, outside the concept. Through the actual thing, both the concept itself and the logical relations and thus the predicates contained in it are posited absolutely, as instantiated in reality. In fact, assuming that things can exist only in the context of real relations with other things which make up the actual world, one might add that through the actuality of a thing, the concept of an entire actual world is absolutely posited. Hence, Kant's T5: "more is posited *through* an existent thing (*durch etwas Existirendes*) than is posited through a merely possible thing" (Ak. 2:75).

Kant's inquiry into whether there is more in existence than in mere possibility is historically significant. For this question is obviously a reference to the treatment of existence in the rationalist school metaphysics. Kant's T4 and T5, which come in the form of answers to the question, suggest that he intends to take issue with the contention that absolute positing of a concept as actually instantiated adds a predicate to its content such that the actual contains more than the merely possible. As I argued in chapter 2, this contention is, mistakenly, derived from Wolff's dictum of existence as the complement of possibility. Interestingly enough, Kant does not bring a direct criticism to Wolff's dictum itself, but complains that it is rather vague: "Wolff's definition of existence, that it is a completion of possibility (*Ergänzung der Möglichkeit*) is obviously very indeterminate" (Ak. 2:76).[28] This might be a fair complaint, if one takes the dictum only in the context of Wolff's ontology, which leaves it unclear whether the dictum entails that the actual contains more than the merely possible. However, as I showed, Wolff's cosmology provides a fuller theory of existence, which does not entail that existence adds a specific predicate to mere possibility in the sense Kant's T4 rejects, but claims, in line with Kant's T5, that through existence the concept of the thing is posited, with all its determinations, as instantiated in the context of the actual world.

On the other hand, Kant's criticism targets Baumgarten directly. In chapter 2, I argued that Baumgarten's account of existence in *Metaphysica* misinterprets Wolff's dictum to entail that existence is to be construed at the level of internal determinations, as 'the collection of affections' or as 'a mode' in its own right.[29] Baumgarten also misconstrues Wolff's principle of complete determination as a principle of existence. For while Wolff takes complete (internal) determinacy to be a necessary condition of existence, Baumgarten takes it to be a sufficient condition of existence.[30] This, of course, ties back to Baumgarten's interpretation of Wolff's dictum. Since essence or internal possibility (in contingent things) determines only essentials and attributes, existence must belong to what is not determined by the essence. So, on Baumgarten's account, what existence adds to mere possibility is that additional mode or layer of modes (i.e., contingent determinations) that is left indeterminate by the essence. This addition is precisely what is more in the actual than the merely possible. Kant is then quite accurate when he writes,

> Baumgarten introduces the concept of thoroughgoing internal determination, and maintains that it is this which is more in existence than in mere possibility, for it completes that which is left indeterminate by the predicates inhering in or issuing from the essence. (*OPA*, Ak. 2:76)

Baumgarten's definition of existence as a specific mode flatly contradicts Kant's T1. But here, in the context of the question of whether the actual contains more than the merely possible, Kant focuses more on Baumgarten's identification of existence with complete determinacy, or more precisely, with what is left indeterminate by the essence of a thing. Kant raises at least two major objections to this approach. His first

[28] One might think that Kant is not very attentive to Wolff's original text here. On this point, Honnefelder (1990), 368 notes that what Kant criticizes in the *OPA* is the conception of existence as the "*Ergänzung*" of possibility, whereas Wolff does not understand existence as "*Ergänzung*" but as "*Erfüllung*."

[29] See Meta. §§55, 134, 257.

[30] See Meta. §54.

objection is just what T4 states: "the difference between what is a real thing and a merely possible thing never lies in the connection of that thing with all the predicates which can be thought in it" (Ak. 2:76). The predicates that an actual S instantiates and the predicates that are thought in the concept of a merely possible S are one and the same set of predicates, because existence simply is not a predicate that can be contained in the concept of anything. Secondly, Kant points out that complete determinacy is neither the sufficient condition, nor the distinguishing mark of existence, because, he holds, there is a certain sense in which indeterminacy or incomplete determinacy is impossible and cannot belong to the merely possible as much as it cannot belong to the actual:

For such indeterminacy is forbidden by the law of excluded middle which maintains that there is no intermediate between two predicates which contradict each other. It is for example impossible that a man should not have a certain stature, position in time, age, location in space, and so forth. (Ak. 2:76)

The principle of excluded middle maintains that in any given pair of contradictory predicates (P1&–P1; P2&–P2;...) one necessarily applies to a subject S and there is no third option for S to remain indeterminate with respect to any such pair. A subject that is not determined with respect to some contradictory pairs cannot exist as such even if it does not violate the principle of contradiction. The principle of excluded middle here is in fact a logical correlate of the principle of complete determination and therefore applies to individuals. So, Kant is not talking about absolute or internal possibilities that are determined only by whatever predicates are contained in the definitions of things (i.e., essentials and attributes). Instead, he has in mind here Leibnizian possibilia (R1), represented by an infinite intellect through their complete concepts as fully determined with respect to all the predicates they can have in their respective worlds. Individual possibilia are determined with respect not only to modes but even also to extrinsic or relative predicates such as spatio-temporal and causal determinations, and are still merely possible. On this basis, Kant criticizes Crusius for regarding "the *somewhere* and *somewhen* as belonging to the unmistakable determinations of existence" (Ak. 2:76).[31] Therefore, from an omniscient point of view, complete determinacy applies to merely possible and actual individuals alike and thus cannot be taken as a distinguishing mark of the latter. Even if one can construct the concept of a thing with utter determinacy, whether such a thing actually exists is an altogether distinct question that cannot be settled on the basis of the content of that concept. In order to answer the question of existence, one has to go outside of the concept and establish the availability of an object that instantiates it.

[31] Kant refers to Crusius's definition of existence here: "Daher ist dies Existenz dasjenige Praedicat eines Dinges, vermoege dessen es auch ausserhalb der Gedanke irgendwo und zu irgend einer Zeit anzutreffen ist" (*Entwurf*, § 46). It is important to note that Kant's claim that spatio-temporal determinations can be contained in the complete concepts of merely possible individuals and thus cannot be regarded as distinguishing mark of existence does not necessarily invalidate Crusius's definition of existence. For Crusius does not claim that determination with respect to space and time is a sufficient condition of existence but that existence is what makes something empirically available in space and time "outside of thought" (*ausserhalb der Gedanke*), which is entirely compatible with Kant's theses on existence.

Therefore, the modality (actuality or mere possibility) of a thing is separate from the content of its conceptual representation and thus from what is posited in it. Whether something is merely possible or actual is an extra-conceptual matter and is about how that content as a whole is posited, i.e., whether it is posited relatively, as a set of merely logical relations, or absolutely, as actually instantiated. Positing is a cognitive act performed by the cognitive subject on the basis of her cognition. Thus, although Kant neither articulates nor exploits it, the crucial implication here is that modality is about how the cognitive subject relates to the concept of an object as a whole, against the background of her epistemic access to the reality outside of her faculty of conceptual representation. This point, only implied in Kant's precritical discussions of existence, will become a central issue in his critical theory of modality.

3.5. The Relevance of Kant's Theses on Existence to the Ontological Argument

While Kant does not offer a refutation of the ontological argument in the *OPA*, a few points on the question of the relevance of his theses on existence to the argument can still be made at this stage. In the third section of the *OPA*, Kant makes a sweeping claim that his T1 is damning for any version of the argument that infers the existence of God from his possibility:

> If the argument is to proceed from the concept of the merely possible as a ground to existence as a consequence, then the same existence must be discoverable in the concept by means of analysis, for the only way in which it is possible to derive a consequence from a concept of the possible is by logical analysis. But then existence would have to be contained in the possible as a predicate. But since... this is never the case, it is obvious that a proof of the truth we are examining is not possible in this manner. (*OPA*, Ak. 2:156)

Kant's confidence seems to rely on two convictions: i) The sort of argument that proceeds from the concept of a possible being to the actual existence of that being necessarily requires that existence be a predicate in the first place, and ii) The ontotheological tradition has produced only this sort of argument thus far. First of all, the latter is supported by the account of the history of ontotheology I offered in chapters 1 and 2. I argued that what motivates the very idea of an ontological argument is the contention that God's existence is logically entailed by his essence, expressed by his mere absolute possibility or concept. Kant's first claim is that such logical entailment can only be the case if existence is contained in the concept of God as a predicate. I have demonstrated earlier that all historical versions of the ontological argument do indeed fit into a two-step logical framework, where the first step is to introduce existence into the concept of God, either explicitly as a distinct predicate, e.g., perfection (Descartes), reality (Wolff, Baumgarten), or more implicitly, under the guise of another predicate, e.g., greatness (Anselm), self-sufficiency (Leibniz), which entails existence. But it is still worth asking why Kant thinks that existence must in principle be treated as a predicate contained in the concept of God in order to secure the logical entailment in question.

Kant's account in the first section of the *OPA* states that existence is not a predicate of objects, i.e., a first-order predicate, but a predicate of concepts, i.e., a second-order predicate, specifically, the predicate of a concept's being instantiated by an actual object. What primarily follows from this is that existence is an extra-conceptual matter strictly separated from the intension of concepts of objects, and thus, the intension of a concept, no matter what predicates are thought to be contained in it, can never logically entail its actual instantiation. More directly put, no first-order predicate can entail anything about existence, because existence is a second-order predicate. If God's existence were to be entailed by a first-order predicate or the totality of first-order predicates contained in the concept of God, it would itself also have to be a first-order predicate contained in the concept of God and thus discoverable by its logical analysis.[32] Hence, Kant's confidence that T1 undermines any argument resting on the validity of the entailment of the existence of God by the intension of the concept of God. This, of course, is a rough idea of an objection that Kant does not care to elaborate in the *OPA*, presumably because he thinks such elaboration is superfluous and has no bearing on his more major and positive project of constructing 'the only possible argument.'

3.6. The Novelty of Kant's Theses: Revisionist or Revolutionary?

We can now raise the question of whether or to what extent the conception of existence Kant offers in the *OPA* is historically novel. I showed in chapter 2 that anticipations of Kant's negative conception of existence can be found in Leibniz and Wolff. I argued that Leibniz's mature view is that in the case of non-divine things existence is an extrinsic property, which indirectly supervenes on the essence but does not inhere in it. Kant's allusions to the Leibnizian idea that God's creation of the actual world does not add any new predicate to its essential constitution in God's mind in formulating his negative theses (T1 and T4) suggests an awareness of this anticipation. I also argued that despite the prevalent misinterpretation in the literature, Wolff's doctrine of *complementum possibilitatis* is not incompatible with T1, but on the contrary, when understood in the right context, conveys a conception of existence anticipating not only T1, but also T4.[33]

Therefore, Kant's T1 is not really historically revolutionary. What may be considered novel is Kant's recognition of the relevance of this thesis to the ontological argument. For contrary to Leibniz and Wolff, who anticipate Kant on that existence is not a predicate of non-divine things but still hold that existence is a predicate contained in or entailed by God's essence, Kant takes T1 as pointing to a negative

[32] Forgie (2008), 130.

[33] One could also argue that Suárez anticipates Kant on T1 and T4, since his view that essence and existence cannot be separated in actuality entails that existence is not among the determinations that exhaust the reality (*res*) or the quiddity of a thing, and thus, does not add anything further to the merely possible. This has not gone unnoticed in the literature. Seigfried (1972) presents a plausible account of the anticipation, where he most notably claims "It is, therefore, not revolutionary at all when Kant insists that a thing 'can never contain more in its actuality than is contained in its complete possibility.' (B 287)" (511).

aspect of existence in general, applying to God as well as non-divine beings. While Kant maintains that T1 is devastating to the ontological argument, as it is constructed by its previous proponents, he obviously does not think that the ontotheological project is thereby doomed. For moving from this diagnosis, Kant formulates an alternative ontological argument, which purports to prove the existence of God *a priori* without treating existence as a predicate and thus is immune to his own objection based on T1. Therefore, Kant's negative conception of existence cannot be understood as marking a revolutionary break from the tradition. Instead, it should be understood as serving Kant's rather revisionist agenda regarding ontotheology.

I propose that the novelty of Kant's account of existence lies rather in his positive theses (T2, T3, T5). For they display the spark of a historically revolutionary shift in understanding not only existence but modality in general, even though the Kant of the *OPA* does not yet seem fully cognizant of what he has discovered and thus does not develop what is really novel in these theses. Kant's most important discovery here is that existence involves a relation to the cognitive subject. To be fair, Kant's overall reflection on the concept of existence in the *OPA* is constructed as primarily aiming to provide an answer to the question of what it means to assert the existence of something (from the viewpoint of a cognitive subject) rather than that of what it means to exist for something in isolation from existence-assertions. However, the positive theses imply the radical idea that existence does not express the way something itself is, but the way in which something is related to the cognitive subject and her representation of that thing. T2 states that existence is a property of the 'thought' or conceptual representation which one [the cognitive subject] has of the thing rather than the thing itself. T3 states that existence indicates a cognitive act, through which the cognitive subject asserts a connection between her representation of the object and her empirical cognition or perception of that object. T5 states that this very connection, and not anything at the level of the intension of the concept, is what is missing in the mere concept or possibility of a thing and is thus what is added to the latter through existence.

Though the precritical Kant just states these positive theses without exploring the significance of the radical idea implied by them, I will argue in Part III that this radical idea will be the basis of his revolutionary theory of modality in the critical period. First, if the difference between the actual and the merely possible lies in the connection between the concept and the object, but not in the content of that representation, then modality can be understood in terms of the kind of the relation between the conceptual representation and the object. Second, if modality is a feature of the relation between the cognitive subject's representation of an object and her cognition of that object, then the conditions under which this relation can be established or an object can be given to the subject's cognition must also be involved in determining the modal status of an object. In other words, modality requires an epistemological inquiry into the conditions of cognition of objects in general and whether and how a particular object satisfies those conditions. Third, such an account suggests a relative conception of modality, i.e., relative to the conditions of the subject's (empirical) cognition of objects and thus to the realm of objects of experience. Kant's revolution in modality will fundamentally consist in redefining modal notions as different 'modes' in which the relation between the conceptual representation and the object obtains against the background of the subjective conditions of the empirical cognition of objects rather than the 'modes' in which things are in isolation from those conditions. So the radical idea of existence that remains unexplored in Kant's precritical and revisionist agenda constitutes the core of the revolutionary theory of modality that Kant will develop in his critical period.

4

Kant's 'Only Possible Argument', Possibility and Necessity

In the last section of the *OPA*, Kant divides all the arguments for God's existence into two main kinds according to the sources they derive from: those which derive from "the concepts of the understanding of the merely possible" and those which derive from "the empirical concept of the existent" (Ak. 2:156). He calls the first, *a priori* kind of arguments "ontological" (Ak. 2:160), and further divides them into two types according to the route they take to the existence of God: those which proceed "from the possible as a *ground* to the existence of God as a *consequence*" and those which proceed "from the possible as a *consequence* to the divine existence as a *ground*" (Ak. 2:156).

As I argue in the previous chapter, Kant takes his negative conception of existence to be fatal to the previous versions of the ontological argument, because they all proceed from the merely possible as a ground to existence as a consequence, and this route requires them to treat existence as a predicate in one way or another. However, Kant still holds that the ontotheological project of proving God's existence *a priori* is salvageable, if it is so reconstructed that it does not treat existence as a predicate. The overarching argument of the *OPA* is that among all possible types of arguments for God's existence, the only one that has a prospect of succeeding is the one that takes the alternative route "from the possible as a consequence to the divine existence as a ground" (Ak. 2:156). Kant intends what he offers in the *OPA* as a framework of an argument, "a main draft" or a "ground of proof" (*Beweisgrund*) (Ak. 2:66), which can be utilized in constructing specific arguments for God's existence.

An argument thus constructed would be immune to Kant's own objection to the previous versions of the ontological argument. In the two precritical texts Kant formulates this framework of an argument, the *NE* and the *OPA*, he is committed to distinct and even mutually exclusive conceptions of existence. As I showed in the previous chapter, while Kant still accepts the dual conception of existence, mental and real, and endorses the first line of objection to the ontological argument in the *NE*, he strictly rejects the inclusion of existence into the conception of anything and favors the second line of objection in the *OPA*. However, this distinction does not lead to a significant difference between his respective formulations of the framework. For this framework just does not rely on any specific conception of existence. Instead, in both of its main precritical statements, it is constructed upon specific conceptions of possibility and necessity.[1] I suggest that if the precritical Kant's objection to the

[1] So I diverge from Henrich (1960), 181–2 and Reuscher (1977), 18–32 who contest the continuity between the arguments in the *NE* and the *OPA* on the grounds that in the former Kant still endorses the idea that existence is a predicate.

traditional versions of the ontological argument reveals his negative and revisionist views on existence, his alternative ontological argument reveals his revisionist views on possibility and necessity.

This argument has enjoyed a wide interest and diverse reconstructions in the literature.[2] I will provide here what I take to be an accurate reconstruction that does not rely on a particularly controversial or contested premise. Though I will follow the more systematic statement of the argument in the *OPA*, I will also pay attention to its earlier statement in the *NE* to illustrate some of the nuances between the two.[3] I will begin with an explication of the conceptions of possibility and necessity at work in the argument.

4.1. Distinctions in Modality

Kant's argument rests on a distinction between two kinds of modality: logical/formal and real/material. It should be noted, however, that it is not textually obvious that the Kant of the 1750s and early 1760s has a distinct notion of 'real modality.' Both in the *NE* and the *OPA*, the distinction Kant offers is rather between two different elements or conditions of possibility, 'formal or logical element' (*das Formale der Möglichket*) and 'real or material element' (*das Reale der Möglichket*), and not between two kinds of possibility such as 'logical' and 'real possibility.'[4] In fact, Kant does not use the term 'real possibility' (*die reale Moeglichkeit*) in any of his precritical works.[5]

Kant comes to fully articulate 'real modality' and sets it against 'logical modality' in his critical period. Accordingly, logical modality pertains to mere thinking and is more properly predicated of the validity or truth of our conceptual representations of things instead of things themselves. Logical possibility, for instance, is the possibility

[2] England (1930); Henrich (1960); Lamacchia (1969); Laberge (1973); Wood (1978); Schmucker (1980); Sala (1990); Dell'Oro (1994); Fisher and Watkins (1998); Schönfeld (2000); Adams (2000); Logan (2007); Stang (2010, 2016); Chignell (2009, 2012, 2014); Boehm (2012, 2014); Abacı (2014, 2017); Yong (2014).

[3] Kant also briefly repeats the argument later in the *Inquiry* (1764). Schönfeld (2000), 197 claims that the 'germ' of the 'only possible argument' could be traced back to Kant's 1759 essay *Optimism*.

[4] See *NE* (Ak. 1:395); *OPA* (Ak. 2:77–8). The concept of the "real element of possibility" (*das Reale der Moeglichket*) is sometimes conflated with that of "real possibility" (*die reale Moeglichkeit*). See, for example, Logan (2007), 351; Byrne (2007), 42.

[5] The only exception I am aware of is R 4196 (Ak.17:452), dated to 1769–70, where Kant does use the term 'real possibility' (*reale Möglichkeit*) without explaining what he means by it. There is another precritical reflection, however, where Kant formulates an extra-logical conception of possibility. In R 4004 (Ak. 17:382), dated to 1769, he writes: "Any concept, in which lies a predicate that is considered in the subject and does not contradict it, is possible; but not every concept in which there is no contradiction is a possible synthesis, i.e., real relations are not seen through the principle of contradiction." Crusius may have been the first to explicitly define the term 'real possibility' as distinct from logical or 'ideal' possibility in his *Entwurf*, which was first published 1766: "If one considers a possible thing only with regard to the question whether it could be thought, that is, whether what one posits (*setzt*) in a concept could be combined (*verbinden*) together without contradiction (*Wiederspruch*), then it is called ideal possibility or possibility merely in thought (*die ideale Moeglichkeit, oder die blosse Moeglichkeit in den Gedanken*). If one, however, . . . can establish that there actually are also sufficient causes available for that thought thing, which only under that condition can come into actuality, then such is called real possibility or possibility outside of thought (*die reale Moeglichkeit, oder die Moeglichkeit ausser den Gedanken*)" (§56).

of thinking a thing through a concept or a proposition without logical contradiction.[6] Logical necessity expresses the necessity of the connection of concepts within a proposition such that the negation of this connection yields a logical contradiction.[7] Real modality, on the other hand, is the metaphysical notion of modality and pertains to the existence of the thing in question, i.e. whether the thing or its state can exist, and if it actually exists, whether it does so necessarily. The underlying motivation of Kant's emphasis on the distinction is that the inferences from logical to real modality are not warranted. Kant's critical references to the real/logical distinction in modality oftentimes figure in the form of a warning against the rationalists, who supposedly succumb to the inferences from logical to real modality.

What gets less attention is that Kant maintains another distinction in modality, which was also used by Kant's rationalist predecessors: relative and absolute modality. In the most general terms, the relative (real) possibility of something is its possibility (of existence) considered with regard to a certain realm of being and its conditions. The absolute (real) possibility of something is its possibility (of existence) considered only in itself, that is, internally and without regard to any specific realm of being and its conditions.[8] On the other hand, while the relative (real) necessity of something is its necessity (of existence) in a specific realm and relative to the conditions of being in that realm, the absolute (real) necessity of something would be its necessity (of existence) in itself, without regard to the conditions of any particular realm of being. Similarly, absolute logical modalities would concern the logical possibility or necessity of concepts or propositions in themselves as opposed to relative logical modalities, which concern whether a proposition is logically possible (i.e., logically compatible with) or logically necessary (i.e., logically entailed by) in relation to a certain other proposition or set of propositions.

As we will see in Part III, the critical Kant restricts real modalities to those relative to the domain of experience and its conditions, which, I will argue, underlies his revolutionary approach to modality in general. The precritical Kant's alternative ontological argument, however, operates at the level of absolute real possibility and necessity.

First of all, Kant presents his argument as an alternative to the previous versions of the ontological argument aiming to demonstrate that God exists as a matter of (absolute) logical necessity, or more precisely, that 'God exists' is logically necessary in itself such that its negation is or yields a logical contradiction. Kant points out that given his T1 existential propositions cannot be logically necessary: "Existence is not a predicate at all, nor is the cancellation of existence the negation of a predicate, by means of which something in a thing is cancelled and through which an internal contradiction could arise" (Ak. 2: 81). Accordingly, the existence of anything, God included, can be denied without logical contradiction, or what is the same, nothing exists with logical necessity. Thus, Kant's argument in the *OPA* replaces the idea of

[6] See Bxxvi, A244/B302, A596/B624; R 4801 (17:732), R 5184 (18:111), R 5556 (18:232), R 5565 (18:235), R 5569 (18:235–6), R 5572 (18:237), R 5722 (18:335), R 5772 (18:349–50).

[7] See A226, A593–4/B621–2; R 4033–5 (17:391–2), R 5572 (18:237), R 5569 (18:235–6).

[8] Kant will change his conception of absolute possibility in the *CPR*. See B285, B38, cf. R 4297 (Ak. 17:499), 5181 (Ak. 18:110), 6376 (Ak. 18: 696).

what could be called an *ens logice necessarium*, a logically necessary being, with that of an *ens realiter necessarium*, a really necessary being. This is also clear from his admission that the cancellation of all existence, the state of absolute nothingness, is in fact not logically contradictory in itself.[9] So when he says that "it is absolutely impossible that nothing at all should exist" (Ak. 2:79), he refers to a state of (absolute) real impossibility. Therefore, Kant's argument points to at least one (absolute) real necessity (i.e., God's existence) that is not (absolutely) logically necessary, and one (absolute) real impossibility (i.e., the cancellation of all existence) that is not (absolutely) logically impossible.[10]

4.2. The Novelty of Kant's Conception of Real Modality

In chapter 2, I challenged the claim that the school metaphysics was committed to 'logicism', explaining modal claims exhaustively in terms of formal-logical principles, and thereby, excluding 'real modality' or ultimately reducing it to 'logical modality.' The implication of this claim is that Kant's distinction between logical and real possibility as two distinct and irreducible kinds of modality is historically novel and amounts to a major advancement over the logicist modal metaphysics of rationalists like Leibniz, Wolff, and Baumgarten. As I hope to have demonstrated, however, this claim is too much of a broad brush on the school metaphysics and does not prove tenable in the face of close enough scrutiny of these figures' accounts of modality.

Both Leibniz and Wolff have quite robust and well-articulated notions of real possibility, defined relative to a possible world (Leibniz's R1) or the actual world (Leibniz's R2 and Wolff's 'extrinsic possibility') and irreducible to freedom from logical contradiction. So if the mentioned historical claim should have any weight, it needs to be restricted to absolute modality and revised as the more modest claim that Kant's introduces absolute *real* modality against the school metaphysicians who hold that absolute modality is definable in terms of the principle of logical contradiction.[11] Even with this refinement, the narrative of logicism would still fall short of doing justice to Kant's predecessors. First of all, they do not conflate the real and logical modalities of things considered in themselves or ignore the former entirely. Leibniz, Wolff, and Baumgarten do understand the possibility of a thing considered in itself as the possibility of its existence and thus as primarily a metaphysical notion, which is why, for instance, Wolff and Baumgarten locate their respective accounts of the

[9] See Ak. 2: 78, 79.

[10] I owe this formulation to Stang. See also Stang (2016), 80. Kant's own retroactive reconstruction of the argument in his lectures on religion (Pölitz) leaves no doubt that he construes himself as having employed the notion of real possibility as well as that of real necessity in the original argument: "For in addition to the logical concept of the necessity of a thing... we have yet another rational concept of real necessity... Of course in the logical sense possibility always precedes actuality... Yet we have no concept of real possibility except through existence, and in the case of every possibility which we think *realiter* we always presuppose some existence..." (*Th.Pölitz*, Ak. 28:1036).

[11] Stang (2016), 15 acknowledges that Leibniz, Wolff and Baumgarten have hypothetical or relative modality in connection with what exists, but admittedly focuses only on their absolute (*in se*) notions of modality.

possibility of a thing (*ens*, *Ding*) at the center of their ontologies. However, in all of the three accounts, (absolute) real possibility (of things) end up being coextensive with (absolute) logical possibility (of concepts of things) and thus the latter turns out to be a reliable epistemic indicator of the former. The main issue here is that despite their recognition of the metaphysical character of the possibility of a thing considered in itself, none of the school metaphysicians formulates an extra-logical or metaphysical conception of internal conflict or incompatibility, without which the domain of what is really possible could not be unambiguously demarcated from that of the logically possible.

Leibniz's special place should nevertheless be underlined here. Although Leibniz defines absolute possibility of a thing in terms of the lack of logical contradiction among the predicates that make up the concept of that thing, he also recognizes the 'actualist principle' which introduces a material condition of absolute real possibility: those predicates that make up the material content of the concept of the thing in question must be grounded in actuality. Leibniz holds that absolute possibilities of individual things are ultimately materially grounded in essences, the intentional objects of God's thoughts. So the actuality upon which all absolute possibility is grounded is the actuality of God's mind or his act of thinking. There are two very important points here. First, Leibniz clearly distinguishes between logical and material grounds. While freedom from logical contradiction is the sufficient ground for the absolute logical possibility of the concept of a thing, the absolute real possibility of that thing requires that it be materially grounded in God's act of thinking. Second, despite the clear distinction between the grounds of logical and real possibility, these two conceptions of possibility seem to pick out the same domain of individuals. For Leibniz does not offer any other formal constraint on the contents of God's thoughts than logical consistency, and maintains that God, through his infinite mind, actually contemplates all logically consistent combinations of concept. So in the case of individuals the two conditions or grounds of absolute possibility, material and formal, collapse into one another, and the absolute logical possibility of an individual turns out to be the sufficient condition of its absolute real possibility. So even though this coextensiveness of absolute logical and real possibility does not result from a failure to recognize the distinction between these two kinds of possibility, it does result in a failure to account for those things that are thinkable without logical contradiction but cannot exist in reality, which is what one would expect a metaphysical conception of modality to account for.

The crucial question here is whether Kant's account of real possibility in the *OPA* lives up to this expectation, or more specifically, whether Kant introduces an extra-logical, i.e., 'real', conception of incompatibility, which would enable him to account for logical possibilities that are really impossible. I will try to answer this question in the following section.

4.3. Absolute Real Possibility

In constructing his alternative ontological argument, Kant makes it clear that he is talking about the real possibility of things considered absolutely or internally: "what I shall be discussing here will always be internal or so-called absolute or unconditional possibility and impossibility, and no other" (Ak. 2:78). For the sake economy, I will drop "absolute" as a qualifier of real possibility from here on in this chapter. Now, Kant offers a condition of real possibility, which will turn into the *nervus probandi* of his entire argument. In its most general form, this condition expresses a dependence or grounding relation between real possibility and actuality: "The internal possibility of all things presupposes some existence or other" (Ak. 2:78). At this level of generality, this condition is in fact what is expressed by 'the actualist principle' (AP) that we first observed in Leibniz. Again, like Leibniz, Kant specifies the principle as expressing the grounding condition of the 'real element of possibility', that is, the predicates that make up the content of the concept of a thing, as opposed to the 'formal element', which consist in the logical compatibility or consistency among those predicates.[12]

In this section, I will only present a brief outline of this condition, and provide my more critical discussion of it, especially regarding its justifiability, after my reconstruction of the argument. However, I should still state here the two main ideas I take to underlie Kant's formulation of the condition. First, the mere thinkability of the predicates of a thing both in themselves and together is necessary but not sufficient for the real possibility of that thing. For the predicates, even if they are logically consistent with one another, may just be chimerical ideas that cannot be instantiated in reality. In other words, for something to be really possible, all of its predicates must be really possible or instantiable in reality. Second, a predicate F is really possible or 'given' or 'grounded' in actuality only through the actual instantiation either of F itself or of another, more fundamental predicate F* that grounds the real possibility of F (i.e., such that F cannot be instantiated without the instantiation of F*).[13] Hence, the actualist principle in its more specified form would be the following.

> <u>The Actualist Principle (AP)</u>: Each and every predicate F of any really possible thing must be grounded in something existent that actually instantiates F itself or a more fundamental predicate F*.

With the addition of this principle, real possibility can be defined as the following.

> <u>Real possibility</u>: Something is really possible, if and only if a) all of its predicates are logically consistent with one another, and b) each and every one of its predicates F is grounded in actuality through either the instantiation of F itself or a more fundamental predicate F*.

Kant's argument employs this conception of real possibility as its pivotal premise. Yet before moving further along, this is a good place to pause and examine a certain interpretation of the conception of real possibility that is at work in Kant's argument. This interpretation is indeed very important, since it is also the basis of an original

[12] Ak. 2:78.

[13] Kant usually uses the term 'given' (*gegeben*) instead of 'grounded' in the *OPA*. See Ak. 2:77, 78, 79. Accordingly, the actuality through which the material content of a possibility is 'given' is the (material or real) ground (*Grund*) of that possibility. Here, I prefer the term 'grounding' to designate this relationship between the possible and the actual.

reconstruction of the argument that has attracted considerable attention in the recent literature.

Chignell claims that in addition to the requirements of logical consistency and the givenness of positive real content, what I called above, respectively, the 'logical' and 'material' conditions of real possibility, Kant introduces a third condition of real possibility in the *OPA*: "The predicates of a thing must be really harmonious with one another."[14] 'Real harmony' is Chignell's term for an extra-logical or metaphysical conception of internal compatibility (or lack of that kind of incompatibility). The idea is that some predicates that are not logically contradictory cannot really be co-instantiated in the same subject, and thus, such a subject would be really impossible. I have argued earlier that a metaphysically irreducible conception of real compatibility (or incompatibility) is missing in pre-Kantian accounts of absolute real possibility, and this is why they fail to demarcate the domains of (absolute) real possibility and (absolute) logical possibility even though they recognize the conceptual distinction between the two notions of possibility. So if Chignell's reading is right and the precritical Kant introduces the 'real harmony' condition, then he indeed makes significant headway toward an account of metaphysical possibility that satisfies our basic intuition that some logical possibilities cannot exist in reality. This would also offer some support to the narrative that the precritical Kant's account of real modality is historically novel, even if the claim that the school metaphysicians commit to a logicist account of modality were to be discounted. Second, Chignell claims that the real harmony requirement motivates not only the central premise of Kant's argument that every possibility must be grounded in some actuality but also the transition to the conclusion that the ultimate single ground of all possibility is God. Third, Chignell claims that the issue of real harmony is responsible for Kant's later abandonment of this argument in his critical period.

I disagree with Chignell's second and third claims. I will show that the 'real harmony' condition plays no role in Kant's argument in section 4.5. I will examine Kant's changing attitude toward his argument in his critical period in chapter 8, and argue that this change has rather to do with the shift in his conception of the actualist principle, which, in turn, suggests a more comprehensive and revolutionary transformation in his overall conception of modality. Therefore, in this section I will only engage with Chignell's first claim that the precritical Kant introduces a 'real harmony' condition on real possibility. I hold that the precritical Kant entertains a notion of metaphysical incompatibility, but he does not incorporate it in his account of real possibility in the form of a condition. There is no compelling textual evidence to uphold Chignell's claim that Kant does so.

The primary textual ground Chignell offers for his claim comes from Kant's various discussions of the concept of 'real repugnance' (*Realrepugnanz*), supposedly the opposite of Chignell's 'real harmony.'[15] Yet two questions need to be raised here: i) Is real repugnance the opposite of 'real harmony', that is, is it a kind of metaphysical incompatibility that results in a state of real impossibility? ii) Does Kant consider

[14] Chignell (2009), 174. See also his (2012). For a broader discussion of Chignell's "real harmony" interpretation, see my (2014), Yong (2014), Chignell (2014).

[15] Chignell (2009), 18.

the absence of real repugnance a condition of real possibility in general? I will start with the latter question.

In neither of the two precritical texts, where Kant provides an analysis of the concept of real possibility, does he count absence of real repugnance among the conditions of real possibility. In fact, in the *NE*, Kant does not even mention real repugnance, and in the *OPA*, the issue of real repugnance does not come up in the section devoted to the analysis of the concept of real possibility, i.e. "The necessary distinction in the concept of possibility" (Ak. 2: 77–8), but only after the argument proper and for reasons that are not relevant to the question of how to conceive real possibility.

Kant's precritical interest in the concept of real repugnance does not lie in its purported relation to the concept of possibility. Kant provides his most extensive account of real repugnance in his *Negative Magnitudes* (*NM*), published a few months after the *OPA* in 1763. Kant's account turns on the idea that the predicates of a subject stand in a relation that cannot be reduced to logical consistency or identity, namely a real relation.[16] Kant focuses on the negative variant of this relation: 'real opposition' (*Realentgegensetzung*) or 'real repugnance' (*Realrepugnanz*).[17] Accordingly, two predicates that are logically consistent may still be in real repugnance and 'nullify' (*vernichtigen*) or 'cancel out' (*aufheben*) each other's effects on the subject. Kant's typical example of real repugnance is that of two motive forces acting on the same body in opposite directions.[18] If two forces have equal magnitudes, they nullify each other and the result for the subject is a state of equilibrium or rest.[19]

Logical opposition or contradiction occurs when two logically contradictory predicates or a reality and its negation such as A and not-A are posited in the same subject. On the other hand, in real repugnance, both predicates represent quantifiable positive realities that are logically not derivable from one another by negation or limitation. These positive realities are opposed to each other only in the sense that their respective consequences quantitatively cancel each other out.[20] This leads to an important difference. Logical opposition results in a state of (both logical and real) impossibility and cancellation of the subject altogether: "The consequence of the logical contradiction is nothing at all (*nihil negativum irrepraesentabile*)" (Ak. 2: 171). On the other hand, real opposition results in a state of (both logical and real) possibility. For it is not the subject that is cancelled out but only the opposing predicates or their effects on the subject. To follow Kant's example:

> [A] body which is both in motion and also, in the very same sense, not in motion, is nothing at all... The motive force of a body in one direction and an equal tendency of the same body in

[16] Beck (1969), 398, 451–2 claims that Kant adopts the concept of real relation from Crusius. What is also noteworthy is that Crusius derives a concept of real possibility from that of real relations, as he defines "real possibility" (*die reale Moeglichkeit*) in terms of the availability of sufficient causes for the possibility in question to come into actuality. *Entwurf*, §56.

[17] Kant also uses 'real contradiction' (*realer Widerstreit*). For the occurrences of these terms, see respectively *OPA* (Ak. 2: 86), *NM* (Ak. 2:171), A264/B320, A273/B329.

[18] *NM* (Ak. 2: 86, 171).

[19] Ak. 2: 86. See also Ak. 2:171, 179, A265/B321.

[20] Kant's examples all reflect this quantitative aspect of real opposition: "motive forces" (Ak. 2:171); "active and passive debts" (Ak. 2:172, 174); "distances covered by a ship with the west wind vs. with the east wind" (Ak. 2:173); "pleasure and displeasure" (Ak. 2:180–1); "vice and virtue" (Ak. 2:182).

the opposite direction... are simultaneously possible in one body. The consequence of such opposition is rest, which is something (*repraesentible*). (Ak. 2: 171)[21]

Although Kant here emphasizes its representability or logical possibility, as opposed to a logical impossibility or 'nothing' (*nihil negativum*), the result of real repugnance is also a state of real possibility. For the two opposing motive forces do not disappear but still act on the body, even if their effects are nullified and thus the subject remains at rest. So the two really repugnant predicates are still co-instantiated by the same subject. This means that real repugnance does not constitute a metaphysical incompatibility that would undermine real possibility. It is then not surprising that Kant would not construe freedom from real repugnance as a condition of real possibility.

According to Chignell this is true of only a "species" of real repugnance, which he calls "predicate-cancelling real repugnance."[22] Admitting that predicate-cancelling real repugnance is what is typically presented by most of Kant's examples (from "mechanics, nautical science, empirical psychology, arithmetic"), Chignell claims that these "easy-to-imagine examples," which Kant picks for pedagogical reasons, do not exhaust the scope of the concept of real repugnance as a "genus."[23] Kant also offers, Chignell argues, a "subject-cancelling" kind of real repugnance, where the opposition of predicates "results in a 'canceling' not merely of their respective effects, but of the subject itself *qua* real possibility."[24]

Chignell draws his primary example of 'subject-cancelling' real repugnance from Kant's discussion of 'ceasing to be' in the *NM* (Ak. 2: 190).[25] Kant suggests that in some cases 'ceasing to be' is the result of the original real ground (i.e., the cause of existence) of the thing getting cancelled by another, opposing real ground (i.e., another positive cause). Chignell quickly concludes that Kant has in mind here the sort of metaphysical incompatibility there is between "the predicates *being water,* and *being XYZ* (where 'XYZ' refers, as usual, to some chemical compound other than H_2O)."[26] Yet Kant in fact talks about the ceasing to be of certain states of substances, e.g. "a movement" (of a physical entity), "representations and desires," or "inner accidents" (of the soul) (2: 190–1), and argues that for "a movement" to stop, there must be a positive motive force (acting on the opposite direction), just as the elimination of a "sorrowful thought" requires a positive effort, for instance, to entertain a happy thought (2: 190). These are all cases of 'predicate-cancelling' real

[21] See also Ak. 2:86 and R 3721–3 (Ak. 17:268–9).

[22] In fact, the term 'predicate-cancelling' is misleading, for, as I suggested above, in real repugnance of this type, not the predicates themselves but their effects on the subject are cancelled. For instance, the opposing moving forces still act on the subject even in the state of equilibrium. In order to emphasize that the result of real repugnance is not a simple lack of the opposing predicates, Kant introduces a terminological distinction between two kinds of negation (*Verneinung, negatio*), and calls the result of real repugnance a 'deprivation' or 'privation' (*Beraubung, privatio*) as opposed to the 'lack,' 'defect,' or 'absence' (*Mangel, defectus, absentia*) of the relevant predicates (Ak. 2:177–8).

[23] Chignell (2014), 55–6.

[24] Chignell (2009), 173.

[25] Chignell (2009), 173.

[26] I will not address here Chignell's Putnam-inspired example of 'water that is not H_2O,' since my point exactly is that this example does not correspond to what Kant has in mind in his precritical discussions of real repugnance. My intention here is not to argue against the general modal-theoretical thesis that 'real harmony' could ground an extra-logical, metaphysical conception of possibility, but against Chignell's specific claim that the precritical Kant makes this connection.

repugnance, for they result in the cancellation not of the subject but of the original state of the subject.

'Predicate-cancelling' real repugnance seems to be what Kant has in mind in the *OPA* as well. Since Kant introduces the notion of real repugnance in his exposition of the being that possesses 'supreme reality', the case of the *ens realissimum* should be the most obvious example in the *OPA* in support of Chignell's interpretation. It does really appear that in the case of the *ens realissimum*, the subject of all positive realities, the real repugnance of predicates results in the cancellation not only of the predicates but also of the subject itself:

> In the most real being of all there cannot be any real repugnance or positive conflict (*Realrepugnanz oder positiver Widerstreit*) among its own determinations, for the consequence would be a deprivation or a lack, and that would contradict (*widerspricht*) its supreme reality. Since a conflict such as this (*Widerstreit*) is bound to occur if all realities existed in the most real being as determinations, it follows that they cannot exist in it as determinations. (Ak. 2: 86)[27]

However, this is in fact still a case of 'predicate-cancelling' real repugnance, albeit a tricky one. Kant's point here is that when real repugnance occurs among the positive determinations of the *ens realissimum*, the subject is indeed cancelled, not because of a real repugnance, but because of the logical contradiction that arises between the definition of the subject as having the predicate of supreme reality and the deprivation resulting from the cancellation of the really repugnant predicates. It is logically impossible for the Supreme Being to have the maximal degrees of all positive realities and yet be deprived of some of them. So in the special case of the *ens realissimum*, (predicate-cancelling) real repugnance leads to a logical contradiction. It follows that we cannot claim that the predicates that are in real repugnance here are metaphysically incompatible such that they cannot be co-instantiated (by any subject). We can at best claim that they cannot be co-instantiated by the *ens realissimum*. Thus, the precritical Kant deems the *ens realissimum*, understood in its all-embracing Spinozistic sense, a "conceptual confusion" (Ak., 2: 85), not because the repugnance in it makes the entity 'really impossible' but because it makes the concept logically incoherent.[28]

Following the structure of Kant's argument for the impossibility of the *ens realissimum*, Stang argues that 'predicate-cancelling' real repugnance can be shown to result in a 'subject-cancelling' one.[29] Stang's strategy is based on introducing into the concept of a being possessing really repugnant predicates one of the

[27] For another instance of the same reasoning, see also *NM* (Ak. 2:200–1).

[28] Kant runs a similar reasoning when he discusses "whether the properties of understanding and will are to be found in the Supreme Being as determinations inhering in it, or whether they are to be regarded merely as consequences produced by it in other things," and concludes that God must be a mind having both understanding and will as his own determinations (*OPA*, Ak. 2:89). Chignell (2009), 174 takes Kant's conclusion to rely on a subject-cancelling real repugnance between the predicates of 'being the Supreme Being' and 'being the *natura naturans*' (with no will and understanding). In fact, however, Kant again observes a logical contradiction between the lack of understanding and will and the supremacy of the Supreme Being. Kant suggests that "if the Supreme Being did not possess understanding and will, every other being... with these properties would... have to take precedence over the Supreme Being" (Ak. 2: 88), and such a god that has no understanding and will "would nonetheless be far inferior to what one must needs think when one thinks of a god" (Ak. 2: 89).

[29] Stang (2016), 96–7.

consequences that is cancelled by their conflict. Assuming that A and B are two equal but opposite forces acting upon the same body, Stang maintains that the "concept <body acted upon by forces A and B and no other forces and moved in direction a>" is a "logically consistent but metaphysically impossibly instantiated concept."[30] As clever as Stang's strategy may be, it is doubtful that his example presents a genuine case of metaphysical incompatibility between positive predicates. Stang is right that 'being acted upon by forces A and B and no other force' is a real and not a logical ground for 'not moving in direction of a' (or 'being at rest'). A and B cancel each other out and in the absence of another force, the body in question has no real ground for moving in direction a (or any other direction). Yet again, it is the logical contradiction between this real or causal consequence, 'not moving in direction a', and 'moving in direction a', which ultimately leads to the impossibility of the instantiation of the concept of the body in question. This same point might perhaps be more easily pushed from the viewpoint of Stang's 'logicist' interlocutor. Let us assume that this logicist is a liberal one: she embraces the Leibnizian notion of 'complete concept' but also concedes that causal relations are logically irreducible. The complete concept of an individual body would include all the causal laws of the particular world in which it inhabits. Then, the logicist would still have a strong argument for the claim that 'conforming to causal-dynamical laws', 'being acted upon by A and B and no other forces', on the one hand, and 'moving in direction a', on the other hand, are logically contradictory, and thus the concept of a body that has all these three predicates is in fact logically inconsistent.

So Stang's example essentially reiterates the main point Kant makes in his criticism of the all-inclusive concept of the *ens realissimum*: a real conflict between some of the positive predicates of a subject can lead (via its real effect of mutual predicate cancellation) to a logical contradiction with some other predicates of that subject. But it does not prove the stronger conclusion that the 'real repugnance' that Kant frequently refers to in the *OPA* and the *NM* is about logically compatible predicates that cannot be co-instantiated in any subject regardless of the other predicates involved. This is important, because only if there is such an absolute metaphysical incompatibility between some predicates, should the lack of it (i.e., 'real harmony') be considered a distinct condition of real possibility.

Having said all this, however, one passage that Chignell cites from the *OPA* seems to present a genuine case of 'subject-cancelling' real repugnance. Again in the context of the specification of what predicates God would have to possess, Kant writes "[All realities] can by no means co-exist together as determinations in a subject... The impenetrability of bodies, extension and such like, cannot be attributes of that which has understanding and will" (Ak. 2: 85). Unlike the previous cases, where the impossibility is due to the logical contradiction between the consequence of two really opposing predicates and a third predicate of the subject, in this case the two predicates, 'being extended' and 'thinking', cannot be co-instantiated in any subject. One would then think that freedom from such metaphysical incompatibility should be a condition of real possibility. However, this reading is impugned when Kant goes

[30] Stang (2016), 97.

on to explain the quoted statement. He first concedes that the incompatibility in the case is not a logical contradiction. Yet, again, he explains 'real repugnance' in terms of opposing forces acting on a body and cancelling out each other's effects. He even goes as far as to say that these forces are "really possible in a body" because the result of their opposition is rest, which is "indubitably, possible" as opposed to the "absolute impossibility" which results from a logical contradiction (Ak. 2: 86).

This insensitivity to the nuance between 'subject-cancelling' and 'predicate-cancelling' real repugnance demonstrates either that Kant is not aware of the importance of this distinction for real possibility, or that he does not think that the two are distinct kinds of real repugnance. In either case, the Kant of the *OPA* does not take freedom from 'subject-cancelling' repugnance to be a condition of real possibility in general. I think this observation alone, if correct, casts serious doubt on Chignell's claim that Kant's ontological argument in the *OPA* operates on the notion of real harmony. In section 4.5., I will demonstrate that the argument does not need to and actually does not employ the notion of real harmony.

4.4. Absolute Real Necessity

The literature on Kant's 'only possible argument' has mostly focused on Kant's notion of (absolute) real possibility. This is understandable given the importance of real possibility in the construction of the argument. In fact, however, equally essential for the argument is Kant's treatment of the concept of (absolute) real necessity. Moreover, while, if my opposition to Chignell's 'real harmony' reading is justified, the precritical Kant's notion of (absolute) real possibility cannot be said to go much farther than Leibniz's, he presents an unprecedented account of (absolute) real necessity in the *OPA* and the *NE*.

Kant's account of real necessity is motivated by his criticism of two traditional conceptions of absolutely necessary being (*ens necessarium*) that figured respectively as the products of the cosmological and ontological arguments: 'a being whose ground of existence is contained itself' and 'a being whose non-existence is a logical contradiction.' Kant's criticism of the former is based on his approach to the PSR. In section 2 of the *NE*, Kant distinguishes between two kinds of grounds: "antecedently determining" and "consequentially determining," the former being "the ground of being or becoming" (*ratio essendi*), the latter "the ground of knowing" (*ratio cognoscendi*) (Ak. 1:392). When the issue at hand is the existence of something, the antecedently determining ground is simply the reason *why* that thing exists, the consequentially determining ground, on the other hand, is the ground of our cognition *that* that thing exists. Accordingly, Kant concludes, the antecedently determining ground necessarily precedes that which is determined (being), whereas the consequentially determining ground presupposes that which is determined (cognition).

The cosmological argument operates on the idea that the ultimate sufficient ground of the series of contingent beings in the world cannot be found within the series itself and thus must lie outside the series and in an absolutely necessary being, which, by containing its own sufficient ground within itself, constitutes an end to the infinite regress of the series of grounding. This reasoning involves the conception of the absolutely necessary being as a being that is self-grounding or that exists through itself, an *ens a se*. However, Kant rejects this conception as absurd, for the ground of existence is the antecedently determining ground, and the latter, by definition, precedes that which it determines and thus cannot be identical with it. The very idea of a being, which contains its ground of existence within itself, is incoherent. But since an absolutely necessary being cannot be grounded by another being either, Kant concludes, in the case of such a being, "an antecedently determining ground is completely absent" (Ak. 1:394). So the existence of an absolutely necessary being cannot be explained in reference to a ground, even though it is itself the ground of everything else.[31]

Interestingly enough, the tradition assumes that the existence of a self-grounding being is also logically necessary. As I showed in chapter 2, Leibniz, Wolff, and Baumgarten treat the cosmological and ontological arguments as components of a unified proof of the existence of an absolutely necessary being. This results in an identification of the cosmological conception of the absolutely necessary being as an *ens a se* and the logical conception of the absolutely necessary being as that whose non-existence is a logical contradiction. Kant also rejects this logical conception. Kant's T1 renders the concept of a being whose non-existence is logically contradictory as absurd as that of a being that contains its own ground of existence: "Existence is not a predicate at all, nor is the cancellation of existence the negation of a predicate, by means of which something in a thing is cancelled and through which an internal contradiction could arise" (*OPA*, Ak. 2:81). If existence cannot be included in the intension of the concept of anything, then the negation of existence never leads to a contradiction in that concept or the non-existence of a thing is never logically impossible. Absolutely necessary existence (or absolute real necessity) must therefore be conceived in terms other than those of self-grounding or logical necessity.

Although the precritical Kant is not shy about criticizing previous conceptions of absolutely necessary being, he does not present this criticism as a distinct objection to the ontological argument, which Henrich would categorize under a third, 'critical' line of objection.[32] Furthermore, Kant's critique of the traditional understanding of absolute real necessity does not lead him to give up on the metaphysical project of proving the existence of an absolutely necessary being. Instead, he introduces a new conception of absolute necessary being to play a fundamental part in his alternative ontological proof.

Kant's account of absolute real necessity unfolds in the third reflection of the *OPA*. He starts with a nominal definition of absolute (or internal) necessity in general.

> <u>Absolute necessity in general:</u> "That of which the opposite is impossible in itself is absolutely necessary" (Ak. 2:81).

[31] Schmucker (1980), 65 sees this rejection of the idea of a self-grounding being in the *NE* as an anticipation of Kant's T1 in the *OPA*. Despite its appeal, I do not think that there is sufficient textual evidence for this view. Furthermore, Kant's commitment to the first, Thomistic line of objection to the ontological argument in the *NE* makes it very hard to think that he has T1 in mind at the time.

[32] Henrich (1960), 75–6.

This definition both fits the logical conversion rule between necessity and impossibility (i.e., $\Box P \leftrightarrow \neg\Diamond\neg P$) and is the standard school definition of a necessary being (i.e., "*necessarium est, cuius oppositum est impossibile*").[33] Kant points out that all concepts of absolute or 'internal necessity' have thus far been understood logically, for 'internal impossibility' has been understood as logical impossibility or as self-contradictoriness. However, he also states that "the necessity, for which I am seeking the ultimate foundation, namely, the necessity of existence, is absolute real necessity (*die absolute Realnothwendigkeit*)," which cannot be logically construed (Ak. 2:82), since, given T1, the non-existence of a thing is never logically impossible.

Thus absolute real necessity is to be defined through a non-logical conception of absolute impossibility, namely, absolute real impossibility. This will also connect Kant's account of absolute real necessity to his account of absolute real possibility and preserve the convertibility relation between necessity and possibility across the board. The challenge here is to define absolute real impossibility without reference to a notion of internal conflict. For logical contradiction is out of question, and Kant does not employ 'real repugnance' or any kind of metaphysical internal incompatibility in defining absolute real impossibility either.[34] Kant's account of absolute real impossibility overcomes this challenge by invoking the actualist principle or the material condition of absolute real possibility. However, he first introduces a generic concept of absolute impossibility that would capture both logical and real variants of it.

> <u>Absolute impossibility in general</u>: P is absolutely impossible if and only if it 'eliminates' (*vertilgen*) or 'cancels' (*aufheben*) all possibility in general.

There are two points here. First, 'elimination' or 'cancellation' is used in the sense of entailment of negation: P cancels Q iff $(P \rightarrow \neg Q)$. But here the condition of absolute impossibility is the cancellation of all possibility or everything that is possible: P is absolutely impossible iff $(P \rightarrow \forall(x)\neg\Diamond x)$. So absolute impossibility is defined in terms not of a conflict (logical or otherwise) but of the removal of any possibility whatsoever. Second, since logical possibility is a condition also of real possibility, what is absolutely logically impossible must be absolutely impossible in general. As I suggested earlier, absolute logical impossibility is to be understood in terms of self-contradictoriness.

> <u>Absolute logical impossibility</u>: P is absolutely logically impossible if and only if it is a logical contradiction in itself.

Now, if the absolute logical impossibility of something entails its absolute impossibility in general, we should be able to reframe the former in terms of the cancellation of all possibility as well. But the question arises: In what sense self-contradiction cancels all possibility? Kant suggests that "through the cancellation of the law of contradiction, the ultimate logical ground of all that can be thought, all possibility vanishes" (Ak. 2:82). This is yet another way of formulating impossibility in terms of

[33] *Metaphysica*, §101.

[34] This further undermines Chignell's "real harmony" reading. For if Kant really treated "real harmony" as a condition of real possibility, one would expect him to define absolute real impossibility and/or absolute real necessity in terms of "subject-cancelling real repugnance" or an equivalent of it.

the cancellation of the ground of possibility. The clue to defining absolute real impossibility (and necessity) lies exactly here. If whatever cancels the ground of possibility is impossible, then we can define both logical and real impossibility without appeal to any kind of internal conflict.

Absolute logical impossibility*: P is absolutely logically impossible if and only if it cancels the *ultimate* logical ground of all (logical and real) possibility.

Absolute real impossibility: P is absolutely really impossible if and only if it cancels the *entire* material ground of all real possibility.

While the ultimate logical ground of all (real and logical) possibility is obviously the law of contradiction, it is not so obvious what the ultimate material ground of real possibility would be. We can draw a clue from the definition of real possibility that Kant offers in the second reflection: Something is really possible, if and only if a) its predicates are logically consistent, and b) its predicates are really possible or grounded in actuality. So the cancellation of all real possibility could occur in two ways: a) the cancellation of the law of contradiction, b) the cancellation of all really possible predicates. And the latter occurs when whatever materially grounds all really possible predicates itself gets cancelled. But there are two alternatives here. There may be multiple beings jointly grounding all really possible predicates, in which case,

Absolute real impossibility*: P is absolutely really impossible if and only if it cancels all of the material grounds of all real possibility.

Or there may be a single, ultimate ground of all real possibility that singlehandedly grounds all really possible predicates, in which case,

Absolute real impossibility**: P is absolutely really impossible if and only if it cancels the ultimate material ground of all real possibility.

While all of the three definitions of absolute real impossibility are true, the first one that refers to the 'entire material ground', without specifying the quantity of grounds, is the most generic. Absolute real necessity or absolute necessity of existence can then be defined through the conversion of this conception of absolute real impossibility.

Absolute logical necessity: P is absolutely logically necessary if ¬P cancels the ultimate logical ground of all (logical as well as real) possibility.

Absolute real necessity: P is absolutely really necessary if ¬P cancels the entire material ground of all real possibility.

In both definitions, 'ground' can be replaced with the respective element of possibility, since the cancellation of the ground is impossible exactly because it will result in the cancellation of the element that it grounds. So Kant writes: "Something may be absolutely necessary either when the formal element of all that can be thought is cancelled by means of its opposite... or, alternatively, when its non-existence eliminates the material element... of all that can be thought" (Ak. 2:82).

Since existence is never a matter of logical necessity, absolute logical necessity can never apply to a thing or its existence. So P in the definition is always a proposition. Absolute real necessity, on the other hand, is always a necessity of existence.

However, it can come in both *de re* and *de dicto* forms, applying, respectively, to a thing's existence and to a proposition about existence.

Absolute real necessity (*de re*): Something P is absolutely really necessary (exists with absolute real necessity) if and only if the non-existence of P cancels the entire material ground of all real possibility (= the material element of all real possibility = all really possible predicates).

Absolute real necessity (*de dicto*): A proposition P is absolutely really necessary (true with absolute real necessity) if and only if ¬P cancels the entire material ground of all real possibility (= the material element of all real possibility = all really possible predicates).

When P is a thing (in the *de re* conception), it must be nothing but the ultimate material ground of all real possibility itself, assuming 'ultimate' indicates the end of a regressive series of grounds. This means that anything that is the ultimate material ground of all real possibility (whether a single being or multiple beings) would exist with absolute real necessity. However, when P is a proposition (in the *de dicto* conception), it is the assertion that something (rather than nothing) exists, for the absolute non-existence of all things would cancel all the material element. As we will see, the transition from *de dicto* to *de re* absolute real necessity will be a central move in Kant's 'only possible argument.'

There are two further points I would like to underline before engaging with the argument. First, the definition of absolute real necessity does not entail that this concept actually applies to anything or any proposition. Right before his presentation of his argument, Kant states, "That this concept, however, is not imaginary but something true is apparent from the following consideration" (Ak. 2:82). And 'the following consideration' is Kant's argument itself. So Kant takes his argument to be a confirmation of the meaningfulness or non-emptiness of the concept of absolutely necessary existence. But as I will discuss in chapters 7 and 8, in his critical period Kant abandons the concept of an absolutely necessary being as "unintelligible" and "illusionary" (A593–4/B621–2).

Second, since Kant emphatically rejects the idea that the absolutely necessary being contains its (antecedently determining) ground of existence within itself, this being could only have a consequentially determining ground, that is, a ground for our cognition of its existence. It follows that if we are to have a valid proof of the existence of an absolutely necessary being, it must move from the consequences of the existence of this being rather than the very concept of it. While the absolutely necessary being is the antecedently determining ground of all real possibility, all real possibility is the consequentially determining ground of our cognition of this being. Thus, our only possible cognition of this being must move from all real possibility. This shift in strategy is what distinguishes Kant's ontological argument from the previous versions.

4.5. The Argument

In both of his statements of the argument in the *NE* and in the *OPA*, Kant follows a two-step strategy. He aims to prove first that it is absolutely really necessary (*de dicto*) that something (rather than nothing) exists, and then, second, that there is a unique being, the ultimate material ground of all real possibility, which exists absolutely really necessarily (*de re*). The starting point of the argument is Kant's definition of real possibility:

(1) Something is really possible if and only if a) its predicates are logically consistent with one another; b) all of its predicates are really possible in virtue of being *given* or *grounded* in something existent.

The second condition renders real possibility materially dependent on actuality such that the material content or element of the concept of a really possible thing must be grounded in actuality. The material content may be composed of logically complex predicates that are themselves derived from other predicates through logical operations (negation, disjunction, conjunction etc.). So this material condition must directly apply to the logically simple predicates that cannot be analyzed into further predicates which, again, would need to be grounded. So here is a more precise formulation of this condition or the 'actualist principle':

(2) Each and every (logically simple) predicate F of any really possible thing must be grounded in something existent that actually instantiates either F itself or a more fundamental predicate F* that grounds F. (1b)

A few points can be further articulated regarding the nature of the fundamentality and grounding relationship between predicates, without having to specify, at this stage, in virtue of what a more fundamental predicate grounds the real possibility of a less fundamental one:

(2a) F* is more fundamental than F (or F is derivative of F*) in the sense that the actual instantiation of F* in an existent subject grounds the real possibility of F or the possibility of the instantiation of F. So if the actuality of F* is removed or cancelled, the real possibility of F is also cancelled.

(2b) It follows from (2a) that a positive predicate F is more fundamental than its logical opposite ¬F, and if F is a gradable predicate, the maximum degree of F is more fundamental than any less degree of F. For the cancellation of, for instance, a predicate like 'powerful' would cancel the real possibility of 'not-powerful', and the cancellation of 'omnipotent' would cancel the real possibility of the lesser degrees of 'powerful.'

(2c) It follows from (2), (2a), and (2b) that the set of most fundamental predicates, which grounds all really possible predicates or the material content of all real possibility, must be composed only of positive and logically simple predicates that are either non-gradable or, if they are gradable, maximal.

The conditions of real possibility also provide us with the conditions under which real possibility gets cancelled:

(3) Real possibility gets cancelled when a) there is a logical contradiction between the predicates; b) the predicates are not grounded in actuality. (1, 2)

It is, of course, logically possible that nothing exists, as "it cannot be that the negation of all existence involves an internal contradiction" (Ak. 2:78). But in a state of nothingness, no predicate would be materially grounded and nothing would be really possible. And what leads to such cancellation of all real possibility is exactly what Kant defines as absolutely really impossible:

(4) P is absolutely really impossible if and only if P cancels the entire material ground of all real possibility.

(5) It is absolutely really impossible that nothing at all exists, since in that case no predicate would be grounded and all real possibility would be cancelled. (3, 4)

(6) It is absolutely really necessary that something exists rather than nothing. (5)

Concluding the first step of the argument, (6) is a statement of real necessity *de dicto*, a modal status which Kant attributes to a proposition P the negation of which cancels the entire material ground of all real possibility. As I argued in section4.4, 'something exists' is the only proposition that can be true with absolute real necessity. But does Kant have to assume that it is necessary that something is possible in order to reach this *de dicto* necessity claim about some existence? The worry here is that the state of absolute nothingness, where there is no existence and thus no real possibility whatsoever, is logically possible. If it is (logically) possible that there is no real possibility and thus it is logically possible that no real possibility is grounded, how can we claim that it is necessary that something exists so that all real possibility is grounded?

Some commentators have suggested that Kant must have been committed to something like the conditional 'if something is possible, then it is necessarily possible', one of the axioms of the system of modal logic S5.[35] In fact, at least one of Kant's precritical reflections can be read as committing him to even the stronger claim that 'all possibility is necessary', which entails the axiom of S5: "since possibility in general is certainly necessary, so is what contains its ground" (R 3712, Ak. 17:252). However, it would be more charitable to read the first step of Kant's argument as stating a set of analytic truths about real modality based on his definitions of real possibility and necessity rather than endorsing a modal-logical axiom unfamiliar to his readership.[36] If the non-existence of anything refers to a state of real impossibility (4, 5), then, via the rules of conversion, it is really necessary that something exists, which entails that it is really necessary that there is some real possibility.[37] On the other hand, by the same token, if there is some real possibility, the cancellation of it would also be really impossible. So whatever real possibility there is is indeed really necessary. Of course, since no existence is ever logically necessary, no real possibility is logically necessary either.[38]

[35] Chignell (2009), 165; Wood (1978), 70; Stang (2016), 129.

[36] For instance, when he writes, after admitting that the negation of all existence is logically possible, "on the other hand, to say that there is a possibility and yet nothing real at all (*gar nichts Wirkliches*) is self-contradictory" (Ak. 2:78), Kant is only stating that once we posit a real possibility, we must also posit some existence as its ground.

[37] For similar analyses, see also Schmucker (1980), 76, and Boehm (2012), 298.

[38] Kant will later put "the fact that something is possible" as an official premise of his argument (Ak. 2:91). Sala (1990), 123–4, Wood (1978), 70, and Guyer (2000), 358 all point out the indispensability of this premise for Kant's argument.

The more pressing challenge for Kant is to make the transition from the *de dicto* claim that it is absolutely really necessary that something exists, the conclusion of the first step in (6), to the *de re* claim that there exists an absolutely really necessary being. For (6) warrants only that one or more things exist and whatever exists grounds all real possibility. There are two ways in which all real possibility can be grounded in existence: either distributively, by multiple discrete actual beings, or singlehandedly, by one actual being. The second step of the argument aims to show that the latter must be the case and that this single ultimate ground of all real possibility is in fact an absolutely really necessary (*de re*) being. So the second step starts with the definition of absolute real necessity (*de re*):

(7) Something is absolutely really necessary if and only if the non-existence of it cancels the entire material ground of all real possibility.

This only tells us what an absolutely really necessary being would be, if it existed. But in order to rule out the possibility of multiple beings materially grounding all real possibility, Kant needs to introduce a novel premise. He writes, "All possibility presupposes something actual in and through which all that can be thought is given. Accordingly, there is a certain reality, the cancellation of which would itself cancel all internal possibility" (Ak. 2: 83). While at first sight this passage does not strike one as advancing the actualist principle in (2), it does in fact introduce a substantial new premise. The idea in (2) was that every particular real possibility presupposes the existence of a material ground, which could get us as far as (6). Here, however, the idea is that all real possibility, i.e. the totality of all particular real possibilities, presupposes a common material ground in which all real content, i.e. the totality of all really possible predicates, is actually given. Hence:

(8) All really possible predicates must be grounded in one existing thing. (Premise)

Kant holds that a really possible predicate is grounded in something "either as a determination existing within it or as a consequence arising from it" (Ak. 2:79). In light of (2), which states that every (logically simple) predicate F must be grounded through the actual instantiation of either F itself or a more fundamental predicate F* that grounds F, if one thing singlehandedly grounds all really possible predicates, it must instantiate the complete set of the (most) fundamental predicates. For any given predicate is either a derivative predicate and must be ultimately grounded by a fundamental predicate, or it is a fundamental predicate and must be instantiated by this being.[39] We can thus modify (8):

(9) All really possible predicates must be grounded in one existing thing that instantiates the complete set of the fundamental predicates. (8, 2)

[39] 1. All predicates must be grounded in one being G.
2. Any given predicate is either derivative or fundamental.
3. If a predicate is derivative, then it must be ultimately grounded by a fundamental predicate.
4. If a predicate is fundamental, then it must be instantiated by G.
5. G must instantiate all fundamental predicates.

Such an ultimate material ground of all real possibility would also be an absolutely really necessary being, since its non-existence would eradicate the entire ground of all real possibility.

(10) There exists an absolutely really necessary being. (6, 7, 9)

(11) There exists an absolutely really necessary being (i.e. *ens necessarium*), and this is a being that instantiates all of the fundamental predicates (i.e. *ens realissimum*). (9, 10)

4.6. The Singularity of the Ground

The success of the entire argument depends on whether two of its premises, (2), the actualist principle, and (8), the claim that all real possibility must be grounded in a single actual being, can be justified. I will first discuss the latter.

Kant's transition from the (*de dicto*) necessity of some existence to the existence of a single (*de re*) necessary being would not be possible without (8). But, as many commentators have asked, why should we accept (8)?[40]

Now, the first step of the argument purports to prove that it is absolutely really necessary that some material ground exists so that all really possible predicates are grounded. This presents us with at least two possible scenarios:

i. There is a single ultimate ground, instantiating the complete set of all fundamental predicates.
ii. There are multiple grounds: each one instantiates a subset of all fundamental predicates, but altogether instantiate the complete set of all fundamental predicates.
    a) Each of these grounds is absolutely really necessary.
    b) Each of these grounds is absolutely really contingent.

(8) warrants the first scenario, so Kant needs to explain why the second scenario is impossible in order to justify (8).

Kant indeed offers an argument that there cannot be multiple absolutely really necessary beings grounding all real possibility.

> Suppose that A is one necessary being and that B is another. It follows from our definition that B is only possible in so far as it is given through another ground, A, as the consequence of A. But since, *ex hypothesi*, B is itself necessary, it follows that its possibility is in it as a predicate and not as a consequence of something else; and yet, according to what has just been said, its possibility is in it only as a consequence, and that is self-contradictory. (Ak. 2:84)

The argument relies on the definition of absolutely necessary being provided in (7) as something that grounds all real possibility by instantiating all fundamental predicates. Such a being would not have to depend on anything else for the grounding of its own predicates. None of its predicates would be 'in it as a consequence of something else.' If A and B are both absolutely really necessary and non-identical,

[40] See Fisher and Watkins (1998); Wood (1978); Adams (2000), Yong (2014); Stang (2016).

then each would have to ground the real possibility of the other, which would not be fitting for an absolutely necessary being. Again, if A and B are non-identical, each would have to instantiate a subset of all fundamental predicates. But in that case neither would be grounding all real possibility, and thus, neither would be absolutely really necessary. A being that instantiates only a subset of all fundamental predicates would be absolutely really contingent, for the non-existence of it would not cancel all real possibility: "in the absence of which . . . there is . . . still something possible—the opposite of such an existence is possible in the real sense; and in that same real sense it is also contingent" (Ak. 2:83). On the other hand, if both A and B instantiate the complete set of fundamental predicates, they would have to be identical. Therefore, "it is not possible for several things to be absolutely necessary" (Ak. 2:83).[41]

However, this argument does not really prove (8) but presupposes it. Given the definition of (*de re*) absolute real necessity as that whose non-existence would cancel the entire ground of all real possibility, the uniqueness of the absolutely necessary being follows trivially. However, without (8), this still does not show us that it is necessary that there is such an absolutely really necessary being that grounds all real possibility single-handedly, but that *if* there is indeed such a being, it must be unique. Furthermore, Kant's argument seems to only eliminate one variant of the multiple-grounds scenario: there are multiple absolutely really necessary grounds. But even if we grant him that, he cannot eliminate the second variant of the multiple-grounds scenario without (8): there is not a single absolutely necessary ground but there are multiple absolutely really contingent grounds. In this case, each of these grounds, which are really contingent in themselves would instantiate different subsets of all really possible predicates (fundamental or derivative), constituting an absolutely really necessary totality without any of them being so individually, and thus doing the job of grounding distributively.[42] This scenario is entirely compatible with the rest of Kant's account of real modality and his insistence that there can only be one absolutely really necessary being.

Various creative solutions to this problem have been offered in the literature. Chignell's 'real harmony' account, though, as I argued earlier, short on offering compelling textual ground, might provide one way to make sense of Kant's emphasis on the singularity of the ground of all real possibility. For if real harmony is a condition of real possibility, then the ground of all real possibility must ground not simply all of the individual really possible predicates themselves, but also of their harmony. That requirement warrants the uniqueness of the ground and justifies (8). Let us assume there are three really possible things in the whole metaphysical space having a limited number of predicates. A has predicates F1, F2; B has F1, F3, F4; and C has F1, F5. Without the harmony condition, a committee of grounds, G1 and G2, respectively instantiating F1, F2, F3 and F4, F5, would ground the real possibilities of A, B, and C. But if real harmony is indeed a condition of real possibility, then this

[41] This rules out the view defended by Adams (2000), 433 that Kant is committed to the idea that a being is necessary if its non-existence cancels any possibility. For, based on Kant's argument against the plurality of absolutely really necessary beings, anything whose non-existence does not cancel all real possibility would be absolutely really contingent.

[42] The observation that Kant fails to consider this scenario has been voiced by Wood (1978), 70–1, Adams (2000), 433; Fisher and Watkins (1998), 375, n. 15.

kind of distributive grounding would not work and the real possibilities of B and C would remain ungrounded. In order for the real harmony condition to be satisfied in the cases of B and C, their respective predicates (F1, F3, F4) and (F1, F5) must be shown to be really harmonious, co-instantiable in one subject. So only a single subject that instantiates the complete set of predicates (F1, F2, F3, F4, F5) can ground the real possibility of all A, B, and C at once.

However, this solution has its own problems. It is a rather uncontroversial aspect of Kant's argument that it sets out to prove that the ultimate ground of all real possibility is a supreme being instantiating only and all fundamental predicates or 'perfections' that do not need any other grounds for their real possibility except the actuality of this being. But if the real harmony interpretation is correct, then grounding all real possibility would surely include grounding the real harmony of a vast number of combinations of mundane, derivative predicates. Let us assume that A is really possible and its predicates F1 and F2 are derivative of the fundamental predicates F1* and F2*. The co-instantiation of F1* and F2* in the ultimate ground of all real possibility would materially ground F1 and F2, but there is no reason to think that the co-instantiation of F1* and F2* would ground the real harmony (or the real possibility of the co-instantiation) of F1 and F2, and thus, the real possibility of A. It appears that the real harmony condition requires the unique, ultimate ground of all real possibility to instantiate not only fundamental predicates but also all really harmonious combinations of derivative predicates. The result is a bloated ultimate ground that looks very much like Spinoza's God or nature. What is worse, if, for instance, A (F1, F2) and B (F1, F3, F4) are both really possible and thus really harmonious in themselves, but F1 and F3 are really repugnant (in a subject-cancelling manner), then the ultimate ground of A and B would be instantiating F1 and F3, which would mean that they were not really repugnant. Chignell recognizes this problem and the remedy he offers is even more Spinoza! The ultimate ground or God instantiates attributes, and each attribute instantiates and thus grounds distinct combinations of really harmonious derivative predicates such as (F1, F2) and (F1, F3, F4).[43] This should be an unpleasant picture for Kant, who holds that such a concept of God is a "conceptual confusion" (Ak., 2: 85). Furthermore, as Yong aptly observes, this picture of explanatorily and causally distinct attributes grounding distinct sets of predicates undermines the very claim of a unique ground of all real possibility, and consequently, the real harmony interpretation loses its main selling point: a solution to the plurality problem.[44]

Boehm suggests that (8) could be justified through the PSR, something well within the realm of Kant's resources, even if he may not have explicitly utilized it in the text. The PSR demands an explanatory ground not just for individual (real) possibilities themselves, but also for their interrelations, which are in fact distinct possibilities in their own right. Boehm argues that in this case, only a single being can satisfy the PSR's demand for grounding all possibilities, for "had certain grounds of possibility been scattered in two or more beings, the relation(s) between these beings themselves would have had to be grounded by yet another being—and so regressively *ad infinitum*. But then, not all possibilities would be grounded."[45]

[43] Chignell (2012), 668. [44] Yong (2014). [45] Boehm (2012), 299.

More recently, Yong presented a similar defense of (8). Arguing that Kant's argument is concerned with accounting not only for the individual possibilities, but also for "the possible relations that obtain between those possibilities" or "the totality of possibility" as a single domain unified through the relations, Yong concludes, via the PSR, that only a single being can ground this totality.[46] For, again, if there were to be multiple beings grounding all of the material content, then the interrelations of these beings would require a further ground.[47]

This way of defending (8) invokes two alternatives. First, if the single being is supposed to ground all individual possibilities as well as their relations, then we are faced with an even more bloated ground than Chignell's, instantiating all real fundamental and derivative atomic predicates as well as relational ones. Boehm embraces this Spinozistic result, and concludes that through the 'only possible argument' Kant commits himself, albeit implicitly (but perhaps knowingly), to Spinozism.[48] If we want to resist this result on Kant's behalf, then we can take the Leibnizian route like Yong does: the single being grounds the 'totality' of all real possibility not through instantiating all real predicates but through thinking them and their possible relations in his infinite intellect.[49] I suggested above that Spinozistic approach ultimately fails to stop the pluralist. And as I will argue in the next section, the Leibnizian conception of grounding real possibilities through the mental representations of God is not compatible with Kant's understanding of the actualist principle and his ambitions in the argument.

Having pointed out the issues with this line of attempt to justify (8), something like the PSR must indeed be behind Kant's commitment to (8). In particular, I agree with Boehm that this is confirmed by the fact that the critical Kant comes to construe (8) as resulting from a transcendental illusion that consists in an illicit assumption that human reason is so strongly compelled to derive from the PSR, i.e., "when the conditioned is given, then so is the whole series of conditions..., which is itself unconditioned, also given" (A307/B364).[50] As I will show in chapter 8, the critical Kant holds that it is giving in to this illusion that leads us to erroneously move from a given distributive unity of conditioned discrete individuals to positing a single unconditioned ground that is never given to our cognition. So Kant himself retrospectively admits that (8) is not justifiable, however rationally appealing it may be.

4.7. What Grounds the Actualist Principle?

The recent literature on Kant's argument has tended to focus on (8) and the problem of the singularity of the ground. There is no doubt that this problem jeopardizes the success of the argument, and is quite intriguing with regard to the implications of the requirement of material grounding of possibility. However, the grounding of this grounding requirement itself or the justifiability of the actualist principle (AP) in (2) deserves at least as much attention. For first of all, the AP is more fundamentally

[46] Yong (2014), 39.

[47] Even more recently, Stang (2016), 138–43, has offered a "refinement" of Yong's suggested solution to the plurality objection.

[48] See Boehm (2012), 308.

[49] Yong (2014), 42.

[50] Boehm (2012), 313.

important for the argument—without the AP the argument cannot even get off the ground. Second, even though the critical Kant does indeed abandon (8) as rooted in a transcendental illusion, as I will argue in chapter 8, the main reason he ends up downgrading the argument from an objectively valid demonstration of the necessary existence of God to a subjectively valid demonstration of the necessary assumption of the idea of God lies in the shift in his understanding of the AP. Third, as I will extensively defend in Part III, this particular shift in Kant's understanding of the AP reflects the very core of his critical and revolutionary conception of modality in general. So we can in fact trace a good deal of the development of Kant's modal theory through the fate of the AP in his thought.

It has been argued that the AP derives from the PSR.[51] Kant's awareness of the relation between the two principles is obvious, especially given that his presentation of the argument in the *NE* is situated within his broader discussion of the PSR. The PSR states that every truth must have an explanatory ground. A modal version of the PSR would apply the idea of explanatory need to modal contexts: real possibilities are not 'brute facts' and must also be explainable. However, this connection to the PSR should not be overstated. For the requirement that there be an explanation for real possibility does not dictate what counts as such an explanation. Kant's AP states a rather specific thesis about what that explanation is and does not follow from the modal PSR without his analysis of real possibility: for something to be really possible, its predicates must be grounded in actuality.[52] But what grounds this thesis? Why must the predicates of a really possible thing be grounded in actuality, and in what sense is this grounding explanatory of its real possibility?

The AP can be construed in at least two ways: i) as an epistemological principle about the way human mind represents real possibilities, or ii) as an ontological principle about the way real possibilities are. It appears convenient to interpret the principle as pointing out the rather uncontroversial yet uninteresting fact that the availability of positive predicates in the logical space is a necessary conceptual condition for our thinking of the possibilities of things.[53] Kant does indeed sometimes write as though the AP is all about this sense of givenness of the "material element" or "datum" of possibility that "can be thought" (*denklich*) (Ak. 2:78).[54] Yet,

[51] See, for instance, Chignell (2009), 157–9; Boehm (2012), 293, and (2014), 16.

[52] This problem undermines Boehm's defense of the AP by reference to a modalized version of what Kant calls in the *Critique* the supreme principle of pure reason: "if something is possible, the complete series of the conditions of this possibility—a series which itself exists conditionally—must be given as well." See Boehm (2012), 297–8, and (2014), 24–5. For, again, this principle only dictates that the conditions of any possibility be given without telling us what those conditions are. The AP tells us precisely the latter and the question here is what justifies the AP itself. Thus, if the AP is to be justified, one has to find a non-circular answer to the question of why real possibility must be grounded in actuality in the specific way the AP prescribes.

[53] This is not an unpopular interpretation. For instance, Wood (1978), 66, and Sala (1990), 78, 119 read the principle along exactly these lines. This sense of availability of conceptual content also seems to correspond to what Chignell calls the "content" requirement of possibility, but he rightly states that it is not a sufficient condition of real possibility. See Chignell (2009), 172, and (2012), 644.

[54] This impression is even stronger in the *NE* version of the argument, where Kant appears to motivate the AP as a condition for conceiving (logical) possibilities: "nothing can be conceived unless whatever is real in every possible concept exists" (*NE*, Ak. 1:395).

this is too weak to be what Kant really means. First, on this reading, the principle would serve only as a condition of our thinking, which, however, would not suffice for real possibility, because the availability of positive predicates in the logical space does not warrant that those predicates designate an object that can exist in reality. Second, this interpretation misses the main insight of the AP: the material element of possibility presupposes some existence or another. The availability in question must thus be understood as groundedness in actual existence.[55]

Alternatively, on a stronger epistemological interpretation, the principle would express a condition of our cognition, as opposed to that of mere thinking, of real possibilities. Kant comes very close to offering this interpretation when he raises the question of how one "knows" (*erkennet*) or what makes one "entitled to accept" (*anzunehmen Recht*) that a "fiery body" can exist in reality (Ak. 2:80–1). He suggests that we can know that a fiery body is really possible if we can prove that fire and body are themselves really possible predicates. But in the absence of actual bodies and fires, the real possibility of these predicates can be known only if the more fundamental predicates, which make fire and body really possible, are themselves known to be really possible. This regressive search for an ultimate explanation can be completed only by an appeal to existence. For we can ultimately prove the real possibility of a fiery body, only if we can prove the real possibility of the fundamental predicates such as extension, impenetrability, and force, at which we arrive by analyzing the concept of 'a fiery body.' But since these predicates can no longer be analyzed, their real possibility can be proved by an appeal to their instantiation in existent things. Kant seems to offer a verificationist motivation for this need to appeal to existence by emphasizing that the logical consistency of our thought of a thing will not settle the question of its real possibility, as its predicates may after all be just "empty words" (*leere Wörter)* that do not "signify" (*bezeichnen*) anything that can be instantiated in reality (Ak. 2:81). This last move can indeed be turned into a promising empiricist justification for the AP: our verification of real possibility depends on our actual experience, for only through the latter can we establish that our logically consistent thoughts are not just chimerical but are of objects that can exist in reality.

Nevertheless, Kant does not further develop the strong epistemological interpretation in the *OPA*, perhaps because this interpretation renders the principle unfit for the desired conclusion of the proof. On this interpretation, the principle expresses a subjective condition springing from the particular structure of our modal cognition: Our cognition of real possibility requires our cognition of actuality. Thus, unless another twist is introduced, any necessity issuing from this subjective condition, a contingent fact about human mind, will also be subjective, while the proof aims to establish an objective necessity, the necessary existence of an object.

[55] In his detailed discussion of the AP, Pinder (1969), 161–2, 199–204 suggests that the 'givenness' (*Gegebensein*) of the material element in the weak sense of being available for thought presupposes absolute positing or actual existence. My worry is that Pinder's formulation still understands the necessity of the reference to actual existence stated by the AP as springing from a condition of thought (i.e. matter be given or available to be thought), and, consequently, identifies the case in which nothing is given to thought as the cancellation of all possibility. As I will argue, basing the proof on such an epistemic condition will undermine the existential conclusion that Kant aims to reach from the outset.

Therefore, for the professed aims of the proof, Kant needs an ontological interpretation of the AP as expressing a condition for something to be really possible.[56] At least two historical variants of this interpretation were available to Kant as of the 1760s. The first variant is what we found earlier in Leibniz: for something to be really possible, its material content must be grounded in the intentional content of an actual mind. Note that this ontological interpretation is significantly distinct from the weak epistemological interpretation. While the latter states that the availability of content is a condition for our thinking of real possibilities, the former states that the actual representation of content by an actual mind is a condition of real possibilities themselves. Accordingly, a divine mind, which represents all predicates, is the ultimate material ground of all real possibility and thus necessarily exists. However, as I suggested in my reconstruction, this variant, though advantageous in avoiding the specter of Spinoza haunting Kant's argument, is not what Kant himself takes. Instead, Kant's argument pursues a non-intentional grounding relationship between God and possibilities, in which God grounds all possibility by instantiating the fundamental predicates as his own determinations.[57]

On the second variant, material grounding is construed in terms of a causal relationship: for something to be really possible, its causes must be actual. One might remember here Descartes' principle that the idea of a thing, insofar as its representational content or objective reality is concerned, must be grounded in the actuality of something that can efficiently cause the thing in question. However, as I discussed in chapter 2, while this principle clearly anticipates the AP, Descartes does not formulate it in explicitly modal terms. A much more explicitly modal formulation of a causal interpretation of the AP can be found in Kant's contemporary Crusius: "If one . . . can establish that there actually are . . . sufficient causes available for that thought thing, which only under that condition can come into actuality, then such is called real possibility (*die reale Moeglichkeit*)" (*Entwurf*, §56). As we saw earlier, Kant emphasizes that the relation of real possibility to actuality can take two forms: "either the possibility is given as a determination existing within the real," in which case the really possible thing or predicate is actually instantiated, or "possibility is given as a consequence through another existence" (Ak. 2:79). Not only is this compatible with a causal interpretation of the AP, there is at least one reflection from the 1760s where Kant seems to embrace this interpretation by formulating material grounding in terms of causal grounding. He writes that real possibility obtains when "that which exists is connected in a certain way, in accordance with laws that are already in it by means of its properties, e.g. if there is wood, then a house made out of wood is possible, or also that an effect flows from forces" (R 3809, Ak. 17:300).

However, the grounding relationship in Kant's argument is supposed to obtain both between the ultimate ground and its fundamental predicates, and between the

[56] This interpretation has been defended by Schmucker (1980), 20, 70–5 and Schönfeld (2000), 203.

[57] For this difference between Kant and Leibniz, see Adams (2000), 425–40, 427; Chignell (2009),182. Nachtomy (2012), 965 claims that the difference between Leibniz and Kant on the issue of God's grounding possibilities is not as significant, for, he argues, Leibniz holds that God grounds possibilities in virtue of both being a 'thinking agent' and instantiating as his attributes the 'simple elements' necessary for what can be thought.

fundamental and the derivative predicates. Causality seems too narrow to capture this relationship. It is not clear in what sense a fundamental predicate F* could be conceived as the cause of a derivative predicate F. Furthermore, there is a serious danger of reducing the AP to the principle of causality *per se*. What does it mean to be the cause of a predicate F? On the most natural reading, it would mean causing that predicate to be instantiated. If all real possibility is indeed grounded, then not only the fundamental predicates, but also all the derivative predicates are caused to be actually instantiated. No unactualized real possibility is left, and the AP would just collapse into causality. Stang, the only commentator who once defended a causal interpretation of the grounding requirement prescribed by the AP and offered an accordingly causal reconstruction of Kant's argument, carefully uses the language of 'causal powers' instead of 'causes.'[58] On Stang's reading, the fundamental predicates are to be conceived not simply as causes of the derivative predicates' instantiation but as God's powers to cause the derivative predicates to be instantiated. Since having the power to cause something does not warrant actually causing it, there is room here for unactualized really possible derivative predicates. Yet what does this reading exactly imply about predicate–predicate grounding? In what exact sense is a fundamental predicate of God, say, 'omniscience', a power to cause its derivatives like 'not-wise' or 'somewhat wise' to be instantiated? What about the predicate 'omnipotence' itself? If 'omnipotence' is an essential predicate of God and his infinite power to bring about anything, why need other fundamental predicates to ground all real possibility?[59]

As an alternative to Stang's reading, one might resort here to a Platonic sort of vertical causation between fundamental predicates, each one of which is presumably instantiated by a Form or an infinite, necessary being, and derivative predicates, instantiated by finite, sensible beings. The Forms can be said to causally ground the real possibility of the sensibles in virtue of the fundamental predicates they instantiate, for the sensibles instantiate their (derivative) predicates only through participating in the fundamental predicates of the Forms. This Platonic picture can offer a working model for a causal interpretation of the predicate–predicate grounding in Kant's argument, once God is conceived as the only Form there is, instantiating all the fundamental predicates and thereby constituting the single causal ground of all real possibility.

The common problem with all of these suggestions regarding the grounding relation in the AP is that none of them is Kant's textually explicit position. What particular ontological interpretation he actually employs remains ultimately ambiguous, and therefore the question of how he justifies this principle remains open. I hold that this ambiguity in the *OPA* is not simply a matter of negligence or failure to clearly communicate an idea on Kant's part. Instead, this ambiguity is quite suggestive and reflects Kant's realization of a tension in his precritical conception of the AP,

[58] See Stang (2010, 2016), 112.

[59] Stang (2016), 144–6 has recently abandoned his 'causal powers' interpretation of Kant's argument, and adopted a more nuanced, agnostic position about grounding. Accordingly, Kant holds that it is in fact not the case that God grounds all real possibility through his causal powers, but 'causal grounding' is the only analogy (with human action) through which we can form a concept of how God grounds all real possibility.

which will grow more visible following the publication of the *OPA* and culminate in a radical transformation in his theory of modality in the critical period. The tension turns around the question of whether the AP expresses an ontological condition of the real possibility of things or an epistemological condition of our cognition of the real possibility of things. While Kant does flirt with the latter, the strong epistemological interpretation of the AP in the *OPA*, he still employs an ontological interpretation in order to for his argument to remain true to its professed objective or existential ambitions. As I will argue in Part III, the critical Kant ultimately abandons his ontological conception of the AP, and adopts the strong epistemological interpretation, for which he provides a solid justification on the basis of his critical theory of modality. This critical shift in Kant's understanding of the AP is vastly important. Not only does it explain Kant's eventual downgrading of his 'only possible argument', but also provides us with an excellent resource to observe the revolutionary leap Kant's modal thought takes in the critical period.

which will grow more visible following the publication of [illegible] [illegible]

PART III

Kantian Modality: Critical and Revolutionary

In Part II, I have argued that Kant's views on modal notions in the *OPA*, though they spring out of specific criticisms of the tradition, ultimately display a revisionist character. These criticisms are oriented toward revising the ontological argument instead of bringing a definitive end to the ontotheological project of proving God's existence *a priori*. Kant remains faithful to ontotheology, aiming to keep it alive by offering an alternative to the kind of argument that he thinks cannot possibly succeed. Thus, whatever advancement or criticism that the Kant of the *OPA* may be said to bring to the tradition with regard to modality should be understood from within this central project of the rationalist tradition.

However, this is only a specific symptom of a deeper continuity between the tradition and the Kant of the *OPA* on modality. Even though Kant strongly commits himself to the thesis (T1) that "existence is not a predicate or determination of a thing," he still does not bring a fundamental objection to the broad conception of modal notions as expressing fundamental ontological features, modes of being, of things. There are two major issues here. First, Kant does not yet explore the implications of T1 for possibility and necessity in general, apart from firmly establishing that no existence, including God's, is logically necessary. This may be because he does not yet recognize existence or actuality as a modal notion along with possibility and necessity as he will in the critical period. Second, Kant does not yet develop what I earlier called the "radical" core idea in his positive theses (T2, T3, T5), construing existence in relation to the cognitive subject, as a feature or predicate of the subject's representation of things instead of things themselves. Again, this may be because his main concern in reflecting on the concept of existence is to undermine the previous versions of the ontological argument, and only his negative thesis, T1, is directly relevant to the argument. And for a more developed positive theory of existence, the reference to the cognitive subject in the positive theses requires an epistemology of the structure and conditions of cognition, which the precritical Kant lacks. In Part III, I will argue that the radical core idea in the positive theses on existence grows into a revolutionary theory of modality, which gets fully articulated in the *CPR* and consists in redefining modal notions (possibility, existence, and necessity) as various modes or ways in which the conceptual representations of things are related to the cognition of the subject rather than various modes in which things themselves are. This theory marks a historical break not only from the rationalist school conception of modality but also from the greater metaphysical tradition, going back to the scholastic reflections on the relationship between essence and existence relationship, which, as I briefly discussed in chapter 1, conceives modalities as basic and genuine ontological features of things.

In Part III, I will also discuss the substantive role of Kant's revolutionary theory of modality in his critical philosophy. In chapters 6, 7, and 8, I will show how Kant's new understanding of modality becomes an integral part of his critical system as it is laid out in the *CPR*. More importantly still, I will argue that the shift from Kant's revisionist views to his revolutionary theory of modality is not merely a logical consequence of Kant's overall critical turn in his intellectual development, along with other components of the critical philosophy, but in fact a very important and constitutive part of the critical turn itself. I will show in chapter 5 that the revolutionary shift in Kant's conception of modality starts in the mid to late 1760s, well preceding and thus independent of his formulation of the very idea of a critical turn in philosophy. This shift starts effecting the critical transformation in Kant's understanding not only of rational theology, the special metaphysic of God, but also of metaphysics in general, before the 1770s, and both of Kant's critique and reconstruction of metaphysics and theology in the Transcendental Ideal extend from this transformation and essentially turn around his revolutionary conception of modality. Therefore, this part of the book aims to offer a reading of the critical turn in Kant's thought through the lens of his revolution in modality. The latter, I will argue, is a motor force of the critical turn and the accompanying system of thought.

5

The Revolutionary Shift in Kantian Modality Prior to the *Critique*

In this chapter, I will focus on the transitional phase between the publications of the *OPA* (1763) and the *CPR* (1781), where we can observe the development of Kant's conception of modality in a revolutionary direction and the role it plays in the broader critical turn in Kant's thought even prior to the *CPR*. A thorough textual account of this development, however, is not easy to provide. For Kant did not publish any work between 1763 and 1781, focusing on modality. Worse yet, he did not publish anything throughout the 1770s, the so-called 'silent decade,' after his *Inaugural Dissertation* (1770), where he introduced some of the major elements of transcendental idealism. Yet it is still possible to trace a consistent development in Kant's views on modality from his *Reflexionen* or annotations in his copy of Baumgarten's *Metaphysica*. These notes reveal how Kant begins to reflect on the implications of the aforementioned radical idea in his positive theses for his 'only possible argument,' the ontotheological project, and metaphysics in general.

It would be helpful to remember here Kant's positive theses briefly. T2 states that existence is a property of the conceptual representation which the cognitive subject has of the thing rather than the thing itself. T3 states that existence indicates a cognitive act, through which the cognitive subject asserts a connection between her representation of the object and her cognition. T5 states that this very connection is what is missing in the mere concept or possibility of a thing and is thus what is added to the concept through existence. These theses attribute a relational and subjective character to existence such that existence assertions necessarily involve a reference to the cognitive subject. If existence is a feature of the relation between the subject's conceptual representation of the object and her cognition, then the epistemic conditions under which this relation can be established must also be considered in determining the existence of an object. As we saw in chapter 3, Kant already holds in the *OPA* that this relation is established or verified through *empirical cognition* (Ak. 2:72–3). This radical idea of existence that remains underdeveloped in the *OPA* transforms into a revolutionary theory of modality in Kant's critical period, which primarily consists in redefining modal notions as different 'modes' or manners in which the relation between the conceptual representation of the object and the subjective conditions of the empirical cognition holds rather than the modes of being of things in isolation from those conditions.

Kant's *Reflexionen* between 1763 and 1781 show that the revolution in Kant's conception of modality starts unfolding in the mid-1760s, and allow us to trace the development of three tightly related lines of thought, each one of which extends from an implication of the aforementioned core idea in Kant's positive theses on existence and underlies one aspect of Kant's emergent revolution in modality: (1) relation to cognition; (2) empiricism; (3) subjectivism.[1]

5.1. Relation to Cognition

I noted earlier that one reason why the Kant of the *OPA* fails to recognize the consequences of his negative and positive reflections on existence for possibility and necessity is that he does not yet construe existence under the heading of modality along with possibility and actuality. In his reflections from around 1770 on, Kant starts reformulating T1, T2, and T3 as applying not only to existence but also to possibility and necessity. For instance, in R 4288 (1770–71) he writes, "Possibility, actuality and necessity are neither concepts of things, nor predicates, but positions (*Stellungen*) of objects" (Ak. 17:497), and in R 4371 (1771), "Possibility and existence are *positiones absolutae*" (Ak. 17:522).[2]

In his reflections from 1770s, Kant underscores the relationality of modalities and their status as second-order predicates of our conceptual representations (of objects), expressing their relation to our cognition in general. For instance, in R 5228 (1776–78), he notes, "Possibility, actuality and necessity are . . . not (real) metaphysical predicates . . . Through them we do not posit or cancel (something in) things, but rather the relation of their concepts to the faculty of the mind" (Ak.18:125). It is also possible to observe a certain pattern through which Kant categorizes each modality as the absolute positing of the relation of the object, or more precisely, its conceptual representation to a distinct aspect of our cognitive faculty. In R 4288 (1770–71), he defines possibility as the positing of something as "the object of fiction"; actuality as the positing of something as "the object of sense"; necessity as the positing of something "through reason" (Ak. 17:494). In R 4299 (1770–71?), he writes "The relation (of an object) to perception (*Warnehmung, perceptio*) is existence; to thought, possibility; to thought, insofar as it determines existence: necessity" (Ak. 17:500). Again in R 4302 (1770–75?), he states, "The object of the concept is possible . . . The object of intuition (*Anschauung*) is actual; the object of intuition, which is also the ground of its own concept, is necessary" (Ak. 17:500). In R 4802 (1775–79), he assigns possibility to the understanding, actuality to sense, and necessity to reason (Ak. 17:733). So, in all these, Kant emphasizes the distinct characteristic of each modality in relation to cognition:

(1) Possibility is the relation of the object to the act and/or the faculty of mere thought (the understanding), i.e., the object is merely thought or represented without being given.

[1] My intention here is not to present a chronological periodization of Kant's development culminating in the *CPR*, as, for instance, Benno Erdmann does in his edition of Kant's *Reflexionen* (1992). The three lines of thought I observe are neither in a chronological nor in a logical order, but they represent the three main perspectives Kant entertains in his interpretation of modality in this transitional phase.

[2] See also the later reflections R 5557 (Ak.18:232) and R 5558 (Ak.18:232).

(2) Existence or actuality is the givenness of the object to the act and/or the faculty of perception, i.e., the representation of the object is connected to the perception of it.
(3) Necessity is the givenness of the object to the faculty of reason through its mere thought or concept, i.e., the connection of the representation of the object to perception is determined by a rational inference.

As I will show in chapter 7, in the *CPR*, while retaining this overall pattern, Kant will reinterpret each of these relations in terms of a relation to the background conditions of empirical cognition laid out by his critical epistemology.

5.2. Empiricism

The Kant of the *OPA* is primarily concerned with the question of what it means to be actual, possible, or necessary, without much regard to the epistemological question regarding the conditions of our cognition of modality. This is not surprising given his interest in utilizing his conception of modality in revising the ontological argument. Yet from the mid-1760s on, Kant appears to gradually develop an epistemology of modality with a strong empiricist emphasis. An important step in this direction is to establish the distinction between the ways we cognize logical and real relations. For instance, in an early note from between 1764 and 1766, Kant states explicitly for the first time that "the cognition of logical relations is rational" or "analytical," while "the cognition of real relations is empirical" (R 3756, Ak. 17:284). This reinforces the empiricist implications of Kant's T1 that existence of anything can be denied without logical contradiction and that existence is not cognizable through derivation from logical relations. Kant further bolsters the empiricist aspect of his thesis, underscoring the radical givenness of existence: "Existence cannot be a predicate, because otherwise a thing could be cognized as existing only through a judgment and by means of the understanding. But we cognize the existence of a thing through sensation" (R 3761, Ak. 17:286). Most importantly, he introduces his sixth thesis on existence, which he will detail in the *CPR*:

T6: Existential judgments are never logically necessary (R 3814, Ak. 17:302), but always synthetic (R 5231, Ak. 18:126).

Kant's focus on the cognition of real as opposed to logical relations in the late 1760s facilitates his move toward an empiricist account of our cognition of real possibility. This account makes both a negative and a positive claim. Freedom from contradiction, which suffices to establish the logical possibility of a concept, is not sufficient to test the real possibility of a thing: For instance, in a note from 1769, Kant writes:

> Every concept, in which lies a predicate that is considered in the subject and does not contradict it, is possible; but not every synthesis in which there is no contradiction is possible, i.e., real relations are not seen through the principle of contradiction. (R 4004, Ak. 17:382)[3]

[3] See also R 5177–8 (Ak. 18:109–10).

Now, combined with his earlier statement that real relations are cognized empirically, this negative claim implies that real possibility requires a positive test based on experience. This positive test follows from Kant's reinterpretation of the actualist principle (AP) that the matter of everything really possible must be 'grounded' or 'given' in something actual. I argued in the previous chapter that in the *OPA*, Kant's ontological ambitions require him to adopt an ontological interpretation of this principle, conceiving it as expressing a condition (*ratio essendi*) of real possibility, even though he clearly flirts with a strong epistemological interpretation of the principle as expressing a condition of our cognition (*ratio cognescendi*) of real possibility. From the late 1760s on, we observe both in his *Reflexionen* and logic lectures that Kant starts endorsing this strong epistemological interpretation of the AP. This is the idea that we depend on our actual experience in order to cognize real possibility. For cognizing the real possibility of something means establishing that the predicates that make up the material content of the concept of that thing are not only logically consistent but also instantiable in reality. But only our empirical verification of the actual instantiation of those predicates themselves or of the fundamental predicates from which they derive can license us to establish that they can be instantiated in reality. In other words, for us to cognize the real possibility of something, the material content of that thing must be *given* to us in experience. So as early as 1769, Kant writes: "by means of reason... we can never entirely cognize the [possibility] of objects, because this involves the presupposition of the possibility of matter, which must be given by means of the sense and therefore *a posteriori*" (R 3999, Ak. 17:381). Similarly, in the early 1770s, he writes, "Even for possibility there needs to be something that is given. The first data are not cognized as possible *a priori*" (R 4483, Ak. 17:569), is reported to state, "The first *principia data* and *materiala* of possibility... must necessarily be given from experience and must arise out of it" (*LB*, Ak. 24:90).[4]

Of course, the strong epistemological interpretation of the principle does not exclude an ontological interpretation of the principle as pertaining to the nature of real possibility in isolation from the question of the nature of the cognitive subject. One might even argue that the former stems from the latter: we are epistemically dependent on our cognition of actuality for our cognition of real possibility because real possibility (ontologically) depends on actuality. However, not only does Kant fail to ground the ontological interpretation in the *OPA*, but he later makes it very clear that the material dependence on actuality is a subjective condition due to our cognitive make-up: "that possibility always presupposes something actual is related to the fact that our understanding can only order data but not produce them" (R 5493, Ak. 18:198); "the first subjective condition... of the representation of the possibility of objects is that all matter of representation be given in sensation and objectively in perception" (R 5526, Ak. 18: 208).

Kant's shift toward the strong epistemological interpretation of the AP has extremely important consequences not only for his conception of real possibility but also for his philosophical development overall. There are at least two major ways

[4] See also R 3930 (Ak. 17:352), R 4006 (Ak. 17:383), R 4244 (Ak. 17:477–8), R 4247 (Ak. 17:480).

in which this shift effect Kant's critical turn, one more specific, the other more general. I will examine the former now and leave the latter for the next section. I have so far identified real possibility as pertaining to the existence of things. But real possibility can be defined in terms of a feature of concepts (or propositions) as well, without conflating it with logical possibility. While logical possibility is the possibility of thinking an object (through a concept) with no logical contradiction, real possibility is the possibility of a concept's being instantiated in reality. The latter can also be expressed in terms of a concept's having an application or relation to really possible objects. So Kant writes:

What agrees with the analytic conditions of thinking is logically possible; what [agrees] with the synthetic [conditions of thinking] is really possible (*real moglich*). Logical without real possibility is the empty concept without content, i.e., relation (*Beziehung*) to an object.

(R 4801, Ak. 17:732)

Kant will later call this feature of concepts having an objective content or relation to really possible objects 'objective validity.' Now, the AP states that our cognition of the real possibility of an object is dependent on our actual experience, for only through the latter can we know that the predicates of the object can be instantiated in reality. Given the interdefinability between the real possibility of objects and the objective validity of concepts, we can take the AP to be stating that our cognition of the objective validity of a concept is dependent on our actual experience. We cognize that a concept has application or relation to (really possible) objects only through experience. While the AP, thus construed, is a perfectly plausible principle, it seems redundant when it comes to the question of the objective validity of empirical concepts, which, as derived from experience, would obviously have application to objects. But the AP presents us with a problem when it comes to the pure concepts that are not derived from experience. How are we to cognize the objective validity of these concepts?

Kant's recognition of this problem in his famous 1772 letter to Marcus Herz is considered by many to mark the critical turn in his 'silent decade' and the genesis of the critical project.[5] In this letter, Kant raises a question that he describes as constituting "the key to the whole secret of metaphysics": "What is the ground of the relation of that in us which we call 'representation' to the object?" (*Corr.* Ak.10:130) This is obviously a question of the ground of real possibility, specifically, the real possibility of a concept's application (or relation) to an object, i.e. its objective validity: 'In virtue of what a concept has relation to an object such that the former is an appropriate representation of the latter?' This immediately raises the tightly connected question about the ground of our cognition of this real possibility: 'How do we know that it is really possible for a concept to have relation to an object?.' Kant acknowledges that it is easy to see how 'passive or sensuous' concepts have relations to objects, as they derive from our experience of the objects they represent. But this cannot be said for our 'intellectual' or pure representations that do not derive from

[5] See, for instance, Guyer (1987), 13 and (2006), 32, Wood (2005), 9, Gardner (1999), 27–30, Mensch (2007) 109.

our experience of objects but originate in our pure intellect or understanding. It would surely be equally easy to say that, if our pure concepts, like those of a divine intellect or an *intellectus archetypes*, actually caused their objects to exist. Kant, states:

However, our understanding, through its representations, is neither the cause of the object (save in the case of moral ends), nor is the object the cause of our intellectual representations in the real sense *{in sensu reali}*. Therefore the pure concepts of the understanding must not be abstracted from sense perceptions, nor must they express the reception of representations through the senses; but though they must have their origin in the nature of the soul, they are neither caused by the object nor do they bring the object itself into being. (*Corr.* Ak. 10:130)

Since we cannot deduce the objective validity of the pure concepts of the understanding empirically, we can only verify it *a priori*. This will be the topic of the Transcendental Deduction, and prove crucial to Kant's answer to the overarching question of the *CPR*, 'how is synthetic *a priori* knowledge possible?.' In the Transcendental Deduction, Kant will argue that we know *a priori* that the pure concepts of the understanding or the 'categories' are objectively valid, because even though they do not derive from experience, they necessarily apply to all (and only) objects of experience. I will revisit this issue in chapter 7, but for now, I would like to clarify my position on the relation between Kant's recognition of the problem of the objective validity of pure concepts in the letter to Herz and his theory of possibility. One might argue that it is this recognition which leads to a change in the mode of inquiry regarding real possibility, a change from the ontological inquiry about real possibility (and its grounds) to the transcendental and epistemological inquiry about our *a priori* representation of really possible objects.[6] So on this account, the change in Kant's theory of possibility is a shift in perspective and is a natural consequence of the overall critical turn in his mode of inquiry. However, I hold that, first, the epistemological shift in Kant's understanding of modality precedes the letter to Herz, and is constitutive of Kant's critical turn rather than a consequence of it, and second, this shift in Kant's understanding of modality amounts to something more fundamental than a perspectival shift regarding the mode of inquiry about real possibility.

The Kant of the *OPA* recognizes through his positive theses that existence is a feature of our representation of the object rather than the object itself. As I showed earlier, Kant starts grouping existence with possibility and necessity from 1770 on. His early account of existence and subsequently modal notions as features of our representations will develop into his critical claim that modal categories are "peculiar" in that, unlike all other categories that contribute to the logical content and structure of our representation of an object, they express only the manner in which our representation of an object is related to our cognition (A74/B100; A216). This idea that modalities involve a necessary reference to the cognitive subject requires Kant to redefine them in relation to the conditions of our empirical cognition or experience. Kant starts defining real possibility as the "agreement" (*Übereinstimmung*) of the object with the conditions of empirical cognition or experience from

[6] This seems to be Stang's position, though he acknowledges that the "modal-epistemic" question of the letter to Herz is "elicited by, but not explicitly posed in, *Beweisgrund*." See Stang (2016), 161.

early 1770s on (R 4483, Ak. 17:569; R 4801, Ak. 17:732; R 5165, Ak. 18:108; R 5184, Ak. 18:111). The redefinition of real possibility in terms of agreement with the conditions of experience, which will also appear as the postulate of possibility in the *CPR*, marks a crucial turning point in Kant's treatment of the notion of possibility. For first of all, it implies that real possibility is relativized to the realm of experience. Second, being a really possible object comes to be identified with being an object of possible experience. Third, the conditions of real possibility (of objects) get identified with the conditions of the possible experience or empirical cognition of objects. This vindicates the AP as an epistemological principle regarding our cognition of real possibility relative to the realm of our experience, i.e., our cognition of whether something agrees with the conditions of experience is dependent on our actual experience of objects. As I suggested above, the AP, thus epistemologically conceived, problematizes our cognition of the real possibility of the instantiation (or objective validity) of the categories. Thus, Kant's recognition of the objective validity of the categories as a problem and, in fact, *the* problem of metaphysics and the consequent critical turn in his mode of inquiry are motivated by this new and revolutionary conception of possibility in particular and modality in general as relative to the cognitive subject and the conditions of her empirical cognition. The question is 'how do we know *a priori* that the categories apply to really possible objects?,' and given the restriction of the realm of real possibility to that of experience, the question turns into 'how do we know *a priori* that the categories apply to possible objects of experience?' The latter is the more structured form that Kant's original question in the letter to Herz will take in the Transcendental Deduction.

5.3. Subjectivism

There is also a broader sense in which the shift in Kant's conception of the AP and modality influences his philosophical trajectory. His unfolding revolution in modality requires him to radically transform his ontotheological and ontological commitments, and more generally, the way he conceives metaphysics as a whole. The most immediate and traceable instance of this transformation is the one Kant's 'only possible argument' goes through. In the presentation of the argument in the *OPA*, the absolutely necessary existence of the *ens realissimum* is derived from the necessity of the groundedness of the material element of all real possibility in something actual. Now once the AP is interpreted as expressing a subjective condition of our cognition of real possibilities rather than an objective condition of real possibilities themselves, the necessity of groundedness dictated by the principle amounts to a subjective necessity. Consequently, the necessity of the ultimate material ground of all real possibility is likewise to be understood as a subjective necessity, that is, a necessity arising from the specific constitution of the cognitive subject. This should dramatically change the meaning of the very conclusion of Kant's argument.

It has become quite a popular contention in the literature especially in the last two decades that Kant downgrades his precritical 'only possible argument' in the Ideal of Pure Reason of the *CPR*, from a demonstration of the (objective) necessity of the existence of God to a demonstration of the (subjective) necessity of presupposing the idea of such a being. In chapter 8, I will provide a detailed account of what exactly happens in the Ideal of Pure Reason and how the shift in Kant's conception of the AP plays a pivotal role in his downgrading of the argument. Yet I wish to point out here that Kant realizes the impact of the change in his approach to the AP and modality on his argument and even presents downgraded formulations of the argument well before the *CPR* and fairly soon after the publication of the *OPA*.

As early as 1764, Kant begins having second thoughts about the conceivability of an absolutely (really) necessary being as that whose non-existence would cancel all real possibility, though he still retains it as a boundary concept.[7] Yet Kant's abandonment of the ontological conclusion of the 1763 argument becomes clearer around 1769, when he reframes the necessary being in terms of an idea, a (subjectively) necessary hypothesis of human reason. For instance, in R 4113, he defines three conceptions of God as three different necessary hypotheses of reason: "1. The necessary (*principium*) logical ideal; 2. The necessary hypothesis of natural order; 3. The necessary hypothesis of moral order" (Ak. 17:421). He cashes out the necessary logical ideal as the idea of the highest reality that is necessarily presupposed by all of our concepts of possible things as they are all derived from this ground-concept. He then concludes: "The highest reality is therefore the substratum of our rational cognition of all possibility, in or through which all possibility is given" (Ak. 17:422).[8]

Kant repeatedly reevaluates his argument in this new subjectivist framework as an argument valid only subjectively and demonstrating only the subjective necessity of positing the idea of God, without any objective import as to the existence of this being.[9] In R 4249 (1769–70), for instance, he says: "The conclusions here follow from the possibility of cognition through our reason, not from the given things, and therefore the argument is valid *ad homines*" (Ak. 17:481). In R 4580 (1772–75), he goes on to say that "the necessity of the divine existence is a necessary hypothesis of mere concepts of the possible," but "the absolutely necessary cannot be proven," and adds, "The proof of God's existence is not apodictic, but hypothetical *sub hypothesi logica*" (Ak. 17:600). In a similar vein, in R 4729 (1773–75), he writes, "the concept of all reality as a substratum of reason is necessary for us; but we cannot on that account regard a highest reality as necessary in itself" (Ak. 17:689–90). Again, in R 5508, a later note from 1776–78, he writes, "the subjective conditions of thinking therefore serve very well for convincing *cat anthropon* but not apodictically" (Ak. 18:203).[10]

Thus, it appears that before Kant formulates the need for a critical turn and maps out his major critical doctrines, he already starts leaning toward the conviction that his argument fails as an objectively valid demonstration of the necessary existence of God, but has the merit of subjective validity in demonstrating the (subjective) necessity of the idea of God. This amounts to a revolutionary approach to ontotheology. Kant is no longer in pursuit of revising the ontological argument to salvage ontotheology, conceived as the project of proving God's existence from his mere concept. Instead, he reorients ontotheology toward demonstrating that the idea of

[7] R 3717 (Ak. 17:260). See also R 4007 (Ak. 17:383); R 4033 (Ak. 17:391).

[8] See also R 4244 (Ak. 17:477–8) and R 4253 (Ak.17:482) from 1769–1770.

[9] See R 4568 (Ak. 17:596), R 4585 (Ak. 17:602), R 4587 (Ak. 17:602), R 5492 (Ak. 18:197–8), R 5522 (Ak. 18:206–7), R 5525 (Ak. 18:207–8), R 5527 (Ak. 17:108), R 5569–70 (Ak. 18:235–6).

[10] For more on the same point, see also R 4261 (Ak. 17:486), R 4345 (Ak. 17:514).

God is a subjectively necessary hypothesis, satisfying a certain need of our rational constitution. Thus, what culminates in Kant's doctrine of the Ideal of Pure Reason is in fact an extension of this novel approach to theology that he adopts long before the *CPR*.

This subjectivist shift in Kant's conception of ontotheology cannot be separated from the broader transformation in Kant's conception of rational theology as a whole and metaphysics in general. As I noted earlier, Kant formulates different conceptions of God as different subjectively necessary hypotheses satisfying different needs of reason. We have seen that he formulates the *ens realissimum* as a hypothesis necessary for our cognition of possibilities. In a number of other notes, Kant formulates the conception of the *ens necessarium* as a hypothesis necessary for our comprehension of the series of contingent things in the world.[11] Again from 1769 on, we see him also referring to a "moral proof" (R 4253, Ak. 17:482) and a conception of God serving as a necessary hypothesis for the practical use of our reason (R 4118, Ak. 17:424), as "a *principium* of deeds" (R 4255, Ak. 17:484) necessary "to determine the final ends and hopes of rational beings" (R 4589, Ak. 17:603). Therefore, Kant moves from a conception of positive theology aiming at demonstrating the existence and attributes of God toward a conception of critical theology studying the various ways in which human reason is compelled to assume the existence and attributes of God.

Again, from the mid-1760s on, Kant starts redefining general metaphysics as the study of the cognitive subject, as opposed to the school conception of metaphysics as an *a priori* science of things insofar as they are considered in themselves. For instance, as early as 1764, in his *Inquiry*, he writes, "metaphysics is nothing other than the philosophy of the fundamental principles of our cognition" (Ak. 2:283). His *Reflexionen* from the same time period provide similar definitions of metaphysics such as "not a philosophy about objects . . . but rather about the subject, namely, the laws of its reason" (R 3716, Ak. 17:255); "the philosophy of the concepts of the *intellectus puri*" (R 3930, Ak. 17:352); "a science of the fundamental concepts and principles of human reason" (R 3946, Ak. 17:359); "a science of the laws of pure human reason and thus subjective" (R 3952, Ak. 17:362–3); "a philosophy of pure reason, of the form, of the subject and not the object" (R 4146, Ak. 17:433). Kant emphasizes the rehabilitative and critical aspect of this subjectivist conception of metaphysics in setting the epistemic boundaries of human reason in contradistinction with the traditional idea of metaphysics as a doctrine or science *per se* of producing positive knowledge of things. For instance, in the *Dreams* (1766), he writes, "metaphysics is a science of the limits of human reason." (Ak. 2:368). In R 3964 (1769), he writes "metaphysics is a critique of pure reason and not a doctrine" (Ak. 17:368), using for the first time the phrase 'critique of pure reason.' Again, in R 4284 (1770–71), he writes, "metaphysics is not a science, not scholarship but rather . . . merely a correction of the healthy understanding and reason" (Ak. 17:495). General metaphysics, thus conceived, is no longer an ontology in the traditional sense of the term.

[11] See, for instance, R 4117 (Ak. 17:423), R 4589 (Ak. 17:603).

5.4. A Critical Theory of Modality

Between 1763 and 1781 Kant's conception of modality undergoes a revolution, which consists in redefining modal notions (possibility, existence, and necessity) as expressing different modes in which the representation of an object is related to the subject's cognition. This revolution is critical in essence. For, first of all, it is the precursor of the very theory of modality Kant will present in the *CPR*. Second, the reference to the cognitive subject in this novel conception of modality and the consequent relativization of modality to the conditions of cognition run parallel to the general epistemological mode of inquiry in critical philosophy. Thus, the radical shift in Kant's views on modality is indeed an integral component of the overall critical turn in his philosophy. However, as I demonstrated in this chapter, Kant's revolution in modality is more than a component of his critical turn. The unfolding of the former historically precedes the latter, and stems from Kant's ongoing reflections on the nature of modality in the 1760s. More precisely, the revolution has two roots: first, the radical idea that underlies Kant's positive theses on existence in the *OPA* and his subsequent application of it to modality in general, and second, Kant's abandonment of the ontological interpretation of the AP in favor of its strong epistemological interpretation, the early signs of which are visible again in the *OPA*. Furthermore, this revolutionary shift in Kant's modal thought seems to have motivated his critical turn, both in the narrow sense of instigating the momentous critical question, 'how can we know *a priori* that the categories are objectively valid?,' and in the broader sense of transforming rational theology and metaphysics in general from ontological and positive doctrines regarding the existence and nature of God and of other supersensible objects into epistemological and critical inquiries into the nature of human reason and its (subjectively) necessary presuppositions regarding these objects. In chapters 6 and 7, I will examine how Kant's revolutionary conception of modality will find its most systematic expression in the Analytic of the *CPR*, and in chapter 8, I will discuss the decisive role this revolution plays in the specific transformation of ontotheology in the Dialectic.

6
The Modality of Judgments

It is not obvious that there is a coherent theory of modality in the *CPR*. Yet if there is indeed one, a comprehensive reconstruction of this theory needs to examine at least four contexts in which Kant talks substantively about modal notions and account for the interrelations of Kant's various statements in these contexts:

i) The first systematic discussion of modality appears in the 'Metaphysical Deduction,' where Kant presents the 'modal functions of judgments' from which he derives the categories of modality (A74-6/B100–1).
ii) In the Schematism, Kant provides his first account of real modality by defining the temporal conditions of the empirical application of the categories of modality (A144-5/B184).
iii) In the Postulates of Empirical Thinking in General, he provides his full account of real modality by specifying the complete set of conditions for the empirical application of the categories of modality (A218/B266).
iv) In the Ideal of Pure Reason, Kant applies his critical theory of modality to reframe his 'only possible argument' as one demonstrating the subjective necessity of the idea of God and to formulate his systematic refutation of the traditional variants of the ontological argument.

This chapter aims to discuss (i). Chapter 7 will discuss (ii) and (iii), and chapter 8 (iv).

* * *

Despite the considerable interest it has received in the literature, Kant's discussion of the modality of judgments may still be the least understood aspect of his overall theory of modality.[1] Kant's account in §9 of the Analytic of Concepts is very difficult to make sense of and presents serious interpretive problems. The literature has produced two prominent interpretations of Kant's account in this passage. The first is a family of interpretations understanding the modality of a given judgment in terms of the judger's attitude toward that judgment based on their epistemic or psychological states. The second understands the modality of a judgment in terms of its location in a formal syllogism or rational inference. I hold that both interpretations have compelling textual grounds and different merits in understanding various aspects of the modality of judgments and yet fall short of doing justice to the

[1] See, for example, Wilson (1978); Mattey (1986); Leech (2010); Kannisto (2012); Wolff (1995); Patzig (1976); Buroker (2006); Blecher (2013); Hanna (2016); Klaus Reich (1992); Brandt (1995); Longuenesse (1998); Grünewald (1986); Rosenkoetter (2013), Hebbeler (2015).

original insight of Kant's account: the recognition of the task of the modality of judgments as establishing relations between individual judgments and the whole of the system of cognition, and thus as constructing the system itself. In the following I aim to offer an interpretation which both combines the merits of the two existing strands of interpretation and captures this insight at the same time. I will propose that Kant construes modal functions of judgments as instantiating relative logical modalities and expressing the logical coherence relations between judgments. This interpretation reflects not only what is so fundamental in Kant's particular account of the modality of judgments, but the spirit of what I narrate here as Kant's overall revolutionary program of redefining modality in terms of the relation between the representations of things and the cognitive subject and her cognitive faculties.

6.1. All Judgments Have a Modality

The fact that Kant locates modality in his table of judgments as one of the four types of logical functions of judgments along with 'quality,' 'quantity,' and 'relation' suggests that he takes modality to be a necessary feature of all judgments. The methodology Kant proposes for discovering logical functions of judgment is to "abstract from all content of a judgment, and attend only to the mere form of the understanding in it" (A70/B95). Thus, each of the four types of function applies to all judgments, no matter what their propositional contents may be. In other words, every judgment *must be* determined with respect to 'quality,' 'quantity,' 'relation,' and 'modality,' by taking on one of the three moments listed under each of them. Kant coins the terms 'problematic,' 'assertoric,' and 'apodictic,' for the three moments of modality, respectively correlating with the notions of possibility, actuality, and necessity.[2] Accordingly, any given judgment is either problematic, assertoric, or apodictic.

Kant's conception of the modality of judgments displays some novelty in the history of logic. Though the reference to modality in logic can be traced all the way back to Aristotle,[3] Kant criticizes the ancients for failing to distinguish the modality of things, which can be represented in the content of a judgment, "e.g., The world exists in a necessary way," and the modality of judgments, which concerns the form of a judgment as a whole.[4] Most eighteenth-century logicians do not fare better in providing substantial treatment of the modality of judgments.[5] For instance, there is no mention of it at all in Wolff's own logical corpus. In his *Vernunftlehre*, which Kant adopted as the textbook for his logic lectures, G. F. Meier refers to 'modal judgment' (*iudicium modale*) as a kind of judgment, distinct from pure judgment (*iudicium purum*), and yet does not present modality as a feature of all judgments.[6] Baumgarten, like Meier, takes modal judgments to constitute a certain subset of judgments as distinct from pure judgments, but further classifies modal judgments into four types: necessity, contingency, possibility, impossibility.[7] Among Kant's predecessors, J. H. Lambert is perhaps the only one that anticipates Kant on the modality of judgments to some extent, especially when he offers modality as another

[2] See Lovejoy (1907), 599; Swing (1969), 16–17. [3] See De *Interpretatione*, 22a 15 cf.
[4] *LV* (Ak. 24:935). [5] See Smith (2003 [1918]), 193. [6] *Vernunftlehre*, §309.
[7] *Logica*, §243.

way of classifying judgments, based on the distinctions between the possible, actual, and necessary (and their opposites). However, Lambert still holds that modalities properly belong to ontology, and understands them as pertaining to the contents of judgments, "i.e., A can be B; A is B; A must be B."[8] At any rate, Kant seems to be the first to unambiguously claim that modality is a necessary logical feature of all judgments and thus must belong to the table of the logical functions of judgments. Against Meier and Baumgarten, Kant insists that modality does not make a judgment 'impure,' but is one of the functions that make a representation a judgment: "no judgment is possible without modality" (R 3111, Ak. 16:663). Thus, Kant holds that we modalize all of our judgments or that modality is an indispensable aspect of our acts of judging. Understanding why Kant holds this requires a close scrutiny of his account in §9 of the Analytic of Concepts.

6.2. There is Something "Peculiar" about Modality

Kant begins his exposition of the modality of judgments with a statement of its special character that distinguishes it from other functions of judgments:

> The modality of judgments is a quite special (*besondere*) function of them, which is distinctive in that it contributes (*beiträgt*) nothing to the content (*Inhalte*) of the judgment (for besides quantity, quality, and relation there is nothing more that constitutes the content of a judgment), but rather concerns only the value of the copula (*Wert der Kopula*) in relation to thinking in general (*Denken überhaupt*). (A74/B100)

There are two claims here, both underscoring what is special about modality, one negative and the other positive. I will start with the negative claim.

6.2.1. *No content: Modality does not contribute to the content of the judgment*

Kant's characterization of not contributing to the content of the judgment as a special and exclusive feature of modal functions of judgments might strike one as baffling. For his methodological instruction that we are to discover all logical functions of judgments by 'abstracting' (*abstrahieren*) from all 'content' (*Inhalte*) and attending only to the mere 'form of the understanding' (*Verstandesform*) implies that not only modality but all four types of functions are about the mere form of the judgment. The question therefore is: In what sense do quantity, quality, and relation contribute to the content of the judgment while modality does not? Obviously, one first needs to understand what Kant means by the 'content' of a judgment.

It might be tempting to resort to the matter/form distinction in judgments that Kant underlines in his logic lectures. For instance, in *The Blomberg Logic*, he calls the *termini* of the judgment, i.e., the subject and the predicate, its "matter," and the relation between the two terms that is expressed by the copula *est*, its "form" (Ak. 24:274). However, this formulation seems to align with the claim that modality concerns the value of the copula only in the case of categorical judgments and thus

[8] *Neues Organon*, §137.

is too narrow. It is evident that Kant has a more general conception of the matter/form distinction in mind applicable to judgments that display different functions of relation. In *The Jäsche Logic*, he suggests that in the case of hypothetical judgments, the matter "consists of two judgments that are connected with one another as ground and consequence," and the form is the connection between two constituents of the complex judgment, logically expressed by the operator 'if..., then...' Similarly, in disjunctive judgments, the matter consists of the constituent disjuncts, and the form is the disjunctive connective itself, i.e. "either...or..." (Ak. 9:105–6). So Kant holds that whatever representations are unified in a judgment constitute its matter, and the manner of this unification is its form.[9]

Textually grounded as it may be, this formulation still does not explain the contrast Kant sets between modality, on the one hand, and quality, quantity, and relation, on the other hand. For all functions of judgments are in fact functions of unity and must all be about different ways in which representations are unified in a judgment. But what is so distinctive about modality such that it does not contribute to the 'content' of a judgment while other types of functions exhaustively constitute it?

First of all, Kant's general conception of modality provides an important clue as to what he means by the negative claim. The separation of modality from the 'content' of what it applies to is a central and recurrent theme in Kantian modality. Let us remember that the first, negative thesis (T1) in the *OPA*, 'existence is not a predicate or determination,' means that existence does not contribute to the determination of the content or intension of the concept of anything. As I showed in the previous chapter, Kant starts expanding this idea of separation from conceptual contents to other modal notions in the late 1760s. As I will argue in chapter 7, T1 culminates in the main claim of the Postulates of the Empirical Thinking in the *CPR* that this separation from content is what makes modal categories distinctive in all classes of categories: "The categories of modality have this peculiarity (*Besondere*): as a determination of the object they do not augment the concept to which they are ascribed in the least..." (A219/B266).

In order to see how this 'peculiarity' also applies to the modal functions of judgments, we should first note here a distinction between at least two basic senses of 'judgment' (*Urteil*) in Kant's usage. First, it is the cognitive act of judging (*urteilen*): the act of combining or unifying various representations into a single representation at a given time by a certain cognitive subject (A68/B93). Second, it is the output or object of this cognitive act (and a possible object of other such acts of judging), that is, the judgment proper or the proposition (*Satz*) expressing the unified representation in a certain logical form (A69/B94, A73/B99). I suggest that the 'content of the judgment' (*Inhalte des Urteils*) should be understood as the propositional content expressed in the judgment. So just as the modality of an object is extrinsic to the content of the concept that represents it, the modality of a judgment is extrinsic to its propositional content.[10] It also follows that modality is in fact more of a feature of the cognitive act of judging than the propositional content that is judged.

[9] In further support of this formulation, see also *LV* (Ak. 24:928–9), *JL* (Ak. 9:101).

[10] Swing (1969), 18, aptly formulates this as the sign of Kant's transformation of the "intrapropositional modalities" of traditional logic into "extrapropositional modalities."

When Kant writes that "[modality] is distinctive in that it contributes nothing to the content of the judgment" he means that modality does not have a role in the logical construction of the unity (of the various representations) that makes up the propositional content of a judgment. The parenthetical remark, "besides quantity, quality, and relation nothing constitutes the content of a judgment," suggests that this internal logical construction of a judgment is completely determined by the other functions. Thus, although quantity, quality, and relation are formal functions and are not part of the specific propositional content of a judgment, these formal functions are what give the propositional content a logical structure, a definite form. This is exactly why they are discoverable by attending to the mere form of a judgment. The function of relation determines how the representations are related to one another: in the case of simple categorical judgments, the constituent concepts are related as the subject and the predicate (S is p); and in the case of complex judgments, the constituent judgments are related as ground and consequence (If P1, then P2), or as disjuncts (Either P1 or P2). Quality determines whether the connection between the subject and the predicate is affirmed (S is p), negated (S is not p), or negated while the subject is affirmed as part of the remaining infinite logical space (S is non-p). Quantity determines whether the predicate is related to all (All S's are p), some (Some S's are p) or a single one of the objects (This S is p) falling under the subject-concept. By contrast, modality does not contribute anything to the logical construction of the content, but, as I will explain below, concerns the external logical relations of this completely structured unity.

This is perfectly compatible with Kant's claim that modality does not take part in the complete determination of the concept of an object, and thus, one can still ask whether that object is "merely possible or also actual, or, if it is the latter, whether it is also necessary" (A219, 225). Thus, discovering modality always requires considerations extrinsic to the conceptual or propositional content in question. Brandt points out that the fact that modality is extrinsic to the 'structuring of a judgment' (*Urteilsgebilde*) entails that we cannot discover the modality of a judgment through the 'heuristic procedure' that Kant proposes for discovering logical functions. From here, Brandt even makes the further, quite extreme claim that "modality is completely invisible in judgments."[11]

6.2.2. Modality concerns only the value of the copula in relation to thinking in general

The meaning of Kant's positive claim depends on what he means by 'value of the copula' (*Wert der Kopula*) and 'thinking in general' (*Denken überhaupt*). One view is that the former is the form copula 'is' takes in a modal expression. For instance, Wilson writes, "'the value of the copula' . . . is simply the modification of the copula," namely, "'may be,' 'is (indeed),' and 'must be'."[12] This view also seems to be what motivates the contention that the modality of a given judgment is a matter of what modal operator is at work in that judgment. For instance, Wood characterizes the

[11] Brandt (1995), 62. [12] Wilson (1978), 253.

three modalities of judgments as corresponding to "three kinds of copula" as in "S is possibly P," "S is (actually) P," and "S is necessarily P."[13] Similarly, Guyer identifies problematic, assertoric, and apodictic judgments as judgments with modal operators "might be," "are," and "must be."[14]

However, one must be cautious about modal operators. Brandt's claim that "modality is completely invisible in judgments" may be too strong, but it conveys a valid warning. The modification of the copula is not a definitive indication of the modality of a given judgment, for the actual modality of the judgment does not always find its linguistic expression, i.e., the corresponding modal operator, in the judgment. The modality of a judgment is fundamentally about the logico-linguistic act of judging or asserting a proposition as a whole, whether it is expressed through the modification of the copula or not. Note Kant's own example of a problematic judgment: "The world exists through blind chance" (B100).[15] The risk here is to miss the extra-propositional character of the modality of judgments and consequently, to view problematic, assertoric, and apodictic judgments as amounting to different *kinds of judgment*, which is precisely what Kant criticizes his predecessors of doing. For Kant, modalities are rather different *manners of judging* applicable to identically structured propositional contents. All possible kinds of judgment, understood as propositional contents with logically distinct structures, are completely exhausted by the combinations of other logical functions. Since the function of modality does not partake in the logical construction of the propositional content, differences in modalities do not yield additional kinds of judgments.

I will argue here that the 'value of the copula' should be understood as the truth value that is assigned to the propositional content of the judgment. On Kant's account, the copula represents both the relation between the subject and the predicate, and thus the conceptual unity in the content of the judgment, as well as the relation of that content with other judgments. The truth that determines the modality of a judgment is not the absolute truth of its content in isolation from other judgments, but one that is relative to other judgments. It is crucial to understand what kind of role the truth of the propositional content plays in the modality of the judgment based on the peculiarity claim, and whether the extra-propositional character of modality entails that the modality of a judgment is entirely independent of the truth of its propositional content. Kant's account is fairly nuanced and turns around two conceptions of truth. I will show that the modality of a given judgment is independent of what Kant calls its 'material truth,' which consists in the correspondence of the propositional content to an objective state of affairs in the world, and yet is determined by what he calls the 'formal truth' of a judgment or what can be also called the relative truth of a judgment, which consists in the logical coherence of the judgment with other judgments in a system of knowledge. In light of this conception of the value of the copula as formal truth, it is also fair to add, at this stage, that Kant's emphasis on the 'relation to thinking in general' correlates with his positive (and

[13] Wood (2005), 42. [14] Guyer (2006), 78. See also Korte, Maunu, Aho (2009), 517.

[15] One can find Kant's other formulations of modality as expressed by the modification of copula: "The soul of man may be/is/must be immortal." See *JL* (Ak. 9:109).

revolutionary) conception of modality as expressing the way in which the representation (of an object) is related to the cognitive faculty of the subject. However, a proper account of this connection has to wait until the last section of this chapter.

Therefore, modality is a relative feature of judgments. This makes it possible for the same propositional content to have different modalities in different contexts. The ultimate question here is, of course, what constitutes a different context, or more specifically, what relation of the judgment determines its modality. I will argue below that the context in question is constituted by logical relations (of compatibility, identity, and entailment) of a given judgment with other judgments, and these relations determine its modal status.

The following five points sum up the provisional framework I have offered so far:

1. Modality is an indispensable feature of all judgments.
2. The modality of a judgment is distinct from other functions in that it is extrinsic to its propositional content (i.e., modality is extra-propositional).
3. The modality of a judgment is a relative feature of its propositional content.
4. The modality of a judgment is about the formal truth of its content, which is constituted by its logical relations with other judgments (i.e., modality is interpropositional).
5. The modality of a judgment is about the relation of its content to the cognitive faculty of the subject.

Any viable interpretation of Kant's account of the modality of judgment should be able to accommodate these five points.

6.3. Modality of Judgment as the 'Attitude' of the Judger

Kant's positive account of the modal functions of judgment is based on the sense in which each modal function is correlated with a modal notion:

> Problematic judgments are those in which one regards (*annimmt*) the assertion or denial as merely possible (arbitrary). Assertoric judgments are those in which it is considered (*betrachtet*) actual (true). Apodictic judgments are those in which it is seen (*ansieht*) as necessary. (B100/A75)

Kant's language here is strikingly subjectivist, strongly suggesting a connection between modality and how the judging subject "regards" (*annimmt*), "considers" (*betrachtet*), or "sees" (*ansieht*) the judgment, and thus, understandably, motivates a family of what I call 'attitude' interpretations in the literature. On these interpretations, modal functions should be construed in terms of the judger's attitudes toward their own act of judging and thus toward the truth value of the propositional content, based on the individual's epistemic or psychological state at the time of judging. Most notably, for instance, Mattey claims that "the modality of the judgment reflects the attitude the judger has toward the acts of affirming or negating" and explicitly identifies the problematic, assertoric, and apodictic judging with epistemic modalities or modes of 'holding to be true' (*Fürwahrhalten*) or assent, i.e., opining, believing, and knowing, respectively.[16] Wilson, emphasizing that Kant's account should not be confused with a logic of alethic modalities, argues that "modal functions are more like epistemic operators."[17] Guyer, likewise, claims that Kant's modal functions of judgment are "subjective and epistemic."[18] Broker suggests that modal functions express the pragmatic or "illocutionary force" the speaker attaches to her particular utterance of the propositional content at the time.[19]

The 'attitude' interpretations appear to be favored by Kant's language and they are compatible with at least the provisional points 1, 2, 3 that I offered above. However, they fall short of capturing the core of Kant's account. First, it is hard to motivate the construal of Kant's modal functions of judgment in the Metaphysical Deduction only in terms of the subject's 'modes of assent' or attitudes toward the truth of the propositional content, when Kant actually discusses the latter in an entirely separate context in the *CPR*, i.e., in the Doctrine of Method (A820–31/B848–59).[20] Second, as we will see below, Kant explicitly states that modal functions express logical modalities, i.e., 'logical possibility, actuality, and necessity,' which is what one would naturally expect in the context of the Metaphysical Deduction, where the aim is to discover the logical structure of the understanding in judging. Besides, even if there is a sense in which the modality of a judgment expresses the judger's attitude toward the act of assertion or the truth of the propositional content (as opposed to, say, the modality of the truth of the content *per se*), we still need to raise the more fundamental question of what ultimately determines the attitude of the judger.

6.4. Modality of Judgment as Syllogistic Topology

Following his subjectivist flavored remarks, Kant characterizes modal functions of judgment in connection with (or even parasitic on) the functions of relation.

> The two judgments whose relation constitutes the hypothetical judgment (*antecedens* and *consequens*), as well as those in whose reciprocal relation the disjunctive judgment consists (the members of the division), are all merely problematic. In the above example the proposition "There is a perfect justice" is not said assertorically, but is only thought of as an arbitrary judgment that it is possible that someone might assume, and only the implication is assertoric. Thus, such judgments can be obviously false and yet, if taken problematically, conditions of the cognition of truth. Thus the judgment "The world exists through blind chance" is of only problematic significance in the disjunctive judgment . . . The problematic proposition is therefore that which only expresses logical possibility (which is not objective) . . . The assertoric proposition speaks of logical actuality or truth, as say (*wie etwa*) in a hypothetical syllogism the antecedent in the major premise is problematic, but that in the minor premise assertoric . . . the apodictic proposition thinks of the assertoric one as determined through these laws of the understanding itself (*Gesetze des Verstandes*), thus asserting *a priori*, and in this way expresses logical necessity. (A76/B101)

[16] Mattey (1986), 426. See also Hebbeler (2015), 41.
[17] Wilson (1978), 252.
[18] Guyer (2010), 127.
[19] Buroker (2006), 91. See also Wolff (1995), 125, and Patzig (1976), 62.
[20] Hanna (2016).

This passages motivates another type of interpretation in the literature, according to which the modality of a judgment is independent of the truth of its propositional content and determined solely by its location in a syllogistic reasoning or logical inference. This interpretation, primarily propounded by Longuenesse, I will call the 'syllogistic topology' interpretation.[21] Longuenesse argues that "A proposition is problematic if it is merely a component in another (hypothetical or disjunctive) proposition," and appears in the major premise of a hypothetical or disjunctive syllogism; assertoric "if it holds the position of a minor premise in a hypothetical [or disjunctive] syllogism"; apodictic if it is "asserted under a principle" and thus appears as the conclusion of a syllogism.[22]

If P, then Q. (Both P and Q are problematic, but 'If P, then Q' itself is assertoric)
P. (P is assertoric)
Therefore, Q. (Q is apodictic)

The syllogistic topology interpretation has significant virtues. Most importantly, it takes Kant's separation of modality from propositional content to heart. The modality of Q, for instance, has nothing to do with Q's content, i.e., whether what Q reports can be, is, or must be true of the world, but has only to do with Q's position in the syllogism. Thus, the content of Q might be false and even impossible in itself, and yet Q could still be judged apodictically. Since the modality of Q is solely and sufficiently determined by its place in the syllogism, the judger's psychological, pragmatic, or evidential state at the time of judging become irrelevant in the determination of Q's modality. Thus, this interpretation is not only compatible with the points 1, 2, and 3 in my framework, but presents a non-epistemological and non-psychological reading of 2 and 3. Since the determining context here is that of logical inference, we can also take this interpretation to be offering a broadly logical account, and thus satisfying point 4, though without specifying how modal functions correlate with logical modalities. Therefore, I think, the syllogistic topology interpretation is on the right path toward a full understanding of Kant's conception of the modal functions of judgment.

There are, however, problems with the syllogistic topology interpretation that need to be addressed. First, the textual ground of this interpretation, which I quoted above, is a little dubious. Kant's reference to a hypothetical syllogism seems to be of rather illustrative significance, offering an example, as indicated by "*wie etwa*," of how the modalities of judgments are correlated with the manner in which they enter into logical inferences, but does not express a general rule about how the modality of a judgment is determined. It may be true that any judgment that is a component of a complex (hypothetical or disjunctive) judgment is problematic, but it does not follow from this that every problematic judgment is part of a complex judgment or that being a component of a complex judgment defines what it is to be a problematic judgment. Similarly, being the minor premise or the conclusion of a syllogism warrants a certain modality for a judgment (respectively, assertoric, and apodictic), but it does not follow that all instances of assertoric and apodictic judgments are

[21] Longuenesse (1998). Longuenesse's reading has more recently been defended by Leech (2010).
[22] Longuenesse (1998), 159.

parts of a syllogism, or that assertoricity and apodicticity are defined by these positions. The syllogistic topology interpretation, therefore, seems to mistake a specific example for a general rule or definition.[23] The specific example of syllogism does not explain why we modalize judgments when they enter into such inferential relations with other judgments and why we do so in connection with logical modalities, as Kant claims, but only states how we do modalize judgments in specific syllogistic formations. It is also not clear how this interpretation could make sense of point 5, which emphasizes that the modality of a judgment concerns the relation of its content to the cognitive faculty of the subject, given that the content becomes entirely irrelevant in the special case of the components of complex judgments.

Moreover, this conflation of an example with a rule leads to a narrow understanding of the modality of judgments, which would not be able to explain why we modalize judgments outside of the context of complex judgments or syllogisms. Longuenesse explicitly states, for instance, that "a judgment is problematic... only insofar as it functions as the component of another judgment, which itself is used assetorically or even apodeictically."[24] This implies that individual discrete judgments, whether categorical or complex, can never be problematic or problematically judged. Worse yet, it implies that such discrete individual judgments that do not enter into logical-inferential relations with other judgments do not have modality. This appears to contradict Kant's insistence that all judgments have modality (point 1), and his numerous examples of how we modalize judgments that are not (at least explicitly) parts of complex judgments or syllogisms. The more promising response here would be that there are in fact no discrete judgments and it is a condition of being a judgment to enter into logical-inferential relations with other judgments. I will pursue this line of response below, but as I will try to show, while this response points to a very important aspect of Kant's theory of judgment, it does not exclusively lead us to the kind of syllogistic topology interpretation that Longuenesse offers.

Therefore, what we need is a more general account of modal functions of judgments, which should explain not just *how* we modalize judgments in specific formal syllogistic contexts, but also and more importantly, *why* we modalize our judgments as a structural feature of our acts of judging in the first place. We need to tease out what is most basic about each individual modal function and the underlying rule that determines all of our modalizations, including those in syllogistic contexts. This account should also offer a substantive explanation of the sense in which modal functions express the manners in which a judgment is related to the judger's cognitive faculty, as stressed by point 5.

6.5. Modality of Judgment and Ground of Assertion

Kant correlates problematicity of judgments with the mere possibility of the act of judging (assertion or negation): "Problematic judgments are those in which one

[23] This charge has been specifically leveled against Longuenesse before by Blecher (2013), 9, and Kannisto (2012), 105. For how Kant's reference to syllogisms should be taken as illustrative instead of definitional, see also Grünewald (1986), 36, Paton (1936), vol. 2, p. 361, n.3., Hebbeler (2015), 43.

[24] Longuenesse (1998), 160.

regards the assertion or denial as merely possible" (A74/B100). This signifies that there is no sufficient ground for either the assertion or negation of the judgment in question, i.e., there is no other judgment in the background that entails or contradicts its truth. Thus, Kant suggests: "A problematic judgment is one in which I only consider the relation of two concepts undetermined, but do not posit it" (*LB*, Ak. 24:276). Kant sometimes also identifies this kind of mere entertainment of the relation between the given representations, without actually asserting it, as "judgment" (*Urteil*), as opposed to "proposition" (*Satz*) whereby the assertion is actually made (*LV*, Ak. 24:934; *JL*, Ak. 9:109). However, this distinction is neither explanatory, nor consistently followed by Kant himself.[25] I think that a more radical characterization of problematicity would be more useful: problematic judgments are not real instances of judging! Kant is reported to claim in his logic lectures that when we judge problematically, we do not in fact pass but "suspend" or "postpone" our judgment (*suspensio judicii*); or merely entertain the propositional content with the *mere possibility* of judging it in order explore its implications; or pass a rather "provisional judgment" (*praevia judicii*) with no commitment to its truth. (*LV*, Ak. 24:860–2, 933; *JL*, Ak. 9:66, 74–5, 109).[26]

Problematic judging is thus future oriented. "It serves (like the designation of the false path among the number of all of those one can take) to find the true one" (B100–1). We suspend judgment when there is no sufficient ground either for its assertion or negation but entertain it (as merely possible) in the hope of finding grounds of its truth in our logical investigations, "in order to seek for the grounds of the determining judgment" (*JL*, Ak. 9:74). More specifically yet, the problematic judgment is entertained "in order see how the proposition that I think would stand in connection with another one" (*LV*, Ak. 24:933). So problematically judging a proposition will help us explore its implications and find out how it coheres with other judgments for which we may have sufficient ground for assertion or denial. The direction is future cognition and the aim is to make the merely possible judgment, actual, or assertoric.

Kant characterizes the components of hypothetical and disjunctive judgments as problematic because the component judgments are suspended in virtue of the logical structures of complex judgments. In a hypothetical judgment, what is asserted is the logical relationship between the antecedent and the consequent, i.e., the former implies the other, and since the individual truths of these components remain immaterial to whether the logical relationship actually holds, the determination of their individual truth-values is bypassed. Similarly, in a disjunctive judgment, what is asserted is the logical relationship between disjuncts, i.e., together they exhaust the logical sphere of a cognition, and the individual truths of the disjuncts are irrelevant to whether the disjunction really holds. However, as I suggested earlier, components of complex judgments are genuine examples of problematic judgments but do not

[25] Despite his bold statement in the *Jäsche Logic* that "A problematic proposition is *contradictio in adjecto*" (Ak. 9:604), Kant himself writes "Der problematische Satz..." (B101) in his discussion of the modality of judgments in the *CPR*.

[26] Grünewald (1986), 35 describes the subject's state of mind in problematically judging as an "objective undecidedness" (*objektive Unentschiedenheit*).

offer a rule or definition of problematicity. These examples should not distract us from the fact that the most fundamental and definitive aspect of problematic judgments is the lack of sufficient logical ground for asserting or negating them.

While problematically judging is to suspend the judgment and merely entertain it without assertion or negation, assertorically judging is to assert the judgment *actually*. Since the problematicity of a judgment is constituted by the lack of sufficient logical grounds for its assertion (or negation), we can assume that the assertoricity of a judgment requires sufficient logical grounds for its assertion of its truth. As I will argue, such logical groundedness is exactly what Kant means when he states that "the assertoric proposition speaks of logical actuality or truth" (A75–6/B101). Again, Kant's reference to the premises of a syllogism as assertoric judgments can be misleading, if it is taken as providing a universal rule or definition of assertoricity, excluding the possibility of asserting a judgment outside of a syllogism. In the specific context of a syllogism, the premises are assertorically judged due to the nature of syllogistic reasoning, i.e., the premises must just be asserted for the syllogism to run. But this should not obscure the fact that the assertoricity of a judgment is most fundamentally about its having logical grounds for assertion. Thus, the location of a judgment in a syllogism as a premise is not what makes it assertoric, but that it is logically grounded (in other judgments) is what makes it a premise of a syllogism in the first place.

Apodictic judging is a kind of assertoric judging in the sense of the actuality of assertion, which, again, requires the logical groundedness of the judgment in question. However, as opposed to the simple assertion in the case of (merely) assertoric judging, in apodictic judging the assertion is (logically) necessary, which Kant broadly construes as being determined by laws. In other words, in apodictic judging, the groundedness of the judgment (in other judgments) follows through laws. I will explain what these laws would be below, but it suffices to note for now that a judgment's following from other judgments through laws just means that it can be logically inferred from them. In this sense, apodicticity can well be defined in terms of being the conclusion of a syllogism. This, of course, gives partial credence to the syllogistic topology interpretation. For while syllogistic topology is not really explanatory in the case of problematic and assertoric judgments, it does offer a fundamental explanation of why the conclusion is apodictic, i.e., it follows from its grounds necessarily.

6.6. Modality of Judgment as Relative Logical Modality

Now, three aspects of judgments fundamentally determine their modality: i) lacking sufficient logical grounds (for assertion or negation); ii) having sufficient logical grounds (for assertion or negation); iii) being logically grounded through laws. The challenge here is to explain why Kant correlates these three fundamental aspects with *logical* notions of possibility, actuality, and necessity.

I claimed in chapter 4 that Kant construes logical modality as the modality of thought, expressing what can be and cannot be and must be thought based on the laws and structure of (human) thought, and thus properly pertaining to our conceptual representations (i.e., concepts or propositions) rather than the objects or events that they purport to represent. On the other hand, he takes real modality to be the modality of the existence of what is represented, expressing whether it can, does, or must exist or obtain in reality. Thus, the latter reflects the mode of the relation or correspondence of the content of representation to reality, while the former reflects the mode of the form of representation to the laws of thought. I also suggested that Kant holds that both kinds of modality can be divided into two, depending on how one considers them: absolute and relative.

Absolute logical modality is the logical modality of a conceptual representation in itself, on the basis of the law of contradiction. Logical possibility, for instance, is the possibility of thinking a thing without logical contradiction. So a judgment P is absolutely logically possible if and only if it is logically not contradictory that P (is true). Logical necessity is the necessity of the connection of concepts within a proposition such that the negation of this connection yields a logical contradiction. So a judgment P is absolutely logically necessary if and only if it is logically contradictory that not-P (or P is not true). Relative logical modality would then be the logical modality of a proposition in relation to another proposition or set of background propositions. It is not difficult to see why modal functions cannot be accounted for by absolute logical possibilities. For while absolute logical modalities amount to the modalities of a proposition considered in itself and thus of the truth of its content alone, modal functions of judgments do not concern the truth of the propositional content in itself. Besides, if we interpret modal functions in terms of absolute logical modalities, any logically non-contradictory judgment would be problematic, and all apodictic judgments would be analytic, neither of which seems to accord with Kant's text. So, Kant must have in mind relative logical modalities when he characterizes modal functions of judgments as expressing logical possibility, logical actuality, which he identifies as 'truth,' and logical necessity. I offer the following three preliminary definitions of relative logical modalities.

1. RLP (P,Q): P is logically possible relative to Q iff it is logically possible that both P and Q (are true).
2. RLA (P, Q): P is logically actual (true) relative to Q iff it is the case that if Q, then P.
3. RLN (P, Q): P is logically necessary relative to Q, iff it is logically necessary that if Q, then P.

Most importantly here, the logical modalities of P relative to Q are independent of the absolute logical (or real) modalities of both P and Q, but depend only on the logical relations between them. In other words, relative logical modalities do not depend on whether the propositions involved are true in themselves. Even in the case that both P and Q are false or even (really or logically) impossible in themselves, P can still be logically possible and even necessary relative to Q. This satisfies Kant's emphasis on the peculiarity of modal functions of judgment as not concerning the propositional contents (in themselves).

These definitions need to be further developed. One might first want to reduce logical modalities involved in these definitions to certain laws or principles, as in the case of the definitions of absolute logical modalities. It is worth noting that Kant himself refers to the "laws of the understanding" in the passage wherein he correlates modal functions with logical modalities:

> The problematic proposition is that which only expresses logical possibility... The assertoric proposition speaks of logical actuality... and indicates that the proposition is already bound (*verbunden*) to the understanding according to (*nach*) its laws; the apodictic proposition thinks of the assertoric one as determined (*bestimmt*) through these laws of the understanding (*durch diese Gesetze des Verstandes*) itself, and as thus asserting *a priori*, and in this way expresses logical necessity. (A76/B101)

Kant does not spell out what these laws are in the Metaphysical Deduction. However, provided that the table of the logical functions of judgment, at least in the context of Kant's initial exposition of it, belongs to general logic, which abstracts from the content of cognition, one would think that Kant alludes to formal-logical laws. This would be an accurate conjecture, because Kant does actually specify these formal-logical laws elsewhere. For instance, in a letter to Reinhold (May 19, 1789), Kant writes:

> All judgments must first, as problematic (as mere judgments) insofar as they express possibility, conform to the principle of contradiction; second, as assertoric (*qua* propositions) insofar as they express logical actuality, that is, truth, they must conform to the principle of sufficient reason; third, as apodictic (as certain knowledge), they must conform to the principle of excluded middle. (*Corr.* Ak. 11:45)

Again, in the *Jäsche Logic*, he is similarly reported to suggest:

> Thus we will be able to advance three principles here as universal, merely formal or logical criteria of truth; these are:
>
> 1. *the principle of contradiction* and *identity*, through which the internal possibility of a judgment is determined for *problematic* judgments;
>
> 2. *the principle of sufficient reason*, on which rests the (logical) actuality of a cognition; the fact that it is grounded, as material for *assertoric* judgments;
>
> 3. *the principle of excluded middle*, on which the (logical) necessity of a cognition is grounded... for *apodeictic* judgments. (*JL*, Ak. 9:52–3)[27]

It is rather easy to see how the principle of contradiction governs relative as well as absolute logical possibility. For we have defined relative logical possibility of P with respect to Q as the logical compatibility between P and Q, which can be cashed out in terms of the absence of logical contradiction in their being both true:

> <u>RLP (P, Q)*:</u> P is logically possible relative to Q iff it is not logically contradictory that P and Q.

This also works well with the basic notion of the problematicity of a judgment as the lack of (sufficient) logical grounds for its assertion or negation. P is neither entailed, nor negated by Q, but only logically compatible with it. In fact, the same logical

[27] See also *LDW* (Ak. 9:721).

compatibility relation between P and Q can be expressed without reference to the principle of contradiction, simply in terms of the absence of the negation of P by Q:

RLP (P, Q)**: P is logically possible relative to Q iff it is not the case that (if Q, then not P).

The less obvious correlation here is between (relative) logical actuality and the principle of sufficient reason. In its logical use, Kant identifies PSR as the external and 'formal' criterion of the truth or logical actuality of a judgment in its relation to other judgments, and characterizes it as a judgment's being 'logically grounded.' A judgment is logically grounded, when a) it has grounds, b) when it does not have false consequences.[28] The logical ground-consequence relationship here is captured by logical implication such that G is a ground of C and C is a consequence of G iff (if G, then C). Yet this is a general definition of logical groundedness, and when we specify the logical groundedness of a judgment P in a specific judgment or set of judgments, Q, we arrive at our definition of relative logical actuality or truth.

RLA (P, Q): P is logically actual (true) relative to Q iff it is the case that if Q, then P.

The second condition that the PSR demands for the logical groundedness of a judgment is that it have no false consequences. Let us assume that P has an exhaustive list of logical consequences (R1, R2, R3), such that if P, then R1 and R2 and R3. From this relationship between P and its consequences, Kant draws negative and positive criteria for the relative logical actuality or truth of P: while one false consequence is sufficient to prove the falsity of P (via *modus tollens*), all of the consequences must be true to infer the truth of P.[29] Hence,

Negative: if not R1 or not R2 or not R3, then not P.
Positive: if R1 and R2 and R3, then P.

We can therefore introduce another the definition of relative logical actuality. Where P is the ground of R1, R2, R3, and P does not have any other consequences,

RLA*[P, (R1, R2, R3)]: P is logically actual (or true) relative to the set (R1, R2, R3) iff it is the case that (if not R1 or not R2 or not R3, then not P) AND (if R1 and R2 and R3, then P).

Combining the two conditions, we have a final and complete definition of relative logical actuality. Where Q is the logical ground of P and R is the conjunction (R1 and R2 and R3) of all logical consequences of P,

RLA** [P, Q and (R1, R2, R3)]: P is logically actual relative to Q and R iff (if Q, then P) AND (if not R, then not P) AND (if R, then P).

Finally, Kant takes the principle of excluded middle to apply to exclusive disjunctions between logical contradictories, whereby "from the negation of one contradictory opposite to the affirmation of the other, and from the positing of one to the negation

[28] *LV* (Ak. 24:826–7); *LDW* (Ak. 24:721); *JL* (Ak. 9:51–2).

[29] *LV* (Ak. 24:827); *JL* (Ak. 9:52).

of the other, the *consequentia is valid*" (*JL*, Ak. 9:130). For Kant, then, the principle of excluded middle depends on the principle of contradiction. For P and T are logical contradictories when T is a negation of P, which makes the simultaneous truth as well as falsity of both a logical contradiction, and implies the exclusive disjunction that either P or T, the negation of which would be a logical contradiction. This disjunction can also be converted into a conjunction of two hypothetical judgments: if T, then not P AND if not T, then P. In either case, the inference to the truth value of P from that of T follows with logical necessity, expressible in terms of the principle of contradiction. Thus, there is nothing ultimately implausible in Kant's characterization of an apodictic judgment, which follows from other judgments through inference, in terms of 'logical necessity.' For he means the relative logical necessity and not the absolute logical necessity of that judgment, which relies solely on the content of that judgment in itself and requires that the negation of it be a logical contradiction. Earlier, I offered the following definition of relative logical necessity:

> RLN (P, Q): P is logically necessary relative to Q, iff it is logically necessary that if Q, then P.

First of all, we can express logical necessity in terms of logical contradiction:

> RLN (P, Q)*: P is logically necessary relative to Q, iff it is logically contradictory that it is not the case that if Q, then P (or (Q and not P)).

Second, we can now see that the same notion of relative logical necessity holds between the conclusion of a disjunctive or hypothetical syllogism and its premises, if we take Q to be a complex judgment, in particular, a conjunction of the exclusive disjunction Q1 (either P or T), and the negative Q2 (not T).

> RLN (P, Q1 (either P or T) and Q2 (Not T))**: P is logically necessary relative to Q1 and Q2, iff it is logically contradictory that it is not the case that if Q1 (either P or T) AND Q2 (not T), then P.

6.7. Modality of Judgment as Formal Truth

The relationship between the modality of a judgment and its truth is important here. Neither of the relative logical modalities I formulated above depends on the truth of the judgment, P, or of any of the other, background judgments in themselves. In that sense, relative logical modalities do not concern the propositional contents of judgments through which they purport to represent the states of affairs in the world. Yet, it is not entirely accurate that relative logical modalities (or modal functions they correlate with) have nothing whatsoever with truth. On the contrary, they do express truth-preserving logical relations between judgments.

What the modalities of judgments do and do not express with respect to truth can be best understood with reference to the distinction Kant makes between two notions or criteria of truth: 'material' (or objective) and 'formal' (or subjective). The material truth of a cognition or a judgment consists in its agreement with its object, i.e., that it really relates to its object, and thus is dependent on what the judgment's propositional content purports to represent in the world.[30] Kant often emphasizes that there cannot be a universal criterion of material truth. Since a universal criterion of truth must abstract from all matter, i.e., the objects and their differences, it can only be formal. Accordingly, the first universal-formal criterion of truth is the agreement of the cognition with itself, i.e., its (absolute) logical possibility.[31] Kant also maintains that the second formal criterion of truth is the (logical) PSR, which amounts to the logical groundedness of the cognition or judgment in virtue of having true grounds and not having false consequences.[32] Thus, the kind of truth that the modal functions of judgments concern with must be formal truth, abstracting from all content. It should be noted that even the formal test of truth, if applied to a single judgment, requires the consideration of its content: how else can one determine whether a judgment agrees with itself, i.e., free of contradiction? This, however, is not the case when we apply the formal criteria of truth to a judgment against the background of another judgment or set of judgments. When thus applied, formal truth expresses only interpropositional logical relations without any dependence on the material or formal consideration of the contents in themselves. Accordingly, the possibility (or problematicity) of a judgment expresses its *agreement* or *compatibility* with the background; the actuality (or assertoricity) of a judgment expresses its *groundedness* or *connectedness* in the background; and the necessity (or apodicticity) of a judgment expresses that its connection with the background is due to laws of inference.

So while material truth amounts to the correspondence notion of truth, which is determined by whether a representation corresponds to (or agrees with) its object and is thus dependent on the facts about the world, formal truth is truth as coherence, which is determined by the logical relations between representations alone and is independent of whether the representations actually correspond to the world.[33] It is possible that all the judgments involved in a system might be materially false and yet maintain the right logical coherence relations among themselves and uphold formal truth in relation to one another. Our consideration of formal truth is akin to that of the formal validity of a logical inference: we set aside the question of the material truth of the propositions in themselves and look only at the way the premises and conclusion are related. The question of material truth is rather tied with that of soundness, which only comes after but remains immaterial to formal validity considerations. However, these formal-logical coherence relations between judgments, though they are independent of the particular truths of judgments in themselves, still carry (material) truth-preserving implications. That is, while formal validity of an inference does not warrant the (material) truth of either the premises or the conclusion, it does warrant that the conclusion would be materially true on the condition that the premises are (materially) true. Thus, modal functions should be understood in the framework of our logical investigation of truth in general.

[30] See *LV* (Ak. 24:823); *LB* (Ak. 24:97); *LDW* (Ak. 24:718). [31] See *JL* (Ak. 90:50).

[32] See *JL* (Ak. 9:51).

[33] I owe this observation of the parallel between Kant's material/formal truth distinction and the contemporary distinction between truth as correspondence and truth as coherence to Kannisto (2012), 85.

6.8. Modality of Judgment 'in Relation to Thinking in General'

My account above satisfies points 1 through 4 and does so by also explaining the most fundamental aspects of Kant's modal functions (having no grounds, being grounded, and being grounded through laws) in terms of logical modalities, which lacks in both the attitude and syllogistic topology interpretations. Yet another aspect of Kant's theory of modality of judgments, which is usually entirely neglected in the literature, is its characterization of modal functions with reference to the relation of a judgment to the cognitive faculty of the judger, point 5 in my provisional framework. This point is crucial, because it is what connects Kant's discussion of the modalities of judgment to what is so revolutionary in his overall critical theory of modality: modality is a feature of the relationship between the representation of an object and the cognitive faculty.

There are at least two ways in which Kant explicitly formulates this aspect of the modalities of judgment. First, he ties each modal function with a specific cognitive faculty. In a footnote to his initial statement of modal functions, he writes: "It is just as if in the first case [problematic] thought were a function of the understanding (*Verstand*), in the second [assertoric] of the power of judgment (*Urteiskraft*), and in the third [apodictic], of reason (*Vernunft*)" (B100). This kind of specific connection between modality and cognitive faculty is not an outlier in Kant's treatment of modality. As I will discuss in the next chapter, in his elucidation of the Postulates, Kant explains the peculiar "subjective-syntheticity" of the principles of modality by an appeal to the fact that "they add to the concept of a thing . . . the cognitive power (*Erkenntniskraft*) whence it arises and has its seat" and "do not assert of a concept anything other than the action of the cognitive faculty (*Erkenntnisvermögen*) through which it is generated" (A234/B287). Both in these passages and in a number of his *Reflexionen* from the 1770s, Kant seems to follow a consistent pattern in formulating modal notions in terms of how an object (or more precisely, its representation) is related to different faculties.[34] Accordingly, possibility is the relation of the representation of an object to the faculty of mere thought (the understanding), i.e., the object is merely thought without being given; existence or actuality is the relation of the representation of the object to the faculty of perception, i.e., the objects is given; necessity is the relation of the representation of the object to the faculty of reason, i.e., the object is given through a rational inference. Kant's above-cited footnote applies this approach to the modality of judgments on the level of formal logic, as abstracted from content, and thus, from the question of whether and how objects are given to our cognition. The account I have provided so far is both explanatory of and further confirmed by Kant's approach. For instance, in problematic judging, we merely entertain or think the proposition with respect to its relative logical possibility without making any assertion (or negation) as to its truth. This act of mere thinking is the job of the faculty of thinking, the understanding (A126). In assertoric judging we assert the judgment (as logically grounded) through the power of judgment. In

[34] See R 4288 (Ak. 17:497); R 4299 (Ak. 17:500); R 4302 (Ak. 17:500); R 4802 (Ak. 17:733).

apodictic judging, we infer the judgment from its grounds through laws. This, of course, is the task of reason, the faculty of inferences and principles (A 300). Thus, we employ different cognitive faculties in our different modes of judging.

Secondly, Kant offers a more general formulation when he presents the positive portion of his peculiarity claim: "The modality of judgment . . . rather concerns only the value of the copula in relation to thinking in general" (A74/B100). Now, there are two levels at which this extremely important statement can be interpreted: general logic and transcendental logic. General logic "abstracts from all contents of the cognition of the understanding and of the difference of its object, and has to do with nothing but the mere form of thinking" (A54/B78), or "considers only the logical form in the relation of cognitions to one another, i.e., the form of thinking in general" (A55/B79), or "considers representations . . . merely in respect of the laws according to which the understanding brings them in relation to one another when it thinks" (A56/B80). This is the primary sense of logic in connection with which the Metaphysical Deduction and the table of the logical functions of judgment are meant to be interpreted. For Kant's aim there is to lay down the formal-logical structure of judgmental representations in isolation from the transcendental question of the conditions of the relation between representations and objects. In line with this formal notion of logic, I have interpreted the modal functions of judgment above as concerning the merely logical relations between judgments, which are governed by formal-logical laws (i.e., POC, PSR, PEM) and express formal rather than material truth. Accordingly, from this general logical point of view, 'the value of the copula in relation to thinking in general' signifies the relation of a given judgment to 'the form of thinking in general,' that is, how it coheres with other judgments based on the formal structure and laws of our thinking, in abstraction from all objective content.

Transcendental logic, on the other hand, does not abstract from objective content but concerns "the origin of our cognition of objects" (A55–6/B80). Thus, transcendental logic is a combination of general logic, concerning the mere form of thinking or representation, and a logic of the material content of representation, concerning the laws and conditions of the possible givenness of objects. For the cognition of whatever (sensible) content is given to us always requires the act of thinking, which, in turn, is always carried out through judgments, formally structured by logical functions. In fact, the key to the discovery of the categories is that our concepts of objects must have logical forms corresponding to these functions, which would make it possible to construct judgments out of concepts: "the same function that gives unity to the different representations in a judgment also gives unity to the mere synthesis of different representations in an intuition, which, expressed generally, is called the pure concept of understanding" (A79/B105). Thus, while logical functions of judgment originally belong to general logic, they also constitute a part of transcendental logic.

Now, from the transcendental logical point of view, judgments should be considered with respect also to their material truth, i.e., whether they relate to objects or correspond to the world, or what Kant calls, 'objective validity' (in addition to their formal truth and validity), and, consequently, with respect to the laws governing the ways in which objects are given to us (in addition to the formal-logical laws of thinking in general). Kant's treatment of modality in transcendental logic will be the main topic of the next chapter. But here I would like to suggest that a transcendental logical perspective on modal functions illuminates the ultimate sense in which they express 'the value of the copula in relation to thinking in general.' This should also account for point 5 and explain how modal functions (or relative logical modalities) play an immensely important role in the construction of the system of cognition.

At the level of general logic, modal functions are functions of logical-coherence relations between judgments. The modality of a given judgment is determined by how it coheres with a background, i.e., whether it is merely logically compatible with it, or it is logically grounded in the background, or it is so grounded in virtue of laws of formal logic. Modal functions are then what logically connect judgments in a coherent system of judgments. Of course, at the level of general logic, this system of judgments is abstracted from all objective content, neither restricted, nor determined by the conditions of how objects can be given to our representations. At the level of transcendental logic, however, the system to which every judgment is to be connected is that of all cognition, determined with respect to its objective content as well as the conditions of the givenness of this content. This system is that of a logically, spatio-temporally and causally unified cognition, where 'the value of the copula in relation to thinking in general' gains a new meaning. In the Metaphysical Deduction, Kant defines the function of a judgment as unifying given representations under a single cognition (A68/B93). In the Transcendental Deduction, however, he deems this definition unsatisfactory for transcendental logic, and suggests that the unifying function of judgments consists not merely in combining given representations into a single cognition, but also in connecting this cognition to the unity of a single consciousness (B140–1). Kant argues that this 'transcendental consciousness' or 'apperception' is what makes possible the unification of all representations into a spatio-temporally, causally, and logically coherent cognition, that is, into "one experience" (A108–10). He claims that this highest unity, i.e., the unity of all representations under a single consciousness, is the necessary condition of the relation of individual representations to objects and thus of the objective validity of the unity of various representations under a single cognition in a judgment. Thus, only in virtue of a connection with this highest unity does the conceptual unity in a judgment become an actual cognition. This connection between an individual judgment and the unity of all cognition, which brings objective validity or material truth-value to the former, is the transcendental logical function of the copula in the judgment: "That is the aim of the copula *is* in them: to distinguish the objective unity of given representations from the subjective. For this word designates the relation of the representations to the original apperception" (B141–2). Thus, while at the general logical level the copula 'is' expresses only the conceptual relation between the subject and the predicate, at the transcendental logical level it expresses that this conceptual relation is indeed connected to the materially determined whole of cognition, which, in turn, means that the conceptual relation actually holds in reality, i.e., it is materially true.

Two crucial points follow from this analysis. The transcendental logical task of the modality of judgments, as the function that concerns 'the value of the copula in relation to thinking in general,' is to incorporate individual judgments to the materially determined, spatio-temporally and causally structured whole of cognition. Accordingly, a judgment is problematic if it is merely (logically) compatible with this whole without being connected to it; it is assertoric, if it is connected to or part of this whole; it is apodictic, if its connection to this whole is (logically) entailed by the laws that structure this whole. Hence, Kant calls the three modal functions "so many moments of thinking in general," expressing the progressive stages of the connection of a judgment to the whole of one's cognition:

> Now, since everything here is gradually incorporated into the understanding, so that one first judges something problematically, then assumes it assertorically as true, and finally asserts it to be inseparably connected with the understanding, i.e., asserts it as necessary and apodictic, these three functions of modality can also be called so many moments of thinking in general. (A76)

Since modality is what gradually incorporates any given judgment to the whole, one can say that when considered from the perspective of the whole of cognition, modality is indispensably instrumental in both building and holding the system of cognition together.

Second, it is the transcendental logical task of modal functions which ultimately explains why modality is a necessary feature of all judgments. For mere conceptual unities become actual judgments with objective validity only through their connection to the whole of cognition. In other words, a conceptual unity between a subject and a predicate is a real judgment only insofar as it enters into logical relations with other judgments in the system of all judgments that is this whole. Therefore, there are in fact no discrete individual judgments, and thus, there are no judgments without a modal status. For modal functions are what establish the logical relations between judgments. And since every real judgment must have a determinate logical relation with the whole of all judgments, every real judgment must be determined with respect to its modality. This is why Kant reacts to previous logicians, who take modal judgments to constitute a distinct class or type of judgment: "no judgment is possible without modality" (R 3111, Ak. 16:663).

7

Modal Categories and Kant's Revolution

Kant's extremely concise exposition of the modal functions of judgment sets the stage for his subsequent, detailed presentation of his critical theory of modality. The substance of this theory is constituted by his treatment of modal concepts as applied to the objects of experience. Kant's break from the traditional conception of modality gets solidified by his formulation of modal concepts as categories that are subject to epistemological conditions for their productive employment. The revolutionary implications of this formulation become clear only after the examination of these conditions for the employment of modal categories. It is the task of this chapter to conduct this examination.

7.1. Transition to the Categories of Modality

Kant introduces the table of the logical functions of judgment as the first step of 'the discovery of all pure concepts of the understanding.' This, of course, implies a strong correlation between logical functions and categories, based on the idea that "the same function that gives unity to the different representations in a judgment also gives unity to the mere synthesis of different representations in an intuition, which, expressed generally, is called the pure concept of the understanding" (A79/B104). It seems then both tables concern the same set of functions of unity working at different levels of representations, i.e., judgmental and intuitional. The doctrine of the Transcendental Deduction, which introduces a yet higher level of unity (i.e., the synthetic unity of transcendental apperception) as the necessary condition of the other two unities, also appeals to this one-to-one parallelism between logical functions of judgments and categories. Kant suggests both that "[t]hat action of the understanding... through which the manifold of given representations (whether they be intuitions or concepts) is brought under an apperception in general, is the logical function of judgments," (B143) and that "[a] manifold that is contained in an intuition that I call mine is represented as belonging to the necessary unity of self-consciousness through the synthesis of the understanding, and this takes place by means of the category" (B144).

So while judgments unify concepts (of objects) in certain logical forms under the unity of apperception, categories, spontaneously employed in our acts of judging as higher order concepts, structure the concepts of objects in accordance with the logical forms in which they are unified in judgments, which, in turn, would make possible that these concepts structure the actual intuitions of objects to which they apply in accordance with the same logical forms. In other words, the correlation between two tables is based on the necessary formal accordance between the unity of concepts under a single cognition and the unity of intuitions under those concepts. Apparently assuming that all the rest is obvious, Kant provides a specific explanation only for the correlation between the form of a disjunctive judgment and the category of community (B111–12). We can, however, explain the correlation between modal functions of judgment (problematic, assertoric, apodictic) and the pairs of modal categories in the table of categories (possibility–impossibility, existence–non-existence, necessity–contingency) based on the interpretation of modal functions I offered in the previous chapter.

Let us recall that modal functions of judgment do not manifest themselves as modal judgments of the form 'S is possible/actual/necessary,' asserting the modal status of some object. Thus, the employment of modal categories in modal assertions, though important for Kant's critical theory of modality, does not present us with an answer to the question of how modal categories derive from modal functions. Instead, the answer lies in the fact that in modalizing a judgment, we exercise a capacity to form our representations under the corresponding modal category. In the case of problematic judging, we suspend the judgment due to lack of sufficient grounds for assertion or negation, and entertain the judgment with respect to its mere possibility. But we can only do so, if we have the capacity to entertain the unity of the given manifold of intuitions, say, the unity of the intuition of a particular object with that of a particular property, as merely possible. It is this capacity to conceptualize our intuitions in accordance with the category of possibility which makes it possible for us to judge problematically.[1] In assertorically judging, we actually make the assertion, and conceive the unity of given intuitions as actually obtaining outside of mere thought. In apodictically judging, we conceive that actual assertion and the unity of given intuitions as necessary. But what does this correlation tell us about the modal categories at the most abstract level, prior to our consideration of their application to objects?

Following the deduction of the categories, Kant restricts the use of the categories to the empirical realm:

> [T]he categories do not afford us cognition of things by means of intuition except through their possible application to empirical intuition, i.e., they serve only for the possibility of empirical cognition. This, however, is called experience. The categories consequently have no other use for the cognition of things except insofar as these are taken as objects of possible experience. (B147–8)

As I will discuss extensively in chapter 9, however, this restriction is a qualified one. First of all, this restriction applies to theoretical cognition of objects, leaving the possibility of the use of the categories beyond the empirical realm for the purposes of practical cognition. Second, this restriction does not stem from a fundamental incapacity of our faculty of thinking to represent non-empirical or noumenal things, but from a fact about our faculty of receptivity: objects are presented to our thought only through sensible intuition and thus the categories can thus only have relation to

[1] See Wood (2005), 43; Guyer (2006), 76.

objects through sensible intuition. Yet the categories are pure conceptual means of thinking things in general and have their origins in the understanding prior to and independent of any particular sensible input. So even if without the relation to sensible intuition the categories would be concepts "empty" of any real content and thus "without objective reality" (A51/B75, B148), they would nevertheless still represent determinate forms of thinking: "if... I leave out all intuition, then there still remains the form of thinking, i.e., the way of determining an object for the manifold of a possible intuition" (A254). Each and every category, even when abstracted from the conditions of sensibility, must still have the function of representing a general property, an aspect of a thing in general. So, what are the abstract or 'unschematized' meanings of the modal categories?

In the previous chapter, I argued that modal functions are captured by (relative) logical modalities. If modal categories are correlated with modal functions because they are required in structuring manifolds of intuition in accordance with modal functions, then modal categories as abstracted from all intuition must carry the structure of logical modalities. Accordingly, the category of possibility, in its most abstract sense, must be the notion of logical compatibility or agreement (i.e., lack of logical contradiction). This notion of logical compatibility is more fundamental than both absolute and relative logical possibility. For while absolute logical possibility can be construed as a concept's or proposition's agreement with itself in accordance with the principle of logical contradiction, relative logical possibility is its agreement with another concept or proposition. Kant consistently follows this most fundamental sense of possibility in many places and defines possibility as an *Übereinkommen* (B265), *Übereinstimmung* (R 4298, Ak. 17:499; R 4801, Ak. 17:732; R 5757, Ak. 18:345), *Einstimmung* (R 4304, Ak. 17:501; R 5698, Ak.18:329), or *Zusammenstimmung* (B184; R 5163, Ak. 18:106; R 5573, Ak. 18:237).

As Kant himself often does, the abstract sense of actuality can be defined as 'absolute positing' or assertion. However, 'absolute positing' does not by itself explain how actuality or existence differs from other modal notions. Furthermore, as I will argue in section 7.5., Kant's account of modality in the Postulates defines all modal notions as various kinds of absolute positing. The notion of logical actuality I discussed in the previous chapter offers a more promising abstract sense of actuality: connectedness or groundedness. Actuality is the absolute positing of a representation as connected or grounded in a certain context or domain. However, at the most abstract level, this notion of connection or ground-consequence relationship must still be understood in the sense of logical implication.

As a general principle for the table of categories, Kant states that "the third category always arises from the combination of the first two in its class" (B110). Then he characterizes necessity as "nothing other than the existence that is given by possibility itself." (B111) Thus, as opposed to actuality, which is simply being connected, necessity signifies being connected in accordance with a rule or law.[2]

[2] In various places Kant defines necessity as being posited "in accordance with (*nach*) a rule" (R 4298, Ak. 17:499), "through (*durch*) its idea" (R 4396), or "through reason" (R 4288, Ak 17:497), and being given "through its concept" (R 4019, Ak. 17:387), "insofar as it is thought" (R 4298, Ak. 17:500), or again, as

We can thus conclude that the unschematized forms of modal categories can be understood in terms of three kinds of logical relations:

Possibility: compatibility,
Actuality: connection or groundedness,
Necessity: connection through a rule.

Before moving on to Kant's transition to the schematized senses of modal categories, I would like to briefly address a couple of specific questions regarding the table of categories.

First, why does Kant correlate assertoric judgments with 'existence' (*Dasein*) and not 'actuality' (*Wirklichkeit*)? It is hard to say that Kant employs a well-defined distinction between '*Dasein*' and '*Wirklichkeit*' (and even '*Existenz*'). While his account of modal schemata takes '*Dasein*' as the unschematized category itself and '*Wirklichkeit*' as its schematized form (A145), Kant does not follow this distinction consistently. In fact, since he formulates the assertoricity of a judgment in terms of its 'logical actuality' (*logische Wirklichkeit*), one might think that the unschematized modal category should be '*Wirklichkeit*' instead of '*Dasein*.' Thus, I will assume here that Kant uses 'actuality' and 'existence' interchangeably in the context of the categories.

Second, how is the modal category of existence (and non-existence) different from the qualitative category of reality (and negation)? The concern that the categories of existence and reality are not so different has been raised by a number of scholars. Guyer, for instance, asks: "Isn't what exists just what has reality, and consequently that which does not exist that which is the subject of negation?"[3] Guyer suggests that existence should be definable in terms of reality and therefore does not have an irreducible place in the table of categories. First of all, one should note that Kant's use of the term "*Realität*" in this context follows the Wolffian/Scholastic use of the term "*realitas*" (or *res*), which expresses a mere positive content of things considered on the existentially neutral level of concepts alone.[4] As Kant's first thesis on existence states, existence is radically extrinsic to concepts and does not participate in the determination of the concepts of things. Reality, on the other hand, is part of the determination of the concepts of things, and, in this context, only amounts to the (relative) positing of a determination in relation to the subject-concept in thought. Without the involvement of the category of existence, this determination of the concept of a thing does not make a reference to an actual object outside the thought, and remains on the level of mere thinking. Therefore, while Guyer is right that what exists is what has reality and what does not exist is the subject of negation, it does not follow that the function of the category of existence is reducible to that of reality. In fact, the separation of these two categories, or more generally, the separation

existence determined by its "possibility" (R 4688, Ak. 17:676), "thought" (R 4299, Ak. 17:500), or "concept" (R 4387, Ak. 17:532).

[3] Guyer (2006), 78. See also Smith (2003), 198. It is evident from a number of passages in his translation of the *CPR* that Smith himself does not quite carefully distinguish between the terms "reality" (*Realität*) and "actuality" or "existence" (*Wirklichkeit/Dasein*).

[4] See R 3706 (Ak. 17:240).

of the categories of modality from those of quality is quite an essential point in Kant's overall understanding of modality. I will revisit this issue in sections 7.4 and 7.5.

It will become clearer as we examine Kant's account of the empirical applications of the modal categories that he holds that these categories do not participate in the qualitative determination of the concepts of objects, i.e., what the objects are, but establish and structure the relations between the concepts of objects and the cognitive subject. But Kant has in mind also a more general distinction concerning the way the categories relate to objects. He splits the table of categories into two divisions, the "mathematical" categories (quality and quantity) and the "dynamical" categories (relation and modality), "the first of which is concerned with the objects of intuition (pure as well as empirical), the second of which, however, is directed at the existence of these objects (either in relation to each other or to the understanding)" (B 110). Accordingly, mathematical categories construct objects by determining them with respect to their qualitative and quantitative properties. Dynamical categories, on the other hand, posit objects in the context of external relations, i.e., in relation to other objects (relation) and in relation to the cognitive subject (modality). Yet modality still stands out as special. For as opposed to all other categories, the categories of modality are not constitutive of the objective content of experience, i.e., what the objects are and how they stand in relation to one another, but determine the manner in which the representation of objective content is related to the cognitive subject.

7.2. Transition from Logical to Real Modalities

The modal categories of possibility, existence, and necessity, in their unschematized senses, respectively signify logical relations of compatibility, connection, and connection through a law. There are two points to be made here. First, as pure forms of thought without objective content, modal categories are instances of mere logical modality, expressing modes of thinking or conceptual representation in general. Second, substantial philosophical work is required to understand how the modal categories can be employed as instances of real modality, expressing the modes of the existence of objects or objective contents of our representations. For the logical relations that modal categories signify in their abstract or unschematized senses do not provide them with sufficiently determinate content for their application to objects of sensible intuition. This, of course, is a general problem for all categories. The only content that is assigned to these pure concepts of the understanding, which, as the Transcendental Deduction purports to prove, necessarily apply to all appearances, is the minimal logical content derived from their correlations with the logical functions of judgment. How, with this underdetermined content, they meaningfully apply to appearances remains a question in the Analytic of Concepts.

Kant's overarching program in the Analytic of Principles is to address this crucial question. We should therefore expect the categories to gain more concrete or extra-logical determinations that would make them really applicable to appearances in this part of the *CPR*. The first chapter of the Analytic of Principles, On the Schematism of the Pure Concepts of the Understanding, lays out the minimal extra-logical, i.e., temporal, determinations that are necessary for the applicability of categories to all appearances. The second chapter, the System of all principles of pure understanding, identifies the *a priori* principles that determine the scope of the actual empirical application of categories. Therefore, the two can be understood as two levels of the schematization of the categories, whereby the empirical meaning of each category is defined. Since Kant's explicit premise here is that the empirical use of the categories is their only real use (i.e., application to objects) to yield (theoretical) cognition, it is natural to assume that the empirical definitions of modal categories provided in the Analytic of Principles will be definitions of instances of real modality. I argue here that these definitions, which constitute the substance of Kant's critical account of modality, present relative real modalities in two major senses. First, these are relative real modalities because they express modes of existence in a certain domain, i.e., whether something can be, is, or must be a part of the domain of experience or empirical cognition. Second, Kant interprets these real modalities as expressing the manner in which the representation of something is related to the cognitive subject. The latter, I will argue, is where Kant's revolutionary transformation of modality lies.

7.3. The Transcendental Schemata of the Modal Categories

I briefly stated above what I consider to be the problem that the Schematism chapter aims to solve. However, since it is the very authenticity of the problem which is contested by the unsympathetic readers of this chapter of the *CPR*, I will start with a general justification of this chapter before focusing on the transcendental schemata of modal categories.[5]

Kant's argument for the need for a separate account of schematism relies on a thesis he introduces concerning concept-application: In order for a concept to apply to an object, there must be "homogeneity" between the concept and the object, "the concept must contain that which is represented in the object that is to be subsumed under it" (A137/B176). The subsumption of an object under a concept always requires a partial identity, or better, a partial correspondence between the predicates that are thought in the concept and the properties that are presented in the sensible intuition of the object. However, Kant does not present the requirement of homogeneity as suggesting a general problem of concept-application.[6] Homogeneity is not a problem in the case of empirical concepts or pure mathematical and geometrical concepts. The concept of a dog, for instance, is homogenous with its object because it contains predicates such as animality, four-footedness, having a wet nose and a wagging tail that all correspond to the properties that can be presented in the actual sensible intuition of objects. Similarly, the concept of a circle is homogenous with the object plate, because circularity is a property that is directly encountered in the intuition of a plate. Homogeneity is a real problem only for the categories, because they are pure concepts that do not have sufficiently determinate contents to pick out real objects. Thus it is only the application of categories to appearances which requires a special discussion. Prior to the Schematism chapter, the *CPR* assigns the

[5] See Smith (2003), 334–42; Bennett (1966), 141–52; Prichard (1909), 249; Warnock (1948); Wolff (1963).
[6] See Guyer (1987), 159–60.

categories only logical contents through their correlation with the logical functions of judgment. As such, the categories do not contain predicates that can be encountered in sensible intuition and must be provided with more determinate content in order to become harmonious with the real content that is presented in our sensible intuitions. Or as Kant's formulates, "There must be a third thing, which must stand in homogeneity with the category on the one hand and the appearance on the other..." (A138/B177)

Kant calls this 'third thing,' which fills the gap between a category and an object of experience and thereby makes the two homogenous, a 'transcendental schema.' Yet this is not very instructive as to what exactly a schema in general is. Kant writes, "the schema is always a product of the imagination," but then cautiously adds, "the schema is to be distinguished from an image" (A140/B179). He rather defines a schema in general as "[the] representation of a general procedure of the imagination for providing a concept with its image" (A140/B180). In every application of a concept to an object of sensible intuition, the faculty of imagination constructs an image appropriate for that concept, making possible the recognition of the object as an instance of the concept. The schema is the procedure or rule for constructing the appropriate image for the concept.

One could ask, 'Aren't concepts themselves rules for constructing appropriate images for recognizing their instances?' or 'Doesn't having a certain concept mean knowing how to construct the relevant image of its instance?' It is important to separate the categories from all other types of concepts. Both pure sensible (mathematical and geometrical) concepts and ordinary empirical concepts are themselves rules for their own applications, and to perform these applications does not require further rules but is a matter of "mother-wit," a "hidden art in the depths of human soul" (A133–4/B172–3; A141/B180). The concept of a number, for instance, is the rule for representing a certain multitude in pure imagination, or the concept of a triangle is the rule for constructing an image of a closed figure in space consisting of three-line segments linked end-to-end. The same is true for empirical concepts. "The concept of a dog signifies a rule in accordance with which my imagination can specify the shape of a four-footed animal in general" (A 141). This, as Guyer argues, unequivocally leads to the conclusion that sensible concepts, pure or empirical, are identical to or function as their own schemata.[7] Therefore, contrary to, for instance, what Bennnett holds, Kant's notion of a schema in general does not suggest a general problem of concept-application, but only the specific problem of the application of the categories.[8]

The fundamental heterogeneity between categories and appearances is what makes the application of the former to the latter problematic. The schema of a category, i.e., the 'transcendental schema,' must be tailored to level this heterogeneity: "This mediating representation must be pure (without anything empirical) and yet intellectual on the one hand and sensible on the other" (A138/B177). Now, in order for the categories to be applicable to appearances, their merely logical and thus underdetermined contents must be further determined in accordance with the *a priori*

[7] Guyer (1987), 159. [8] Bennett (1966), 141.

formal conditions under which objects can be given in sensible intuition, i.e., space and time. Yet space is the form of outer sense and "is limited as an *a priori* condition merely to outer intuitions," whereas time, as the form of inner sense directly and of outer sense indirectly, "is the *a priori* formal condition of all appearances in general" (A34). Thus, in addition to its apriority and sensible character, it is the universality of time as a condition of the givenness of all objects in sensible intuition that leads Kant to nominate it as the best fit for the schematization of the categories:

> Time, as the formal condition of the manifold of inner sense, thus of the connection of all representations, contains an *a priori* manifold in pure intuition. Now a transcendental time-determination is homogenous with the category (which constitutes its unity) insofar as it is universal and rests on a rule *a priori*. But it is on the other hand homogenous with the appearance insofar as time is contained in every empirical representation of the manifold. Hence an application of the category to appearances becomes possible by means of the transcendental time-determination which, as the schema of the concept of the understanding, mediates the subsumption of the latter under the former. (A138–9/B177–8)

The schemata of the categories are thus temporal interpretations of whatever logical content the categories derive from the logical functions of judgment. Yet Kant's argument assumes that the logical contents of the categories can be sufficiently expressed in temporal terms with no reference to spatiality. This does not seem to be a warranted assumption. For the main point here is to provide categories with objective reality by finding their sensible expressions, and the logical contents of some categories may require reference to space in their sensible expressions. As a matter of fact, at the very end of the Analytic of Principles, in a remark that he inserts into the second edition of the *CPR*, Kant suggests that the sensible expressions of all three categories of relation, i.e., substance, causality, and community, necessarily require reference to space. For instance, persistence, the sensible expression of the category of substance (and also the schema of substance), can only be given in an outer intuition, "since space alone persistently determines, while time, however, and thus everything that is in inner sense constantly flows" (B291). Similarly, alteration and interdependent coexistence, the intuitions that correspond to the logical senses of the categories of causality and community, can only be exemplified as outer intuitions determined by space (B291–3). The merely temporal schematization of these categories, then, would not make them sufficiently determined for their actual empirical use.

Guyer points out a more fundamental reason for the indispensable involvement of spatiality in the sensible interpretations of categories: time cannot be perceived in itself but can only be represented through spatial relations or as linked to objects in space.[9] This is repeatedly stated by Kant in the Analogies of Experience (B225, 233, 257). It seems then even if Kant defines the schemata of the categories in purely temporal terms, as various transcendental time-determinations, the empirical use of the categories and thus of the corresponding schemata requires reference to space.[10]

[9] See Guyer (2006), 98–9; (1987), 167–9.

[10] Guyer (1987), 168 argues that this is a strategic move on Kant's part in order to provide "a vital premise" for his ultimate argument against the skeptics of external reality in the Refutation of Idealism.

Whatever Kant's motives may be in organizing the Analytic of Principles in the way he does, this organization entails a certain reading of the relationship between its two chapters, the Schematism and the System of all Principles. Namely, the Schematism chapter lays out the minimal extra-logical or real, i.e., temporal, determinations that are necessary for the empirical applicability of categories to appearances, and the System of all Principles identifies the *a priori* principles determining the scope of the actual empirical application of categories. In accordance with this general framework, it is natural to expect the two sets of definitions of real modality, provided respectively by the Schematism and the Postulates, to be at least consistent with one another, as two levels of the schematizations of the modal categories. The problem, however, is that these two sets do not appear to be consistent.[11] In the rest of this section, I will offer a reading of Kant's schemata of modal categories that can account for their consistency not only with the postulates but also with the rest of Kant's modal theory in the *CPR*.

* * *

Kant formulates three individual schemata of modal categories in the following way:

> The schema of possibility is the agreement (*Zusammenstimmung*) of the synthesis of various representations with the conditions of time in general (e.g., since opposites cannot exist in one thing at the same time, they can only exist one after another), thus the determination of the representation of a thing to some time.
>
> The schema of actuality is existence at a determinate time.
>
> The schema of necessity is the existence of an object at all times. (A145/B184)

The schema of possibility is Kant's first attempt in the *CPR* to define real possibility relative to the domain of experience. The unschematized category of possibility signifies mere logical compatibility or agreement. The schematized category of possibility is supposed to signify that something can be an object of experience. In the absence of the object, the real possibility of the object can only be tested through its conceptual representation. The question of the real possibility of the object then becomes that of whether our concept can represent an object of experience, which requires the logical compatibility of the concept with the conditions of experience. Yet since the schema of possibility concerns only whether something can exist in time or be part of a temporal order (i.e., whether something can be represented as being in time), Kant defines it as the compatibility of the conceptual representation of a thing with the conditions of time.

Since the concept of an object is a synthesis or combination of certain representations, i.e., predicates, Kant also construes the question of real possibility here as a question of whether certain predicates can be synthesized in the concept of an object against the background of the conditions of time. His example in the parentheses suggests that the idea of the synthesizability of predicates is a version of the notion of

[11] A number of commentators have pointed out the apparent incongruence between the modal schemata and the postulates. See, for instance, Bennett (1966), 167; Grünewald (1986), 21.

real compatibility or harmony of predicates, i.e., whether they can be instantiated by the same subject in reality, which I discussed in chapter 4 in connection with Kant's precritical account of modality. However, there are two important points here. First, the synthesizability of predicates is neither a matter of their logical consistency with one another, nor a matter of a bare metaphysical compatibility between them (i.e., 'real harmony' à la Chignell). So it does not turn around an absolute consideration of the synthesis in itself. Instead, the synthesis is relativized to the conditions of time. Second, this relation to the conditions of experience is still that of logical compatibility: for a certain set of predicates to be synthesizable, their synthesis must be logically compatible with the conditions of time or the conditions of time must not entail the negation of their synthesis.

The primary question here is whether the logical compatibility with the conditions of time warranties the synthesizability or possibility of the co-instantiation of predicates in an object of experience. Kant's example does not answer this question, for the kind of opposition that is at stake in it is unclear. If the 'opposites' (*Entgegengesetzte*) that cannot simultaneously exist in one thing are merely logical contradictories, then the schema of possibility does not add much to (absolute) logical possibility. If what is at work here is real opposition or repugnance, then we still need more specification. For, as I argued in chapter 4, an empirical instance of real opposition between predicates does not necessarily rule out the possibility of their co-instantiation. Let us recall that two motive forces in opposite directions can coexist in one thing even though they cancel out each other's effects on that thing. Putting Kant's example aside, we can still see that the compatibility with the conditions of time is not a sufficient condition of real possibility relative to the domain of experience. For there are at least geometrical (or spatial) real impossibilities that exemplify representations that are not really synthesizable even if they do not violate the conditions of time, e.g., parallel lines intersecting at one point or a triangle whose one side is longer than the sum total of the other two sides.

Therefore, the schema of possibility suffers from the restriction of schemata to temporal terms. Although it provides the minimal or necessary conditions for real possibility relative to the domain of experience, the schema of possibility fails to provide the category of possibility with sufficient determination for its application to all objects of sensible intuition. It may well be right, as Allison suggests, that "time provides the field within which, and in terms of which, real possibility is determined," but that field of temporality underdetermines the notion of real possibility.[12] Real possibility, if it is to function as an effective filter for all possible objects of experience, must be constructed as a thicker notion containing more than the compatibility with the conditions of time. As I will discuss in section 7.5., Kant presents this thicker notion in the first postulate, as the compatibility with all formal conditions of experience, i.e., space and time and the principles of the understanding (A218/B266). Yet since the compatibility with the conditions of time is implied by the compatibility with formal conditions of experience, the schema of possibility remains at least consistent with the postulate of possibility.

[12] Allison (1983), 190.

The schema of actuality seems unproblematic. The abstract sense of actuality is groundedness or connectedness in a system of ground-consequence relations. To be actual in the empirical sense is to exist in the domain of experience, which, in turn, requires being connected in the thoroughgoing causal nexus of all empirical objects. Kant's schema of actuality as 'existence at a determinate time' captures this empirical sense of actuality, for being an object of experience and being at a determinate point in time reciprocally imply each other. The existence of something at a determinate time, even if its spatial location, if any, is undetermined, is not only a necessary but also a sufficient condition for that thing to be an actual part of the thoroughgoing causal unity of experience. Moreover, the cognition of an object at a determinate time requires an actual perception, which is the material condition of experience that Kant appeals to in the postulate of actuality (A218/B266). Therefore, the schema of actuality as "existence at a determinate time" sufficiently determines the scope of actual objects of experience.

Kant's formulation of the schema of necessity as 'existence at all times' (*das Dasein eines Gegenstandes zu aller Zeit*) seems to be the most problematic one among his schemata of modal categories. First of all, as Kant himself states, his formulation comes to identify the necessity of phenomena with eternity: "*aeternitas [est] necessitas phaenomena*" (A146/B186). This identification is familiar from Kant's precritical texts. Both in the *NE* and the *OPA*, Kant lists infinitude (Ak. 1:395) or eternity (Ak. 2:85) among essential predicates of the absolutely really necessary being, *ens necessarium*, the material ground of all real possibility. Kant's attribution of eternity to the *ens necessarium*, however, is not motivated by a temporal interpretation of necessity. Instead, Kant's reasoning is strictly metaphysical: the absolutely really necessary being, since it is presupposed by all real possibility, cannot be determined in any other way than it actually is and its non-existence is absolutely really impossible, and therefore, it must be immutable and eternal.

Second, as I showed in chapter 5, even though Kant's precritical formulation of the absolutely really necessary being as the material ground of all real possibility is an attempt to revise the traditional conception of the *ens necessarium*, whose existence is contained in its essence as a determination and thus is a logical necessity, as early as the mid-1760s Kant begins expressing concern about the comprehensibility of the notion of an absolutely necessary being in general. As I will discuss in chapter 8, in the Ideal of Pure Reason, he explicitly deems the concept of an absolutely (really) necessary being empirically meaningless, retaining it only as a regulative ideal (A593–4/B621–2). Moreover, as we will see in section 7.5., in the Postulates, he restricts the only notion of real necessity applicable to appearances to that of a relative one, a "hypothetical" necessity concerning the states of things (rather than their existences *per se*) cognized as the effects of other given states (A227–8/B280). Accordingly, everything is in fact relatively or hypothetically really necessary since everything is located somewhere in the thoroughgoing causal nexus of time (and space). Therefore, the only empirically applicable notion of real necessity does not at all entail the 'existence of a thing at all times.' On the contrary, it determines everything within the temporal series of events as causes and effects. Besides, even if we assume that Kant somehow intends the schema of necessity to pertain to absolute (but not relative) real necessity, 'existence at all times' would still be off the mark. For, as Kant's discussion of the Fourth Antinomy demonstrates, the idea of an absolutely really necessary cause being part of the temporal-causal series of appearances and thus *existing in time* is a contradiction (A460/B488). The absolutely really necessary being, if it is to be conceptualized at all, must be thought of as free from both the spatial and temporal conditions of sensibility.

Third, in his explication of the First Analogy as the principle of the persistence of substance, Kant identifies "existence at all times" (*Dasein zu aller Zeit*) or "everlasting existence" (*immerwährendes Dasein*) with substance, or more precisely, with the absolute quantum of substance, not with the modal category of necessity (A185). Thus, if 'existence at all times' is to be the temporal schema of any category, the category of substance seems to be the more suitable candidate.[13]

Perhaps, then, Allison is right in suggesting that "the only way to make sense of Kant's characterization of the schema of necessity is to assume that he does not mean what he appears to mean."[14] In fact, Schneeberger adopts this approach long before Allison's suggestion. He claims that Kant takes "strict universality" to be an indispensable mark of necessity (B3), and interprets "at all times" (*zu aller Zeit*) as "universality expressed through a time-determination."[15] Schneeberger has in mind Kant's second analogy: every alteration or event occurs in accordance with a causal connection. This is the universal principle that the condition under which an empirical occurrence "always and necessarily" (B238, B246) follows is to be found in what precedes it in temporal order. Accordingly, what the schema of necessity entails is not the permanent existence of a thing at all times but the regular occurrence of a type of event under the same given conditions. Schneeberger's interpretation makes the schema of necessity more comprehensible, and, given that Kant's ultimate rendition of real necessity in the Postulates involves the law of causality, its appeal to the second analogy seems apt. Yet, reading 'at all times' as referring to regular occurrences at determinate points in time is an interpretive stretch and dismissive of the explicit reference to eternity and the 'sum total of time' in Kant's text.

Alternatively, Allison himself proposes to follow Paton's wording of Kant's schema of necessity as "existence in relation to the whole of time."[16] Paton's formula makes it possible to construe the schema of necessity as expressing the determination of existence as an effect following from the whole of a causal chain that must exist throughout time up to that point. This might be a viable rendering of Kant's schema of necessity. However, neither Paton nor Allison offers a textual basis for it. They only rely on a fairly arbitrary translation of Kant's somewhat ambiguous phrase '*zu aller Zeit*.' They do not offer a firm philosophical ground for holding that 'existence in relation to the whole of time' is what Kant really means either, but only a general charity in an effort to figure out 'what he ought to have meant.'

I think that there are both textual and genuine philosophical reasons for holding that Kant must have in mind something like what Paton's formula suggests. It is evident that Kant was not entirely at peace with the published formulation of the modal schemata in the Schematism, for he kept entertaining alternative formulations

[13] Bennett (1966), 167 claims that "the schema of necessity is indistinguishable from that of substance!"
[14] Allison (1983), 192. [15] Schneeberger (1952), 90. [16] Paton (1936), Vol 2, 60.

even after the publication of the first edition of the *CPR*. For instance, in a reflection from between 1785 and 1788, Kant writes:

> Possible is what is determined with regard to time in general, thus with regard to the concept of time, wherein consequently the conditions of determination in time lie... Secondly, actual (also possible) is that which is determined in the relative time, thus within some definite time; i.e., that which is determined through and with regard to other things given in time. Finally, a necessary being... is what is determined with regard to the absolute time. (R 5763, Ak. 18:347)

In his own notes on the marginalia of his own copy of the *CPR*, Kant attempts at yet another schematization of modal categories:

> What can be thought indeterminately in time at all is possible.
>
> What is in time determinately is actual.
>
> What is determined by the concept of time itself is necessary. (Ak. 23:32).

Both sets of alternative schemata follow a pattern that reflects the essence of Kant's general treatment of real modality in the *CPR*: each modality expresses a certain way in which the representation of the object in question is related to the unity of experience, against the background of its conditions. Given this general approach, the transcendental schemata of modality, as interpretations of modal categories from the viewpoint of the temporal dimension of human experience, should concern the nature of the relation of the representation of an object to the whole of time. Possibility, then, should be defined as logical compatibility with the conditions or laws that make the whole itself possible; actuality as being connected to, or a part of, that whole; necessity as being a part of the whole in virtue of its very laws. Indeed, both sets define possibility as representability of something in time, which, of course, requires a logical compatibility between the concept of the object or event in question and the laws of time in general. Actuality, on the other hand, is defined by both sets as being at a determinate location in time, which, as the first set elaborates, requires being causally connected to other temporal objects or events. And both sets offer a schema of necessity that involves a reference to the whole of time. For by 'determination with regard to the absolute time,' Kant means being at a definite time, i.e., being actual, as determined by the totality of all time-determinations, which altogether constitute the whole causal nexus of objects in time. The second set reinforces the same idea: necessity is being at a definite point in time (i.e., being actual) in virtue of time itself, that is, the whole of temporal relations. In other words, necessity is actuality, being connected to the causal nexus of all things in time, where this connectedness is entailed by the whole of time, i.e., the entire content as well as the formal structure of time. This rendition not only makes all three modal schemata entirely consistent with the postulates, but also helps us make sense of Kant's remarks that the schemata of modal categories contain "*time itself*, as the correlate of the determination of whether and how an object belongs to time," and that they concern "the sum total of time" (*Zeitinbegriff*) (A145/B185), a statement found "obscure" or simply neglected by some scholars.[17]

[17] See Guyer (1987), 174.

The only worry here might be that the distinction between the schemata of necessity and actuality gets blurred. For in virtue of the law of causality and its universal validity for all temporal representations, anything that exists at determinate time exists as determined by a causal relation, which is itself determined by the rest of the causal chain that spans throughout all time. Causality is the very law that governs the relations between temporal objects. So if actuality is being connected to the temporal series of all things, then being thus connected is the same as being causally determined by the whole of the series. The categories of necessity and actuality, when applied to temporal objects, end up being coextensive. However, this consequence should not register as a handicap but a virtue of the interpretation. For coextensiveness with actuality reflects the essential character of the notion of relative (empirical) real necessity (i.e., 'hypothetical necessity') that Kant will develop in the Postulates: "Everything that happens is hypothetically necessary" (A228).

7.4. The Principles in General

The transcendental schemata define the real senses of the categories in terms of the conditions of time, which are only minimally necessary for the application of the categories to objects of experience. The second chapter of the Analytic of Principles, System of all principles of pure understanding, goes one step further in determining the real senses of the categories by laying out the principles governing their empirical use.

The principles that govern the empirical use of the categories would also determine *a priori* the form of possible experience in general and, consequently, the form of all the possible objects of experience. Such principles, which, on the one hand, follow from the logical senses of the categories, and on the other hand, contain the conditions of sensibility, are the more concrete items of our synthetic *a priori* cognition of objects. Only through the mediation of these principles does the understanding perform its real legislative role, for these principles are nothing but the highest-order laws that the understanding prescribes for nature, the sum total of all appearances. Kant construes the empirical laws of nature as subordinated to these principles: "without exception all laws of nature stand under higher principles of the understanding, as they apply the latter to particular cases of appearance" (A159). Corresponding to the categories of quantity, quality, relation, and modality, Kant divides the principles into four groups, Axioms of intuition, Anticipations of perception, Analogies of experience, and Postulates of empirical thinking in general (A161). He further divides these four groups into two, and calls the Axioms and Anticipations 'mathematical,' and the Analogies and Postulates 'dynamical principles' (A162).

Let us remember from chapter 6 that Kant applies this mathematical/dynamical distinction also to the logical functions of judgment. Unfortunately, he does not do a good job of explaining this distinction. He grounds the distinction at the level of principles on the idea that the mathematical principles (Axioms and Anticipations) pertain "merely to the intuition [of an appearance]," whereas the dynamical principles (Analogies and Postulates) pertain to "the existence of an appearance in general" (A160). He also adds the qualifications that the mathematical principles are "unconditionally necessary, i.e., apodictic" (A160) and "are capable of intuitive certainty" (A162), the dynamical principles are necessary "only under the condition of empirical thinking in an experience, thus only mediately and indirectly" (A160) and are "capable only of a discursive certainty" (A162). What are we to make of these statements?

I suggest that the ultimate import of this distinction is that the mathematical principles concern the construction of the individual objects of experience themselves, whereas the dynamical principles concern the relations of objects to one another and to the whole of experience. More specifically, the principle following from the axioms of intuition is that all objects that can be represented in intuitions have extensive magnitudes. This, of course, is a consequence of the structure of space-time as an extended continuum, in which all objects of intuition must occupy a certain region. The anticipations of perception prescribe the principle that the real or material content in all objects of empirical intuition has intensive magnitudes. That is to say, the qualities we perceive in an object of empirical intuition always come in degrees of influence on sense. The analogies of experience rely on the fundamental principle that the objects given in empirical intuition can only be represented as related to one another through necessary connections in time. The individual analogies are the means of the understanding to bring all appearances together under the necessary temporal unity of apperception and thus to unify experience into a single whole. And finally, the postulates of empirical thinking, the principles defining the real senses of the modal categories relative to the domain of experience, express the manners in which objects of empirical intuition are related to the whole of experience, i.e., whether an object is possibly, actually, or necessarily part of that whole.

Kant uses more useful terminology to capture the distinction between the mathematical and dynamical principles in his general remarks on the analogies: the former are constitutive of our objects, whereas the latter are valid of objects merely regulatively (A180). The mathematical principles are directly constitutive of objects, for they determine *a priori* what absolute properties the objects necessarily instantiate in their possible representation in our mode of intuition, i.e., extension and intension. The dynamical ones, on the other hand, prescribe, at best, what relative or relational properties objects necessarily instantiate in their possible representation in intuition, either in relation to one another or in relation to the whole of experience.

We could now make more sense of Kant's aforementioned remarks about the distinction. The mathematical principles pertain to "the intuition [of an appearance]" and the dynamical principles pertain to "the existence (*Dasein*) of an appearance in general" (A160).[18] Both of the mathematical principles, that all intuitions are extensive magnitudes and that the material content in all empirical intuitions come in degrees of intensity, directly define the ways objects are necessarily represented as individual appearances in intuitions. The reason why Kant would state that the

[18] See also R 4758 (Ak. 17:705–6). We should also remember that in the Metaphysical Deduction, Kant grounds the mathematical/dynamical distinction with respect to the categories upon an almost identical description of functional difference. He states that the categories of quantity and quality are "concerned with objects of intuition (pure as well as empirical)," and the categories of relation and modality are "directed at the existence (*Existenz*) of these objects (either in relation to each other or to the understanding)" (B110).

dynamical principles are about the existence of objects of intuition is more complicated and seems contrary to the natural expectation that the existence or modal status of objects should be the business only of the postulates rather than the dynamical principles in general. As I will argue in more detail below, one implication of the doctrine of the Postulates is that the actuality of an object is a matter of its being part of the thoroughgoing connection of experience, that is, its positedness as an object in relation to other objects and to the whole of experience. And such determination of the position of an object in this thoroughgoing unity of experience requires not only the postulates but also the analogies as principles regulating our representation of the interconnections of objects.

This also makes clearer why Kant deems the mathematical principles unconditionally necessary and the dynamical principles conditionally necessary. For the extensive and intensive magnitudes are absolute properties that necessarily belong to all possible individual objects of intuition, whether or not they are actually given in intuition. In this sense, the mathematical principles are valid even at a modally neutral level. The existence or actual givenness of an object, however, is contingent in itself. Yet, as Kant insists, the analogies and postulates, as expressing relational properties, apply to objects of intuition necessarily only "under the condition of empirical thinking in an experience" (A160), that is, under the condition of the involvement of a cognitive subject or an apperception that is to unify all the individual objects of empirical intuition into a single experience, a coherent whole of cognition.[19] And finally, it only naturally follows that the mathematical principles are "capable of intuitive certainty" (A162), for they concern the structure of intuitive representation and are provable directly in experience; and the dynamical principles are "capable only of discursive certainty" (A162), for as regulative principles of unification, they presuppose the involvement of (empirical) thinking, which, in turn, is always discursive. As a system, then, all the principles of the understanding carry a "universal" and "complete" certainty (B200): All objects of possible experience are both absolutely and relationally determined *a priori* through this system of principles.

7.5. The Postulates of Modality

In the framework of Kant's system of principles, the postulates, as the full schemata of modal categories, serve to regulate the relations between our representations of objects and the unity of experience as a coherent whole. In the following, I will argue that the doctrine of the Postulates is the ultimate culmination of Kant's critical and revolutionary theory of modality.

Kant formulates 'the postulates of empirical thinking in general' in the following way:

1. Whatever agrees (*übereinkommt*) with the formal conditions of experience (in accordance with intuition and concepts) is possible.

[19] This is perfectly consistent with the conclusion of my account of Kant's modal functions of judgment that modality is a necessary feature not of the content of an individual judgment considered in itself, but of its relation with other judgments and the whole of the system of judgments.

2. That which is connected (*zusammenhängt*) with the material conditions of experience (of sensation) is actual.
3. That whose connection (*Zusammenhang*) with the actual is determined (*bestimmt*) in accordance with general conditions of experience is (exists) necessarily. (A218/B266)

These formulations, Kant suggests, are "the definitions of concepts of possibility, actuality, and necessity in their empirical use" or the full schemata of modal categories (A219/B266). He also claims that since these formulations delineate the domain in which things can be given to us, i.e., the domain of human experience, they are "at the same time restrictions of *all* categories to merely empirical use, without any permission and allowance for their transcendental use" (my emphasis, A219/B266). This is in fact a strong statement vindicating my claim that modality has an extraordinarily important role in the shaping of Kant's critical philosophy. The restriction of modal categories to their empirical use is not simply a consequence of Kant's general critical restriction of theoretical cognition to the objects of possible experience. In fact, the latter follows from Kant's revolutionary transformation of modalities from fundamental ontological properties of things themselves into features of our representation of them in relation to our cognitive make-up.

Thus, the postulates present real modalities relative to the empirical domain or empirical interpretations of modal categories. In each postulate the crucial term, expressing the kind of relation between the item whose modality is in question and the background against which the modality of the item is defined, corresponds to the logical relation each respective modal category signifies in its unschematized form. More specifically, the notions of 'agreement' (*Übereinkommen*), 'connection' (*Zusammenhang*), and 'connection determined (*bestimmt*) by (general conditions of experience)' that constitute the logical structures of the postulates are the same as the logical notions of 'compatibility,' 'connectedness or groundedness,' and 'connection entailed by a rule.' So what makes the postulates instances of real modality (relative to the domain of experience) is the content that Kant assigns to the background in their definitions. Since the general background is the domain of experience itself, what can be, is and must be part of this domain is determined by various aspects of the conditions of experience. I will discuss the details of each postulate and the corresponding background conditions below, but let me first underline the logical structures of the postulates. When F represents the set of the formal conditions of experience, M the set of material conditions of experience, and G the set of general conditions of experience, the postulates can be formalized in the following way:

1. PP: Something (P) is really possible relative to the domain of experience iff it is not logically contradictory that (P AND F), or, what is the same, it is not the case that (if F, then not P).
2. PA: Something (P) is really actual relative to the domain of experience iff (if M, then P).
3. PN: Something (P) is really necessary relative to the domain of experience iff P (is actual) AND if G, then P.

7.5.1. Postulate of possibility

The postulate of possibility defines Kant's full-fledged notion of real possibility relative to the domain of experience. As an instance of real possibility, it expresses the possibility of a thing's being in the domain of experience. However, since the notion of logical compatibility implies a relation between two conceptual or propositional representations, the test of real possibility that the postulate offers requires the involvement of the concept of the object or event whose possibility is in question. In other words, the possibility of a thing's being in the domain of experience can be tested through the possibility of its concept's relating to or being instantiated by an object of experience. Kant makes this switch from 'object' talk to 'concept' talk in his elucidation of the postulate, when he writes, for instance, that "the postulate of possibility of *things* requires that their *concept* agree with the formal conditions of experience in general" (A220/B267, my emphases). This is also perfectly in line with the schema of possibility as "the agreement (*Zusammenstimmung*) of the synthesis of various representations with the conditions of time in general" (B184). So, the postulate of possibility states that something is a possible object of experience if the synthesis of the predicates that constitute the concept of that thing is compatible with or not negated by the formal conditions of experience. Again, the same idea can be expressed as the representability of the predicates in question as co-instantiated (by an object or event) in experience against the background of the formal conditions of experience.

Kant does not directly spell out what he considers to be the formal conditions of experience, though he characterizes them as constituting "the objective form of experience," which "contains all synthesis that is requisite for the cognition of objects" (A220). Thus, formal conditions of experience must be the *a priori* formal elements of cognition that determine the matter, i.e., the object, of experience: the form of sensible intuition, i.e., space and time, and the form of thinking, i.e., the categories and the principles of the understanding setting the rules for their empirical application. Something is then really possible if and only if it can be represented as given in space and time, having extensive and intensive magnitudes, and conforming to the universal laws of experience expressed by the analogies, i.e., the conservation of the quantum of substance, causation, and interaction, or what is the same, the synthesis of its predicates is logically compatible with or is not negated by any of these conditions. So,

> PP: Something (P) is really possible relative to the domain of experience iff it is not the case that (if F, then not P).

This would have been a rather straightforward account of real possibility relative to the domain of experience. However, some of Kant's examples of what appear to be real impossibilities suggest that there is more to his account of (empirical) real possibility. Kant draws his examples from the concepts of things that are imaginatively constructed out of the material provided by actual experience: "a substance that was persistently present in space yet without filling it . . . , or a special power of our mind to intuit the future . . . , or finally, a faculty of our mind to stand in a community of thoughts with other men" (A222/BB270). He insists that these concepts would remain "nothing but figments of the brain" since the instruction of experience is omitted in their construction. But it is not clear what formal conditions of experience these concepts violate or logically contradict. What is even more puzzling is Kant's claim that these fictitious cases lack objective reality, because their possibility "cannot be grounded in experience and its known laws (*deren bekannte Gesetze*)" (A223). One might interpret 'known laws of experience' as 'formal conditions of experience,' since the latter include laws regarding the pure structure of space and time and the principles of the understanding that are themselves fundamental *a priori* laws to which all particular laws of nature must conform. Yet since it is not evident at all what precisely constitutes a contradiction between these fundamental *a priori* laws and the fictional cases Kant mentions, it seems also tempting to read 'known laws of experience' as particular empirical laws of nature (e.g., laws of mechanics, neurology, psychology), which could be shown to specifically negate each of the cases in question.

Even though particular empirical laws of nature must conform to the *a priori* formal conditions of experience, Kant certainly take the former and the latter to be presenting distinct sets of laws. In the "Introduction" to the *Critique of the Power of Judgment* he suggests that all particular empirical laws of nature form a systematic unity under the universal transcendental laws that make the experience of nature into a thoroughly interconnected whole of cognition (Ak. 5:179–86). Yet he also makes it clear that such systematic unity of the manifold of empirical laws is only a regulative presupposition of our power of judgment and functions merely as a guideline for our empirical research into nature. In the *CPR*, just after boldly stating in the Transcendental Deduction that "the understanding is itself the source of the laws of nature, and thus of the formal unity of nature," he cautiously adds: "to be sure, empirical laws, as such can by no means derive their origin from the pure understanding" (A127).[20] The relationship between the principles of the understanding and the particular empirical laws of nature is one of conformity of the matter to the *a priori* form. Through its pure laws or principles, the understanding confers a formal unity on nature, which makes the lawfulness (and thereby the necessity) of all empirical laws possible. Therefore, Kant cannot plausibly include the particular empirical laws of nature in the 'formal conditions' or the 'objective form' of experience in general. The formal conditions of experience are exhausted by the conditions that are brought by the structure of space and time, i.e., sensible formal conditions of experience, and the principles of the understanding, i.e., intellectual formal conditions of experience.

This means that reading 'known laws of experience' as particular empirical laws of nature leads one to the contention that Kant has at least two distinct notions of real possibility in the Postulates, one requiring agreement with the formal structure of experience as explicitly prescribed by the postulate of possibility itself, the other requiring agreement also with all the empirical laws of nature and thus delineating a narrower domain. Despite this *prima facie* undesirable consequence, this reading is not unpopular in the literature. I will now be content with raising this problem of the

[20] See also A234/B287.

duplication of real possibility and examine it in 7.6, in connection with my discussion of another puzzling remark Kant makes about the extensions of modal categories.

7.5.2. Postulate of actuality

The postulate of actuality defines actuality within the domain of experience as being 'connected to the material conditions of experience.' The postulate itself does not state what the 'material conditions of experience' are, and it is not immediately obvious what it means to be 'connected' to a set of conditions either. In his elucidation of the postulate of actuality, Kant first identifies the material condition of experience as perception: for us to cognize something as an actual object of experience, it must be given to us through a connection with perception or sensible intuition. By specifying a connection with perception as the mark of existence, the postulate provides Kant's precritical account of existence with an epistemological basis and a strong empiricist direction. "In the mere concept of a thing no characteristic of its existence can be encountered at all" (A225/B273), that is, existence is not a predicate or a determination that could be contained in the concept of anything. Even if we could form the complete concept of a thing with all its determinations, the existence of that thing would still not be found in its concept. This is so, because existence "[has to do] only with the question whether such a thing is given to us in such a way that the perception of it could in any case precede the concept" (A225/B273). In other words, existence is about whether the concept of an object is connected to the perception of that object. It is important to recognize again that rigorously speaking, the predicate of existence operates at the level of concepts and not objects. That is, existence expresses a predicate or determination not of the object but of its conceptual representation, i.e., that its concept, with all its intension, is posited as related to an object outside that concept and located in the context of the thoroughgoing connection of objects of experience. The locative connotation of 'positing' finds its full expression here.

Kant emphasizes that the perception the connection with which is required by the postulate of actuality does not have to be "the immediate perception of the object itself the existence of which is to be cognized" (A225/B273). A 'real connection' between the object whose immediate perception we lack and another object whose perception we have can also verify the existence of the former. And by a 'real connection,' Kant means any connection between objects regulated by the analogies of experience: "One can also cognize the existence of the thing prior to the perception of it . . . if only it is connected with some perceptions in accordance with the principles of their empirical connection (the analogies)" (A225/B273).

The analogies provide *a priori* laws for the connections of objects and unify experience into a coherent, causally and spatio-temporally connected whole. Through the immediate perception of an object, we simply posit or locate it in the thoroughgoing connection of the actual objects of experience. In some cases, the immediate perception might lack due to a spatial and/or temporal distance of the cognitive subject from the actual object. In certain other cases, the immediate perception of the object is impossible due to the constitution of our perceptual apparatus, as in the case of "a magnetic matter penetrating all bodies" (A226/B273). Yet in the absence of an immediate perception of an object, we can still posit it as an object of possible perception in the context of the unity of experience by way of inference from its connection to another object, of which we have an actual perception, through the *a priori* laws prescribed by the analogies.[21]

The postulate of actuality thus offers two ways in which something can be connected to the material conditions of experience, i.e., an actual perception, through which its actuality is cognized. First, there is an immediate perception of the object:

> PA(i): Something (P) is actual relative to the domain of experience iff there is an immediate perception of P.

Second, there is a mediate connection of the object to an actual perception through the analogies (A):

> PA(ii): Something (P) is actual relative to the domain of experience iff there is a perception of something (Q) AND if (Q AND A), then P.[22]

7.5.3. Postulate of necessity

Kant makes clear that the notion of necessity that the postulate defines "pertains to material necessity in existence, not merely formal and logical necessity in the connection of concepts" (A226). Thus, like existence, real necessity relative to the domain of experience is not a predicate or determination that can be contained in the concept of an object but an extra-conceptual fact that can only be given through a connection with an actual perception: "the necessity of existence can thus never be cognized from concepts but rather always only from connection with that which is perceived, in accordance with the general laws of experience" (A227).

This also gives us an idea of how to interpret 'general conditions' (*allgemeinen Bedingungen*) of experience mentioned in the postulate of necessity. It is natural to think that the general conditions of experience are a combination of formal and material conditions of experience that are respectively mentioned in the postulates of possibility and actuality. Accordingly, something is really necessary if it is actual and thus is connected to the material conditions of experience, i.e., an actual perception, and this connection is determined or entailed by the form as well as the matter of experience. In other words, something is really necessary if its existence is entailed by the combination of the formal conditions of experience, i.e., space and time and the principles of the understanding, and the thoroughgoing nexus of things that constitutes the material content of experience up to that point in time.

> P(N): Something (P) is necessary relative to the domain of experience iff P (is actual) AND if (F AND M), then P.

[21] The analogies, however, are only universal and formal principles, and as Kant's example of the magnetic matter suggests, in particular cases of our 'comparatively *a priori*' cognition of existence, we employ the relevant specific empirical laws of nature (e.g., specific causal laws of magnetism). As we discover empirical laws of nature, we further extend the reach of our cognition of existence.

[22] It is important to note here that P is not logically grounded in or connected to Q. That is, P's existence or actuality is not included in the concept of Q or logically entailed by Q's existence. P's existence is entailed by Q's existence only on the condition that A (the set of analogies), which is not contained in the concept of either P or Q, holds.

I suggested earlier that the ultimate implication of the postulate of actuality is that cognizing the actuality of something is positing it as located in the thoroughgoing nexus of the objects of experience. Since these connections are regulated by the *a priori* laws of relations, i.e., the analogies, we can cognize the existence of a member in the nexus of experience from another already given member 'comparatively *a priori*.' The postulate of necessity defines real necessity as a matter of cognizing the actuality of something as a necessary consequence of both the actual content and form of the nexus, that is, the members of the nexus and the analogies prescribing how those members must be connected to one another. In other words, real necessity relative to the empirical domain is the necessity of ground-consequence relations between phenomena as dictated by the very form of experience. Since the relevant kind of relation here is that of ground and consequence, Kant specifies the relevant analogy as the second one: "Necessity concerns only the relations of appearances in accordance with the dynamical law of causality, and the possibility grounded upon it of inferring *a priori* from some given existence (cause) to another existence (the effect)" (B280/A228). Kant also recognizes the consequence of defining necessity as the necessity of causal connections: "It is not the existence of things (substances) but of their state of which alone we can cognize the necessity... since [the existence of things as substances] can never be regarded as empirical effects." (A227/B280) So necessary existence pertains to the alterations of the states of substances and thus to events. We can ascribe it only to the state of a thing, rather than to the thing itself, following from another temporally preceding state of that thing in accordance with the dynamical law of causality.

It is noteworthy that the notion of real necessity defined by the postulate neither explains nor meaningfully applies to the law of causality itself (or any other formal condition of experience in general). For the law of causality itself is not in any way given through the perception of actual alterations, but it is a necessary condition to which all alterations must conform. In other words, the law of causality, along with other formal conditions of experience, is necessary for the very possibility of the domain of experience, and not so because it exists within that domain. This kind of transcendental necessity that belongs to the formal conditions of experience cannot be explained in reference to another law or ground within the domain of experience. But the notion of real necessity offered by the postulate is the immanent and relative necessity of whatever actual alteration that is entailed by the law of causality and the causal nexus of things within the domain of experience. The postulate of necessity thus underscores that every actual alteration is necessary, not in itself, but from a causal point of view: "Everything that happens is hypothetically necessary; that is a principle that subjects alteration in the world to a law" (A228/B280). This leads to a picture of a world where "'nothing happens through a mere accident' (*in mundo non datur casus*)" but also "'no necessity... is blind, but is rather conditioned, consequently comprehensible necessity' (*non datur fatum*)" (A228/B280–1). Through the postulate of necessity, Kant thus restricts real necessity to that of conditioned necessity of phenomena in the domain of experience, where nothing is absolutely really necessary (either in the sense of not being conditioned by anything else or in the sense of being the condition of everything else), and yet everything is necessary in relation to its cause and the law of causality.

7.6. Are Real Modalities Coextensive?

Following his elucidation of the postulates, Kant makes some puzzling remarks. He appears to reject the contention that the extension of possibility is greater than that of actuality, which, in turn, is greater than that of necessity (A230/B283). Further, he states that "the actual adds nothing to the possible" (B284). This leads to the fairly common view in the literature that Kant endorses what I will call here 'the coextensiveness thesis': all modal categories, in their empirical applications, are coextensive, that is, real possibility, actuality, and real necessity relative to the domain of experience all pick out the same set of objects in this domain.[23] The coextensiveness thesis comes with striking modal commitments. If the thesis is correct, then there are no mere real *possibilia* in the domain of experience. All real possibilities actualize, and all actualities are really necessary. Whatever is not actual is really impossible, and whatever is really possible is really necessary. The modal distinctions introduced by the postulates collapse.

I diverge from the common view in two important ways.[24] First, Kant remains carefully agnostic as to whether the extension of real possibility is greater than that of actuality, without excluding unactualized real possibilities. He can, at most, be said to be committed to a weaker version of the coextensiveness thesis: what we can assert to be really possible is coextensive with what we cognize to be actual. Second, the real intention behind Kant's remarks is not to introduce a positive ontological thesis about the extensions of modal concepts, but rather to criticize what Kant takes to be the prevalent way of understanding modalities among his rationalist predecessors, i.e., as various determinations that enter the intensions of concepts of things. Kant's criticism reveals the rich connection between the postulates and his precritical theses on existence. Accordingly, (i) the doctrine of the Postulates extends the negative thesis on existence to all modal notions (i.e., Modal categories do not add any further determinations to the concepts of the objects to which they are ascribed); (ii) the doctrine of the Postulates transforms what I earlier called the core, radical idea in the positive theses into a revolutionary conception of modality (i.e., Modalities express the ways in which concepts of objects are related to the cognitive subject).

Kant's brief discussion appended to the elucidation of the postulates (A230–2/B283–5) sets the stage so as to suggest that he will provide a positive argument for the coextensiveness of modal categories when he raises the questions of "Whether the field of possibility is greater than the field that contains everything actual, and whether the latter is in turn greater than the set of which is necessary" (A230). However, this impression is weakened by his immediate explanation of what these questions mean to ask. Emphasizing that these questions, though appropriate, "fall under the jurisdiction of reason alone," Kant writes:

> For they mean, roughly, to ask whether all things, as appearances, belong together in the sum total and the context of a single experience, of which each given perception is a part which

[23] See, for instance, Smith (2003), 392; Paton (1936), 339–40; Cassirer (1954), 207; Wolff (1963), 298; Hartnack (1967), 86–7; Dicker (2004), 80–2; Stang (2011) 446, 454; and Pape (1966), 221.

[24] For a complete exposition of my view, see my (2016).

therefore could not be combined with any other appearances, or whether my perceptions could belong to more than one possible experience (in their general connection). (A230/B283)

7.6.1. *Are 'other forms of experience' really possible?*

When this passage is read in light of his following elucidation, Kant seems to be asking two distinct versions of that question without making the distinction explicit. First, we have reason to think that Kant is asking whether a different form of experience or an experience formally different from our actual experience is possible.[25] For he writes:

> Even were they possible, we could still not conceive of and make comprehensible other forms of intuition (than space and time) or other forms of understanding (than the discursive form of thinking or that of cognition through concepts); and even if we could, they would still not belong to experience, as the sole cognition in which objects are given to us. (B283)

Kant refrains from making a definitive statement as to the modal status of other forms of experience and remains agnostic. He cannot be asking the rather uninteresting question of whether the concept of a different form of experience, i.e., a non-spatio-temporal form of intuition or a non-discursive form of thinking, is logically possible. The answer would be obviously in the affirmative. Yet the question is in fact that of the real possibility of a different form of experience. The first postulate is intended to define the real possibility of being within the domain of our experience. However, given these conditions of real possibility, it is not clear why Kant cannot simply rule out other forms of experience as really impossible. For surely, the forms of intuition other than spatio-temporal and the forms of understanding other than discursive would disagree with the actual form of our experience, which consists in spatio-temporal intuition and discursive understanding.[26] So Kant must have a different worry in mind.

The first postulate implies that Kant takes the specific form of our experience or subjectivity to generate a specific notion of the real possibility of objects. But a notion of possibility that issues from a certain form of experience cannot meaningfully apply to another form of experience consisting in different conditions under which objects are given to and thought by the subject. For on this account, different forms of experience amount to different ways of delimiting the metaphysical space into distinct domains of really possible objects, i.e., objects conformable to different forms of givenness and thinking. Raising the question of the real possibility of a certain form of experience against the background of the actual form of our experience would, then, be committing a category mistake.

The question of the real possibility of a different form of experience can be meaningfully raised only against the background of the entire metaphysical space that is not limited or conditioned by the particular form of our experience. In other words, the question should rather be formulated as that of the absolute real possibility of a different form of experience. But the notion of real possibility provided by

[25] For instance, Guyer (1998), 305 holds that this is what Kant is asking at A230/B283.

[26] Stang (2011), 455 raises a similar point, and suggests that Kant's agnosticism about the possibility of other forms of experience is unjustified.

the first postulate expresses the real possibility of things relative to the actual form of our experience, and cannot be utilized in cognizing the absolute real possibility of things. It is indeed both one of the overarching conclusions of the Transcendental Deduction (B147–8) and the special emphasis of the Postulates (A219) that the only legitimate cognition-yielding use of the categories (including the modal categories) is their applications within the domain of experience and their 'transcendental use' in cognizing things 'in general and in themselves' without regard to the conditions of our mode of experience would remain illegitimate.[27] In this sense, Kant's agnosticism about the (absolute) real possibility of other forms of experience is not surprising.

Of course, this restriction of categories to empirical use does not rule out our ability to entertain the *thought* of things that do not conform to the formal conditions of our experience. Since the categories themselves, in their unschematized forms, are not bound by the sensible conditions of the givenness of objects (i.e., in space and time) and can extend further than what is bound by those conditions, we do in fact have the conceptual apparatus to *think* whatever disagrees with our form of intuition, be it a noumenon or a different form of experience. However, this still does not give us the license to determine the (absolute) real possibility of things that are not intuitable by us. For the unschematized categories, without relation to sensible intuition, are "mere forms of thought without objective reality" (B148).[28] Once again, therefore, the right epistemic attitude toward the real modality of things in themselves and altogether different forms of experience should be agnostic instead of negative or positive.[29]

7.6.2. Are other 'experienceable worlds' really possible?

While the postulate of possibility neither posits nor rules out the (absolute) real possibility of other forms of experience, it does place a formal constraint upon our possibility considerations regarding the content of experience. If we are to entertain the real possibility of other worlds than the world that constitutes the actual material content of our experience, we are constrained not only by logical conditions of thought, but also by the formal conditions of our experience. Thus, the 'Kantian real possible worlds' must be not only logically consistent in themselves but also compatible with the actual form of our experience. To borrow Hanna's term, they must be "(humanly) experienceable worlds," i.e., the subset of all logically possible domains that are also intuitable in space and time and thinkable through the categories.[30] Interestingly enough, however, Kant seems to remain agnostic also about the real possibility of other (experienceable) worlds.

Following the last quote above, in addition to the question of the real possibility of other forms of experience, Kant gives us reason to think that he is also raising the quite different question of whether a different material content of experience is really

[27] See A219, A246/B303, A255, A239/B298.

[28] See A254–56/B310–12.

[29] For Kant's other agnostic statements about the real modality of things in themselves and other forms of experience, see also B72, *Prolegomena* (Ak. 4:531).

[30] Hanna (2001), 243. Kant himself does not cash out his modal theory in terms of possible worlds semantics, and in fact, except in a few precritical texts (*NE*, Ak. 1:414; *Optimism*, Ak. 2:29–35; *OPA*, Ak. 2:72.), he does not engage with the talk of possible worlds.

possible. For instance, he writes, "Whether other perceptions than those which in general belong to our entire possible experience and therefore an entirely different field of matter can obtain cannot be decided by the understanding" (A231). Now, the possibility that is at stake here is exactly the real possibility of other experienceable worlds, which formally agree with but materially diverge from the actual world of our experience.[31] However, Kant's agnosticism in this case does not appear evidently justified, and in fact, appears even contrary to the most natural reading of the first postulate. For the pure subjective form of our experience underdetermines its material content, and if real possibility requires nothing more than compatibility with the form of our experience, there seems to be no reason why we should not be licensed to attribute such possibility to 'an entirely different field of matter.' In other words, if the first postulate sets forth the sufficient conditions of real possibility, then alternative worlds (and their individual members) that can be represented spatio-temporally, as having extensive and intensive magnitudes (i.e., Axioms and Anticipations), and as conforming to the universal principles of substance, causality, and interaction (i.e., Analogies) should be deemed *really possible.*

Moreover, the doctrines of the Aesthetic and the Analytic do not require that the totality of the material content that is actually given to our perception be necessarily fixed so as to exclude the possibility of alternate totalities, but only that the ways in which the individuals in this totality are given to our perception and the ways in which our perceptions are ordered in their interconnections are determined *a priori* by the form of experience.[32] In fact, claiming that the actual content of experience is necessary in the former way would be ascribing real necessity to the noumenal ground of the sum total of appearances, which would be a transcendental use of the category of necessity. The illegitimacy of the latter is also underscored by the third postulate, which restricts the ascription of real necessity to the existence of effects the causes of which are given to our perception, and thus, to the existence, not even of things, as substances, but of their states. Thus, the kind of causal determinism that Kant holds in virtue of the third postulate pertains to the order of the given material content of experience and does not by itself exclude the real possibility of a still causally ordered but 'entirely different field of matter.'

7.6.3. *The problem of the duplication of real possibility*

There is then a puzzling disparity between what the postulate of possibility officially states and Kant's modal conservatism about what to positively allow as really

[31] Of course, other worlds, when conceived, à la David Lewis, as real entities causally and spatio-temporally disconnected from our world, will never be part of the thoroughgoing unity of our actual experience. Thus, in that sense, other worlds, even if they agree with the formal conditions of experience, are not really possible. Dicker (2004), 81 and Wolff (1963), 298, for instance, take this shortcut in excluding the real possibility of other worlds. But Kant's question here is not whether we can have the experience of such other worlds, but whether, given the formal structure of our mode of experiencing things, a different configuration of still spatio-temporally and causally ordered things could have constituted the world of our experience.

[32] On the contrary, Kant seems to assume the contingency of the given content of experience on a fundamental level when he limits necessary judgments to *a priori* ones, which draw on the form of experience, and declares the content of experience itself as never giving its judgments, i.e., empirical judgments, true necessity, and universality (B3–4).

possible. As I noted in section 7.5., this disparity is also suggested by Kant's elucidation of the postulate of possibility. His first example of a really impossible object, "a figure that is enclosed between two straight lines" (B268), is easy to make sense of. Although there is not a logical contradiction in the concept of such a figure, the construction of it in (Euclidean) space is impossible. Yet it is not easy to see how Kant's second set of examples fail the test of real possibility: "a substance that was persistently present in space yet without filling it"; "a special fundamental power of our mind to intuit the future"; "a faculty of our mind to stand in a community of our thoughts with other men" (A222–3/B270). For while he unambiguously refers to the violation of 'the conditions of space and its determinations' in the first case, he does not explain which formal conditions of experience these objects violate, but only states that their possibility "cannot be grounded in experience and its known laws [*bekannte Gesetze*]" (A223). As I suggested earlier, one might want to read the expression 'the known laws of experience' as referring to the empirical laws of nature.[33] However, neither does Kant specify which empirical law of nature is violated by the cases in question, nor does the postulate of possibility itself seem to require us to include the particular empirical laws of nature in our real possibility considerations. More importantly yet, Kant never actually says that these objects are impossible, but that "there would be no indications at all" or "grounds" of their possibility and that they are "fictions" (*Erdichtungen*) or "nothing but figments of the brain" (A222/B270). Thus, he ultimately remains agnostic with respect to the real possibility of fictionally constructed objects, which do not seem to violate the formal conditions of experience.

Attributing the coextensiveness thesis to Kant appears to be a convenient interpretive option here. For if Kant really is committed to the coextensiveness of real possibility and actuality, then we do not have to worry about accounting for his exclusion of non-actual worlds or objects that are not incompatible with the formal conditions of experience from the extension of really possibility. For then all real mere *possibilia* would be real impossibilities. Yet the problem is that the postulate of possibility itself does not support this option. If anything, the formalism of the postulate appears to allow the real possibility of alternative experienceable worlds and fictional objects.

As I mentioned earlier, there is a certain trend in the literature that seeks to solve this interpretive difficulty by reading into the Postulates a second notion of (relative) real possibility. Recognizing that Kant's official postulate of possibility does not support the coextensiveness thesis, some commentators have claimed that Kant must have in mind an additional, narrower notion of real possibility, which requires compatibility not only with the form of experience but also with the actual matter of experience. For instance, Kemp Smith claims,

[33] There is at least one considerable support for this reading. In his metaphysics lectures, Kant is reported to introduce a notion of "physical possibility" as pertaining to "that which does not conflict with the laws of experience," and given the example provided for physical impossibility ('a large palace being built in four weeks'), one could argue that 'the laws of experience' must refer to particular laws of physics (*MM*, Ak. 29:812).

> [Possibility] is employed in a much more narrower sense to signify that which can have "objective reality"... The possibility of the objectively real rests upon a fulfillment of a threefold condition: (1) that it agree with the formal conditions of experience; (2) that it stand in connection with the material of sensuous conditions of experience; (3) that it follow with necessity upon some preceding state in accordance with the principle of causality, and so form part of a causally necessitated order of nature.[34]

Similarly, Stang argues that while the postulate itself conveys a notion of "formal possibility," which requires compatibility with the actual form of experience, Kant also implicitly employs, as a dual of the notion of real necessity invoked by the postulate of necessity, a notion of "empirical-causal possibility," which requires the compatibility of the alteration or event in question with "actual natural laws and the past history of the empirical world up until time t."[35] This further constraint on real possibility dramatically narrows its extension. It deems really impossible other worlds that completely or partly diverge from the actual history of the world and fictional objects or events that were not and will not be part of the actual content of our experience. Moreover, the deterministic character of the causal laws of nature will also make this-worldly counterfactuals and alternative futures really impossible. Accordingly, events such as Germany's winning WWII or the 2018 FIFA World Cup being hosted by a country other than Russia are all really impossible, because, even though they do not violate any of the formal conditions of experience, they are not compatible with the actual causal context of the world. Nothing that does not actualize is really possible and whatever is really possible actualizes. And since there are no non-actual real possibilities, actuality, real possibility, and real necessity are all coextensive.[36]

This approach has at least three drawbacks. First of all, it attributes to Kant something that he does not explicitly commit to. It could be argued, of course, that the second, narrower notion of (relative) real possibility is implicit in the text, but it is nevertheless a forced reading to assume that Kant would leave such an immensely important point only implicit at the heart of his most rigorous exposition of his modal theory.[37]

[34] Smith (2003), 392.

[35] See Stang (2016), 215; Stang (2011), 453. For further examples of this view, see also Wood (1978), 45, and Pape (1966), 220–1. Stang's derivation of an "empirical-causal" notion of possibility from Kant's postulate of necessity by using modal logical rules of conversion (i.e., "necessarily *p* if and only if not possibly not-*p*") is certainly an interesting suggestion. And we have textual ground to believe that Kant adheres to the interdefinability of logical possibility and logical necessity (*ML2*, Ak. 28:557). Yet the Postulates chapter displays no sign that Kant intends interdefinability between real modalities. For the postulates define the real modalities separately, not only in terms of different relations, i.e., agreeing with, being connected with, having a connection with the actual that is determined in accordance with, with different sets of conditions of experience, i.e., formal, material, general, but also in terms of relations to different cognitive faculties, i.e., understanding, power of judgment, and reason (A219). I will discuss this further below.

[36] As Dicker (2004), 81 notes, the idea here is similar to what Lovejoy (1936), 52, 240 calls "the principle of plenitude," i.e., that whatever is possible must at some point in time be actualized such that the actual world contains the maximum number of things and events that it can.

[37] In reference to the notion of real possibility defined by the postulate, Kant uses the term 'formal possibility' at least once: "All appearances as possible experiences, therefore, lie *a priori* in the understanding, and receive their formal possibility from it, just as they lie in the sensibility as mere intuitions, and are only possible through the latter as far as their form is concerned" (A127). Yet this is not sufficient evidence for his introduction of a separate, non-formal notion of real possibility in the Postulates.

Second, it is not clear whether this approach leaves room for a meaningful distinction between possibility and actuality. The postulates maintain a conceptual distinction between possibility and actuality by assigning them distinct types of logical relations, 'compatibility' and 'connection,' respectively. Now, the notion of 'empirical-causal possibility' attributed to Kant by Kemp Smith and Stang would undermine the distinction between possibility and actuality even on a conceptual level. Provided that the causal laws of nature sufficiently determine the effects of actually given causes, being 'compatible' with the actual content of experience in accordance with the causal laws of nature is nothing but being causally 'connected' with the actual content of experience. For if there is only one event compatible with a given state of things in accordance with the relevant causal laws of nature, then the former is also entailed by the latter. Thus, actuality is indeed built into the very notion of 'empirical-causal possibility,' in which case the coextensiveness of possibility and actuality follows trivially.

Finally, and most importantly, this approach overlooks the blatant textual fact that Kant does not deny the real possibilities of divergent experienceable worlds and fictional objects, but remains agnostic about them. 'Empirical-causal possibility,' by excluding these items as really impossible on account of their being incompatible with the actual content of experience, conflicts with this considered agnosticism of Kant's.

7.6.4. Kant's critical conception of the 'actualist principle'

Rather than duplicating the notion of real possibility, I propose that the solution to the apparent disparity in the text lies in recognizing that Kant employs one notion of real possibility in the Postulates but introduces a distinction between the conditions of real possibility and the conditions of our cognition of real possibility. Kant's overall elucidation of the postulate of possibility underscores that it is one thing for something to be really possible, and another for us to know or prove that something is really possible, which makes it only natural to talk about additional conditions to be met in the latter case due to the specific constitution of our cognition. Thus, what might appear to be two distinct notions of real possibility is in fact a distinction between the *ratio essendi* and the *ratio cognoscendi* of real possibility.

The postulate of possibility defines real possibility in terms of agreement with the sensible and intellectual formal conditions of experience. However, the postulate itself does not explain how exactly we know that an object agrees with these formal conditions. Kant's agnosticism about the real possibility of the spirits, psychic, and telepathic forces implies that he does not consider the absence of a salient conflict with formal conditions of experience in our conceptual representation of an object to be a sufficient condition for our cognition of its real possibility. Kant seems to think that conceptual representation is not a reliable apparatus for us to decisively cognize the compatibility in question. Accordingly, we need more than the postulate itself to positively apply the notion of real possibility to objects.

The crucial point here is that even though real possibility is the possibility of the *object* as opposed to the logical possibility of the *concept*, in the absence of the givenness of the object in, or through a connection with, an actual perception, which would establish the actuality of the object and obviate the question of its possibility, our cognition of real possibility must still rest on a consideration of the mere concept of the object. That is to say, when we are inquiring into real possibility, we are in fact asking whether the synthesis of predicates in a concept can be instantiated by an object of experience. Kant calls this positive feature of a concept its 'objective reality,' and argues that what distinguishes the concept of a really possible object from a merely logically possible concept is the objective reality of the former (B268).

Kant points out two ways in which we can cognize the objective reality of a concept: either the concept is an empirical concept the synthesis in which is "borrowed from experience," or the concept is a pure concept but one related to experience as an "*a priori* condition, on which experience in general (its form) rests" (A220). Empirical concepts are those that we acquire a posteriori, i.e., through reflection on our actual experience, and thus it does not take a special effort to prove that they represent possible objects of experience. The objective reality of pure concepts, on the other hand, requires "the explanation of the way in which [they] *can* relate to objects [of experience] *a priori*," which Kant calls "transcendental deduction" (A85). For instance, Kant states that only from the fact that the categories of relation (substance, causality, and community) "express *a priori* the relations of the perceptions in every experience does one cognize their objective reality" (A221/B269). So we can verify that a concept can be instantiated in experience, either when it is actually instantiated in experience, i.e., when it is the concept of an actual object of experience, or when it is necessarily instantiated in experience, i.e., when it is a formal condition of experience of objects in general.[38]

This sheds some light on Kant's reluctance to grant real possibility to spirits, precognition, and telepathy. We cannot provide a transcendental deduction of these concepts, which seem to be chosen by Kant as fictional analogues of the categories of relation, since they do not express the relations of perception in every experience, or more generally, are not among the formal conditions of experience. But aren't these concepts somehow 'borrowed from experience" and thus, can't we cognize their objective reality a posteriori? Not quite:

> If we wanted to make entirely new concepts of substances, of forces, and of interactions from the material that perception offers us, without borrowing the example of their connection from experience itself, then one would end up nothing but figments of the brain, for the possibility of which there would be no indications at all, since in their case one did not accept experience as instructress nor borrow these concepts from it. (A222)

Therefore, a concept's being 'borrowed from experience' consists in the empirical exemplification not just of the individual predicates that constitute that concept but also of their synthesis within that concept. It is in this sense that we depend on the instruction of actual experience for our cognition of the objective reality of the concept, or what is the same thing, the real possibility of the objective content of

[38] We can verify the objective reality of geometrical figures by constructing them in pure intuition. In this special case, the verification is *a priori*, but the corresponding concepts are not among the formal conditions of experience, even though they are related to a formal condition of experience, i.e., space (A224/B272).

the concept. The example of actual experience sets the boundaries of our positive knowledge of real possibility, and we are to remain ignorant about the real possibility of the alleged objects of those concepts whose constructions do not rely on experience.

The idea that real possibility depends on actuality first appears in what I earlier called the 'actualist principle' and is formulated in the following way:

> The Actualist Principle (AP): Each and every predicate F of any really possible thing must be grounded in something existent that actually instantiates F itself or a more fundamental predicate F*.

I demonstrated in chapter 4 that the AP forms the crux of Kant's precritical argument for the absolutely real necessity of a unique ultimate ground of all real possibility in his *OPA*. However, I also claimed that Kant fails to offer a substantive justification for the AP. In the context of his argument, which sets out to prove the existence of the absolutely really necessary being, Kant employs the AP as an ontological principle regarding real possibilities of things themselves. Yet he also flirts with what I called the strong version of the epistemological interpretation of the AP as expressing a condition of our cognition of real possibilities of things. For instance, he claims that one knows that a 'fiery body' can exist in reality, only if one can prove that fire and body, or more fundamental predicates that make them really possible such as extension, impenetrability, and force, can exist in reality, which ultimately requires an appeal to their actual instantiation, i.e., one's experience of their actuality (*OPA*, Ak. 2:80–1).

The critical Kant's warning against making entirely new concepts of things without borrowing their examples from experience, and offering the "instruction of experience" as a condition of cognizing the real possibility of objects in his elucidation of the postulate of possibility (A222) suggests that he incorporates the strong epistemological interpretation of the AP into his critical account of real possibility. Two points need attention in the critical version. First, Kant now unambiguously understands the principle epistemologically, as stating a condition not of real possibility *per se* but of our cognition of real possibility: we are, due to our cognitive make-up, dependent on actual experience to cognize the real possibility of objects. This is why the postulate of possibility itself does not make any reference to this dependence. The latter becomes an issue only in Kant's elucidation of how we apply the postulate in cognizing real possibility.

Second, while the precritical Kant takes the AP to require the actual instantiation or groundedness of the individual predicates of the concept of the thing in question, the critical Kant's emphasis on the role of experience in verifying the real possibility of the 'connection' or 'synthesis' of the predicates suggests that he now takes the AP to require our cognition not only of the actual instantiation (or groundedness) of each individual predicate but also of their co-instantiation in an actual object of experience. Because only then do we know that the synthesis of the predicates in the concept denotes a possible object of experience. Kant's worry here is clear: even if there is no logical contradiction in our imaginative combination of a certain set of predicates, and even if we know, either through past perception or a transcendental deduction, that those individual predicates are separately instantiated in actual experience, the synthesis of the predicates may still fail to represent an object that can be given in experience. For instance, the predicates in the concept of a spirit, i.e., persistence in space and not filling space, may be thought as instantiated respectively by material and thinking substances, but we have no experiential evidence that these can be co-instantiated in the same substance.

It is plausible to think here that Kant has in mind the idea that logically consistent predicates may stand in a non-logical, i.e., real or metaphysical, opposition with one another. This is a familiar idea. Kant refers to a "real opposition" (*Realentgegensetzung*), "real repugnance" (*Realrepugnanz*), or "real contradiction" (*realer Widerstreit*) both in his precritical texts and in the *CPR*.[39] I argued in chapter 4, against Chignell, that the examples provided in these texts suggest that Kant understands 'real opposition' between predicates to result in the cancellation of their effects on the subject, and thus in a state of equilibrium and real possibility. But Kant's reservation regarding the real possibility of spirits, precognition, and telepathy seems to be motivated by what Chignell calls the 'subject-cancelling' type of real opposition, the result of which is the cancellation and thus the real impossibility not just of the effects of the predicates on the subject but also of their synthesis in the subject. What also follows is that Kant thinks that unlike freedom from logical contradiction, we cannot establish freedom from real opposition by means of the mere thought or conceptual representation of the predicates; we depend on the actual experience of their co-instantiation. Without the latter, we cannot rule out 'subject-cancelling' real opposition and are bound to remain agnostic about the real possibility of the objects in question.

Kant thus refuses to grant real possibility to the alleged objects of concepts that have been 'made' or 'invented' by us without the guidance of actual experience. Such concepts are a special class of "empty" concepts denoting mere "thought-entities" (*entia rationis, Gedankendinge*) "which one thinks without contradiction, to be sure, but also without any example from experience even being thought" (B347). Yet Kant also refrains from labeling these thought-entities as really impossible. He unambiguously states that these entities "cannot be counted among the possibilities although they must not on that ground be asserted to be impossible" (B347), and he contrasts them with really and logically impossible objects (*nihil negativum*) such as a "rectilinear figure with two sides" (A292/B348). In refusing to grant real possibility to spirits and psychic and telepathic forces, Kant does not appeal to an impossibility (logical or real) but to a lack of positive ground, stating that "there would be no indications at all [*gar keine Kennzeichen*]" (A222) of their real possibility or that their real possibility is "entirely groundless [*ganz grundlos*]" (A223).

So what Kant introduces here is an epistemic rather than an ontological predicament. The question is whether this epistemic predicament leads us toward a weaker version of the coextensiveness thesis that could also accommodate Kant's agnosticism, according to which the sphere of what we can assert to be really possible is restricted to that of what we cognize to be actual.

[39] See respectively *OPA*, Ak. 2:86; *NM* Ak. 2:171; A264/B320, A273/B329.

7.6.5. The weak coextensiveness thesis: Types and particulars

It is possible to observe both in the *OPA* (1763) and in the *Blomberg Logic*, which reflects his views in the early 1770s, that Kant remains adherent to the Leibnizian notion of a complete concept representing an individual in its complete determinacy.[40] However, in his critical period logic lectures, from the 1780s on, Kant explicitly abandons the idea of a singular concept representing an individual with complete determinacy. He repeatedly emphasizes that the abstract/concrete or universal/singular distinction does not apply to concepts themselves, as they are all abstract and universal representations, but only to the use of concepts.[41] He holds that the idea of a lowest concept representing a single object (a *conceptus infimus*), that is, of a concept completely determined with respect to all possible predicates, is only a negative and exclusive limit in the logical subordination of concepts and thus can never be reached.[42] Accordingly, every concept, as to its content, is only a partial or partially determinate representation of the object.

This means that real possibility *qua* mere real possibility, i.e., in the absence of actual perception, will always be considered with reference to an incomplete (conceptual) representation of the object in question. This also informs the exemplification requirement prescribed by the AP: in the absence of the actual perception of an object, we affirm the real possibility of that object if the synthesis of the predicates in its concept is actually exemplified in actual experience. If concepts are only partial representations of objects, then exemplification need not express a complete identity between that whose real possibility is sought and the actual example that grounds that possibility. The co-instantiation of a certain set of predicates in actuality, even if the actual example is a particular object or event that is itself completely determinate, is sufficient to establish the real possibility of the *type* of object or event that is represented by that set of predicates. For instance, the existing world record in a branch of athletics gives us sufficient ground to establish the real possibility of a human being's equaling the world record, if not improving it. The actual observation of one particular narwhal would be sufficient for establishing the real possibility of the species of whales with horn-like tusks. Therefore, at least on the level of types, the exemplification requirement restricts what we can positively cognize as really possible to what we do actually cognize, that is, the weaker coextensiveness holds. We verify that a type of thing is really possible only if that type is exemplified by an actual object or event in experience.

Where does this leave us on the question of the real possibility of alternative worlds? Our experience of the actual world cannot guide us in determining the real possibility of the experienceable worlds that are composed of entirely different types of objects. This is what Kant means when he emphasizes that we cannot decide the real possibility of "an entirely different field of matter" (A231). On the other hand, it appears as though we can attribute real possibility to the worlds that conform to the formal conditions of experience and that are sufficiently similar, though not identical,

[40] See *OPA* (Ak. 2:72–3); *LB* (Ak. 24:257).
[41] See *LV* (Ak. 24:908); *JL* (Ak. 9:1, 99–100).
[42] *JL* (Ak. 9:97).

to the actual world such that they are composed of objects whose types are exemplified by the particular objects of the actual world. Then the actual world presents a general archetype or schema for the material contents of other really possible worlds that are formally identical to it. There is, however, a further analysis that we have to carry out regarding particular objects in order to do full justice to Kant's modal agnosticism.

The strongest indication of the real possibility of something is its actuality, which can only be cognized through a connection with a perception. And the actual object is completely determined with respect to the whole of empirical reality. Yet as was stressed earlier, when we lack an actual perception of the object or a connection with it, our real possibility considerations must rest only on the mere concept of the object, which can never represent the latter in its complete determination. Only empirical intuition or perception can represent the qualitative and quantitative determinations, i.e., intensive and extensive magnitudes, and causal and spatio-temporal determinations required for the individuation of an object in the thoroughgoing nexus of empirical reality. Rigorously speaking, then, *mere* real possibility, when it comes to apply to a particular object, is bound to remain an incomplete possibility.[43] The problem again is epistemic: our discursive understanding, in its capacity to think conceptually, is not able to capture the "complete possibility" of particular objects. In fact, Kant introduces the notion of "complete (*vollständige*) possibility" as a correlate to the Leibnizian idea of a completely determinate concept of an individual.[44] And just like the complete concept of an individual itself, Kant holds that the notion of complete possibility is an unattainable and merely regulative idea of reason: "We can never have insight into complete possibility, because we cannot determine a concept in all detail thoroughly and *in concreto*" (R 6298, Ak. 18:565).

Therefore, first of all, we cannot posit the real possibility of non-actual particulars that are completely determined in the spatio-temporal and causal context of an experienceable world. Fictional characters and counterfactual events are never defined well enough so as to be completely determinate, but even if they were, we would still remain ignorant about their real possibility. In the case of particulars, we would have sufficient evidence to posit the real possibility only of the ones to which we have a perceptual connection. The same restriction applies to whole worlds too. The only completely determinate spatio-temporal and causal unity whose real possibility we can posit is the actual world of our experience. What is actually given in experience does not license us to posit the real possibility of more than a single particular experienceable world. Thus, with regard to the real possibility of particular contents of whole experiences, we are restricted by the unity of our actual apperception. This is what Kant means when he writes "[T]hat another series of

[43] To my knowledge, H. W. Cassirer is the only author to pay some attention to the distinction between the possibility of types and that of particulars in the context of Kant's postulate of possibility. Cassirer (1954), 202–3 claims that Kant distinguishes between "the notion of possibility, as applied to particular objects of experience and the notion of possibility, as applied to objects of experience in general." He also adds, without any explanation, that the former notion of possibility "plays no part in Kant's enquiry into the principles of modality."

[44] A234/B287; R 5590 (Ak. 18:242).

appearances in thoroughgoing connection with that which is given to me in perception, thus more than a single all-encompassing experience, is possible, cannot be inferred from that which is given" (A231–2).

So not just the form but also the matter of our actual experience constrains our positive cognition of real possibility concerning particulars. In other words, in the case of a particular object, our cognition of its 'agreement with the formal conditions of experience' is only possible through our cognition of its 'connection with the material conditions of experience.' For the very particularity of the object requires not just that it be thought generally as a spatio-temporal and causally determined object whose type is exemplified in the actual world but also that it be specifically located in the spatio-temporal and causal context of the actual world, which, in turn, requires an actual perception of it or of what is connected to it through the law of causality. Our inability to conceptually represent objects as thus completely determined and individuated without their giveness to us in perception is what motivates Kant's emphasis on the AP. Thus, in the final analysis, on the level of particular objects too, Kant holds the (weak) coextensiveness of what we can cognize as really possible and what we cognize as actual. This still does not mean that he endorses the (strong) coextensiveness of the real possibility and the actuality of particulars. Kant's position is unambiguously agnostic: he does not deny the real possibilities of non-actual particulars, but points out our inability to posit them. His critical actualism works only as a condition on our modal cognition and does not exclude mere *possibilia* as really impossible.

Kant's remarks about the extensions of possibility and actuality should not be understood as entailing the positive ontological thesis that all real modalities are coextensive but the negative epistemological claim that we cannot legitimately extend what is really possible beyond what is given to us in actuality because of a certain incapacity of our discursivity. His account is still based on a single notion of real possibility, but takes into consideration the distinction between the conditions of real possibility and the conditions of our cognition of real possibility. The postulate of possibility alone sets the general conditions of the possibility of being in the domain of experience. Yet it is Kant's elucidation of the postulate that teaches us that the full schema for our application of the category of possibility is the matter of our actual experience.

7.7. The Real Target of Kant's Remarks and His Revolution in Modality

Why does Kant append such a perplexing discussion to his elucidation of the postulates? It might be said that Kant intends to issue a critical warning against extending possibility beyond actuality without regard to the epistemic conditions of our cognition of modality. However, he has much more in mind. The real target of his remarks is a certain metaphysical way of thinking about modalities that he attributes to his rationalist predecessors, namely conceiving of modalities as various real determinations of objects and thus in terms of the contents or intensions of the concepts of objects. For it is this conception of modalities which, Kant thinks, motivates the contention that the extension of possibility is greater than that of actuality, which, in turn, is greater than that of necessity. In fact, Kant's critique of this conception of modality reveals what exactly is so historically revolutionary in his critical theory of modality.

As I noted earlier, the critical Kant maintains that concepts are universal and incomplete representations of individual things. Let us also recall it follows from this view of concepts that our cognition of possibility *qua* mere possibility is always that of an incomplete possibility, and 'complete possibility' is a merely regulative and unattainable idea of reason paralleling the equally unattainable idea of a 'complete concept.' Kant's critical view of concepts underscores a significant divide between him and his rationalist predecessors, who subscribe to the idea of complete concepts of individual things.[45]

With this tension in mind, we can now look at another part of Kant's discussion of the extensions of modal categories. In the midst of his formulations of the questions of extension, Kant points to the 'poverty' of the series of inferences through which it is concluded that the realm of possibility is greater than that of actuality:

> 'Everything actual is possible'—from this it follows naturally, in accordance with the logical rules of conversion, the merely particular proposition, 'Something possible is actual,' which then seems to mean as much as 'Much is possible that is not actual.' It certainly looks as if one could increase the number of that which is possible beyond that of the actual, since *something must be added to the former to constitute the latter*. But I do not acknowledge this *addition* to the possible (A231/B284, my emphases).[46]

Kant's rejection of 'this addition' is not at all trivial. It follows from his negative account of existence, which, as we saw in chapter 3, he first offers in his precritical polemics with the tradition. Especially pertinent here is his T4, "Nothing more is posited in an existent thing than is posited in a merely possible thing" (*OPA*, Ak. 2:75), which he will reiterate in his refutation of the ontological argument (A599–600/B627–8). Kant's recurrent talk of the actual not 'adding' anything to or not 'containing' anything more than the possible might appear to be lending textual support for the view that Kant endorses the (strong) coextensiveness thesis: if the actual does not contain more than the possible, the two concepts must be coextensive. In fact, however, the coextensiveness thesis is beside Kant's point. Kant does not name any figure, but it is obvious to his late eighteenth-century German audience that he is alluding to Wolff's dictum that existence is the 'complement of possibility,' which I discussed in chapter 2.

Let us remember that the reasoning that Wolff runs in reaching this dictum is very similar to the series of inferences Kant critically reconstructs above:

[45] See Leibniz's *Discourse on Metaphysics* §8; Wolff's *Ontologia* §§225, 226, and Baumgarten's *Metaphysica* §§53, 148, 151.

[46] Kant continues the quoted passage with the proposition that "For that which would have to be added to the former would be impossible." This unfortunate proposition is a failed attempt on Kant's part to quickly block the preceding inference on a purely formal level and might lead to a gross misunderstanding of what he intends to get across. As I explain below, Kant's rejection of the 'addition' relies on more substantive grounds regarding the way modalities should be conceived.

(1) While everything actual is also (absolutely) possible, not everything that is (absolutely) possible is actual.
(2) Then, mere possibility is not the sufficient ground of actual existence. The actual existence of a thing requires something more than possibility.
(3) The possible must be complemented by something more from outside in order to exist.
(4) What we call existence is this complement of (absolute) possibility (*complementum possibilitatis, Erfüllung der Möglichkeit*).[47]

Wolff also holds that whatever exists is completely determinate, and that only what is so determinate exists, and thereby establishes a necessary connection between complete determination and actual existence.[48] Baumgarten interprets Wolff's dictum and his principle that the actual is always completely determinate to suggest that existence is an internal but contingent determination in its own right, a 'mode' that complements the essence (*complementum essentiae*) or internal possibility of a finite thing, which by itself is underdetermined or incomplete.[49] Kant emphatically complains about this interpretive move by Baumgarten already in *OPA*:

> Baumgarten introduces the concept of thoroughgoing internal determination, and maintains that it is this which is more in existence than in mere possibility, for it completes that which is left indeterminate by the predicates inhering in or issuing from the essence (Ak. 2:76).

Thus, this view locates the difference between the actual and the merely possible at the level of the determinacy of the concept of the object: while the former is completely determined with respect to all possible predicates, the latter is not. Once modal distinctions are thus defined in terms of the determinacy of the intensions of the concepts, the conclusion that the concept of the actual in general has a narrower extension than the concept of the merely possible appears to follow naturally. And this construal of modal distinctions is exactly what Kant refuses here. If modality has nothing to do with intension, then the question of extensions of modal categories is motivated by a false premise.

7.7.1. *The peculiarity of modality*

Kant's revolutionary accomplishment in the Postulates primarily consists in replacing the construal of modalities in terms of *what* determinations are represented *in* the concepts of objects with the construal of modalities in terms of *how* the concepts of objects are related to the cognitive subject and the conditions of cognition. Kant puts forward this revolutionary approach right in the beginning of the Postulates by emphasizing that there is something 'peculiar' about modality as opposed to all other categories of the understanding:

> The categories of modality have this peculiarity (*Besondere*): as a determination of the object they do not augment the concept to which they are ascribed in the least, but rather express only the relation to the faculty of cognition. If the concept of a thing is already entirely complete, I can still

[47] *DM*, §§14–15; *Ontologia*, §§170–174. [48] *Ontologia*, §226. [49] Metaphysica, §§55, 134.

ask about this object whether it is merely possible, or also actual, or if it is the latter, whether it is also necessary? No further determinations in the object itself are hereby thought. (A219)

Unlike ordinary predicates that add further determinations to the concept of an object and narrow down its extension, modal categories do not bring any further determinations to the conceptual representation of the object or change the extension of it. I take Kant here to be extending his T1 that existence is not a predicate or a determination that can take part in the intension of the concept of anything. Kant now holds that real possibility and real necessity share this peculiarity with existence.

It is for the rhetorical purpose of highlighting this peculiar nature of modal categories that Kant appeals to the idea of a completely determinate concept in this context. Even were we to have a Leibnizian divine intellect with the capacity to represent an object in its complete determination with respect to all possible predicates, the modal status of that object would still remain an open question, because (real) modality is an extra-conceptual matter and cannot be settled by mere conceptual thought. In other words, mere concepts are neutral with respect to (real) modality; from the viewpoint of determination or conceptual content, the putative complete (real) possibility of a thing could not be distinguished from its actuality and necessity.[50] Kant's insistence that the actual does not contain more than or 'add' anything more to the possible (T4) and his famous example of the hundred dollars (A599–600/B627–8), which I will examine more closely in the next chapter, should also be understood in the vein of this critical rhetoric. The difference between the merely (but completely) possible and the actual hundred dollars does not lie in the contents of their corresponding conceptual representations. Instead, the difference lies outside the concepts, in the ways the concepts are related to the cognitive faculties of the subject and the conditions of cognition. It is Kant's main aim in the Postulates to give flesh to the latter, positive claim in the framework of his critical theory of cognition.

7.7.2. Kant's radical idea

The negative aspect of Kant's peculiarity claim is that modal notions, as strictly separated from the intensions of concepts of objects, are not properties of objects expressing the ways they are. This negative aspect is here complemented by the positive claim that modality is in fact a feature of our conceptual representations of objects, expressing how they are related to our cognition. Again, this is an extension of what I earlier called the core radical idea in Kant's precritical positive account of existence, expressed most clearly by his T2: "Existence is a predicate not so much of the thing itself as of the thought which one has of the thing" (*OPA*, Ak. 2:72). In other words, modal categories, when they apply to objects, are in fact second-order or concept-level predicates. This is in fact why in his more rigorous formulations of modality, Kant switches from 'object' talk to 'concept' or 'representation' talk. The latter is not a gesture in favor of the phenomenalist view that all objects are in fact our representations. Instead, it underlines the exceptionality of modal categories: unlike the other categories, modal categories do not even purport to express features of objects themselves but of our representations of them in relation to our cognition in general.

[50] See R 5772 (Ak. 18:349–50); R 5784 (Ak. 18:354); A234/B287.

I argued in chapter 3 that the precritical Kant does not yet realize the immense importance of this radical idea for modality in general, despite its original expression by T2. However, the critical Kant builds a new conception of modality on this very idea. After stating that modality is an extra-conceptual matter separate from the intension even of the complete concept of an object, Kant explains what it means to ask whether that object is merely possible, actual, or necessary:

No further determinations in the object itself are hereby thought; rather, it is only asked: how is the object itself (together with all its determinations) related to the understanding and its empirical use, to the empirical power of judgment and to reason (in its application to experience)? (A219)

Kant here extends his T3 that existence is absolute positing (*OPA*, Ak. 2:73) to modality in general. There are two important aspects of T3. First, it suggests that existence, instead of picking out a certain determination in the concept of an object, expresses an extra-conceptual fact about the concept as a whole, i.e., with whatever determinations contained in its intension. Second, this expression is carried out through a cognitive act, i.e., positing or asserting the concept as corresponding to an actual object outside thought. Now, Kant maintains, just like existence, (real) possibility and (real) necessity are also peculiar kinds of 'predicates' that posit the concept of an object as a whole, with all its determinations, with regard to an extra-conceptual fact. Thus, all three types of modal assertions are in fact different instances of absolute positing.

Yet, this is not a simple extension of T3. In the *OPA*, T3 aims at explaining what it means to make an existence assertion in a proposition, leaving the questions of what the epistemic conditions of absolute positing are and what kind of cognitive act this is unanswered. The doctrine of the Postulates provides substantive answers to these questions. The initial statement that I cited above makes it clear that to posit something as possible, actual, or necessary is to posit it in relation to different cognitive faculties. The postulates themselves identify the conditions of these different instances of absolute positing with different conditions of experience. But the key to a better grasp of what these really mean is yet another entirely novel thesis about modal assertions that Kant introduces for the first time in the Postulates: modal assertions constitute a special type of synthetic judgments.

7.7.3. The "subjective-syntheticity" of modality

There are two ways in which Kant defines syntheticity of judgments, one negative, the other positive. According to the negative definition of syntheticity, a judgment is synthetic if "[the predicate] *B* lies entirely outside the [subject] concept *A*, though to be sure it stands in connection with it" (A7/B10). Kant generalizes his precritical T1, existence is not a determination or predicate that can be contained in the intension of the concept of anything, to all modal categories in the Postulates. This makes all categorical modal propositions, e.g., 'S is (really) possible/actual/necessary,' fit the negative definition of syntheticity. That is, all modal predicates lie outside the subject-concept, though they may be connected with it. The propositions of actuality and (real) necessity require connection with actual perception, and the propositions of (real) possibility require consideration of the formal conditions of experience, which are not contained in the concept of any object. In each case, the modal status of what the subject-concept represents cannot be reached by means of an analysis of this concept, and consequently, the truth of the proposition is dependent on extra-conceptual facts about the relationship between the concept, and the cognitive subject against the background of the conditions of experience.

According to Kant's positive definition of syntheticity, synthetic judgments are "judgments of amplification" because they "add to the concept of the subject a predicate that was not thought in it at all" (A7/B11). However, none of the modal categories is the kind of predicate that could add any further determination to the concept of anything. A modal judgment does not state anything new regarding *what* the object of the subject-concept is. In what sense, then, are modal judgments synthetic? What kind of 'synthesis' or 'addition' do they carry out? The answer to this question is crucial in understanding the essence of Kant's revolutionary perspective on modality. The answer lies in the peculiar kind of syntheticity Kant attributes to modal judgments:

> The principles of modality are not... objective-synthetic, since the predicates of possibility, actuality and necessity do not in the least augment the concept of which they are asserted in such a way as to add something to the representation of the object. But since they are nevertheless always synthetic, they are so only subjectively, i.e., they add to the concept of a thing (the real), about which they do not otherwise say anything, the cognitive power whence it arises and has its seat. (A234/B286)

Accordingly, objectively synthetic judgments are those in which the synthesis operates on the level of the intensions of the conceptual representation of the object and thus contributes to the determination of that representation. This is not the case in modal judgments. Principles of modality have a distinct place in the grand picture of Kant's 'system of all principles of pure understanding.' For they are, first of all, 'dynamical principles' and thus distinct from the 'mathematical' principles, i.e., Axioms and Anticipations, which are constitutive of objects in that they determine *a priori* what absolute properties objects necessarily display in their possible representation in intuition. Second, the principles of modality are also distinct from the other set of dynamical principles, the Analogies of Experience, for while both regulate the ways objects are located in experience, the latter still concern interrelations of objects and the former concern the relations of the representations of objects to the systematic whole of experience. Thus, in the ascription of a modal category to the concept of an object, neither an inner determination of the object nor its (external) relation to another object is directly expressed. Modal judgments are subjective in that rather than stating anything about what the representation of the object contains, they express how that representation is related to the cognition of the subject.

More specifically yet, Kant correlates the three kinds of modal assertions with three distinct cognitive faculties and their specific cognitive acts that are involved in these modal assertions. In a possibility assertion, e.g., 'S is possible,' the understanding establishes the compatibility between the concept of S and the formal conditions of experience. In an existence assertion, e.g., 'S exists,' the power of judgment simply connects the concept of S with an actual perception and thus with an actual object. In a necessity assertion, e.g., 'S exists necessarily,' reason connects the concept of S with an actual perception and its object through a causal inference. Just like the case with the correlation Kant proposed earlier between the modalities of judgment and the same respective faculties (A76/B101), in formulating this correlation, Kant seems to have in mind the idea of a modal progress. In the possibility assertion, the concept is set forth, without its object being given. In the second, the object is simply given through perception and subsumed under the concept. In the third, the givenness of the object and its consequent subsumption under the concept is inferred by means of a rule.[51] Therefore, in all three kinds of modal assertion, there is an act of subjective synthesis carried out by the faculty of cognition. Unlike the ordinary, objective synthesis, this subjective synthesis does not 'amplify' the concept of the object by adding a further determination to it, but it does amplify our empirical cognition in general by adding to it a possible, actual or necessary object.[52] So, there is certainly an addition, though a different sort, involved in modal assertions. In refusing the construal of existence as an addition to the merely possible Kant writes:

I do not acknowledge this addition to the possible.... All that can be added to my understanding is something beyond agreement with the formal conditions of experience, namely connection with some perception or other. (B284)

At this point, Kant reinterprets his precritical theses, T4 and T5, "Nothing more is posited in an existent thing than is posited in a merely possible thing" (*OPA*, Ak. 2:75) but "More is posited through an existent thing than is posited through a merely possible thing" (Ak. 2:75), in light of his now fully unfolded revolutionary account of modality, according to which the 'addition' in question or modal difference must be construed in terms of the difference in the manner of the relation to cognition:

Through the actuality of a thing I certainly posit more than possibility, *but not in the thing*: for that can never contain more in actuality than what was contained in its *complete possibility*. But while possibility was merely a positing of a thing in relation to the understanding (to its empirical use), actuality is at the same time in connection with perception.

(A234–5/B287, emphases added)

This is no small addition. Real possibility posits the object, insofar as it is represented by its concept, as compatible with the general form of experience. But through the addition of a connection with a perception, the object is posited or located in the context of actual experience as a whole. And when this connection with perception is established through a causal inference, the object is posited in the context of actual experience as determined by its thoroughgoing causal unity; that is, the entire actual world or cosmology of objects is added to it. Therefore, the assignment of each modal status to an object, rather than contributing to the representation of the object itself, contributes to our overall construction of a system of empirical cognition against the background of the structure and conditions of experience. The principles of modality

[51] See also R 4802 (Ak. 17:733).

[52] See R 5558 (Ak. 18:232): "The propositions of modalities, e.g., A is possible (real) (actual), are synthetic propositions. But nothing is added to the concept of the thing but to thinking in general." See also A601/B629.

are only subjectively synthetic because the differences between them do not lie in what they posit *in* the concepts of objects but in *how* they posit the concepts of objects in relation to the unity of our experience.

* * *

In Part II, I argued that Kant's negative account of existence, particularly his well-known T1, through which he rejects the conception of existence as a predicate containable in the concept of an object, does not present what is really radical in his engagement with the subject in the *OPA*. For one thing, the idea that existence is a specific determination of an actual object was not universally espoused even among the proponents of the ontological argument, nor was Kant the first to object to it. I suggested that Kant's radical idea is instead rooted in his positive account of existence in the *OPA*: existence is a predicate of our representation of things rather than things themselves (T2) and is inseparably tied up with a cognitive act, i.e., positing, performed by a cognitive subject (T3). It is this very idea which constitutes the core of what I call here Kant's revolutionary theory of modality, according to which modal notions no longer express modes of being of things but modes in which the representations of things can be related to the cognition of the subject.

This revolution in Kant's understanding of modality, which, I argued in chapter 5, starts unfolding as early as 1760s through his growing emphasis on the relation to cognition (5.1.), empiricism (5.2.), and subjectivism (5.3.), culminates in the doctrine of the Postulates. Kant pursues a heavy and multidimensional agenda regarding modality in the Postulates. He restricts real modality to the particular domain of empirical objects. Since this domain is delineated by the subjective conditions of human empirical cognition or experience, being part of this domain requires standing in relation to those conditions. The postulates define being possible, actual, and necessary in this domain as different ways in which the representation of an object relates to the conditions of experience. This amounts to redefining modality as a feature of our representations of things, and thus as fundamentally relative to our cognitive constitution. This redefinition transforms the question regarding the modality of things: it is no longer a question of what determinations a thing itself has and thus of what the intension of our concept of it contains, but how that concept relates to our cognition. The absolute (real) modality of things, in isolation from the conditions of their relation to our cognition, though still thinkable, remains beyond our modal cognition. In the next chapter, I will show how this implication of Kant's revolutionary theory of modality will radicalize his critique of traditional ontotheology.

8

Kant's Radical Critique of Ontotheology

I have shown so far that Kant's precritical and revisionist views on modality develop into a revolutionary theory of modality in his critical period. My aim in this chapter is to demonstrate how this revolutionary theory of modality radicalizes his critique of ontotheology in the Transcendental Dialectic of the *CPR*. I will examine this critique in two parts. First, I will focus on section two of the Ideal of Pure Reason, where Kant provides a subtle critique of his own precritical 'only possible argument.' Second, I will discuss his refutation of the traditional ontological argument in section four of the Ideal. I will argue that both steps of Kant's critique of ontotheology are motivated by the principles of his revolutionary theory of modality laid out in the Analytic. What makes this critique radical, as opposed to Kant's precritical revisionist critique, is that it aims to demonstrate the impossibility of ontotheology and reframe it in terms of a subjectively valid procedure of pure reason. Such downgrading of a claim about the existence of a supersensible being to one about the nature of human reason is central to Kant's critique of rational theology and metaphysis in general. Thus, I will argue, Kant's revolutionary theory of modality plays an essential role in his overall critical transformation of metaphysics.

8.1. The Fate of the 'Only Possible Argument' after Kant's Modal Revolution

Kant adopts a quite perplexing attitude toward his precrtical alternative ontological argument in his critical period. In the Ideal of Pure Reason, he explicitly rules out the very possibility of any theoretical proof of God's existence, providing refutations of the three traditional proofs (i.e. ontological, cosmological, and physico-theological) as "the only three kinds of proof there are" (A590/B618), and makes no explicit reference to his own 'only possible argument.'[1] The question of what then happens to the 'only possible argument' in the critical period has attracted scholarly attention in the last few decades as much as the argument itself. The interest in this question is well justified, not only because of the genuine curiosity of the case, but also because

[1] Kant mentions "the only possible argument" (*einzig mögliche Beweisgrund*) twice in the Ideal (A625/B653 and A630/B658), but somewhat confusingly, in both occasions he refers to the rather classical ontological argument as the 'only possible' one.

tracing the fate of the argument in the critical period proves fruitful in understanding the intricate grounds and structure of Kant's critique of rational theology.

Despite its economical appeal, it would be too hasty to conclude that this apparent omission in the *CPR* suggests a silent abandonment of the argument on Kant's part. For one thing, Kant continues to express sympathy for the rational appeal of the argument, stating that even if it cannot demonstrate the (objective) necessity of the existence of God, it establishes the (subjective) necessity of assuming the idea of such a being.[2] Moreover, it is widely held in the literature that the section two of the Ideal involves an implicit restatement of the argument, somewhat preserving its logical structure but downgrading its status to the demonstration of the subjective necessity of an idea of reason.[3]

The main interpretive debate turns around the questions of what shift in Kant's thought between 1763 and 1781 leads to this downgrading, and whether and how this shift follows from Kant's critical doctrines. The more mainstream view, propounded by, for instance, Henrich, England, and Wood, is that the downgrading is due to the restrictions on theoretical cognition imposed by transcendental idealism.[4] Watkins and Fisher claim that Kant's "weakening" of the conclusion of the argument is based on the distinction between the respective objects of the understanding and reason, which does not depend on transcendental idealism.[5] Schmucker and Logan claim that the evolution of the argument into the argument of the Ideal is brought about by an empiricist shift in Kant's thought prior to and independent of his major critical doctrines.[6] Chignell argues that Kant's "undoing" of his original argument is due to his realization that the argument relies on the unwarranted assumption that the *ens realissimum* itself is really possible.[7] Boehm argues that Kant rejects the argument and "transforms" its conclusion into a regulative ideal, because of his commitment to the doctrine of transcendental illusion.[8]

There are elements of truth in all of these accounts. However, the story behind Kant's downgrading of his argument is more historically and textually complex than has been portrayed in the literature. What is conspicuously problematic is that despite the popularity of the thesis that the Ideal involves a transformed version of the argument, an account of how exactly this transformation happens within the framework of the extremely convoluted text and the doctrine of the Ideal has not been offered.[9] Without such an account, the transformation thesis itself cannot be fully substantiated, nor can the role of the doctrine of transcendental illusion in the transformation of the argument be understood. Worse yet, the main question of what specifically leads to Kant's rejection of the objective validity of the argument cannot

[2] See *Th.Pölitz* (Ak. 28:1034); see also R 6293 (Ak. 18:561).

[3] See Sala (1990), 237–55; Schmucker (1980), 495–500; Fisher and Watkins (1998); Wood (1978), 71–9; Logan (2007), 360–1; Henrich (1960), 141–3; England (1930), chapter 6; Dell'Oro (1994), 164–72; Guyer (2006), 145–7; Boehm (2012), 312–17.

[4] See Henrich (1960), 143; Wood (1978), 73–9; England (1930), chapters 5–6.

[5] See Fisher and Watkins (1998), 370, 393.

[6] See Schmucker (1972), 496; Logan (2007), 347.

[7] See Chignell (2009) 190.

[8] See Boehm (2012), 313.

[9] Grier (2001), ch. 7 and Allison (2004), ch. 14 have produced exceptionally detailed commentaries on this chapter of the *CPR*. Unfortunately, however, except for Grier's brief discussion (2001, 248–50), neither work offers a substantive account of Kant's treatment of the argument in the Ideal.

be fully answered. For the Ideal is the only text where Kant, albeit indirectly and almost cryptically, provides an answer to this question.

In this section, I will offer an account of the transformation of the argument in the Ideal in its historical context.[10] First, I will locate Kant's dual perspective on the argument (as subjectively valid but lacking objective validity) in the broader shift in his conception of metaphysics in general and rational theology in particular, from an ontological to an epistemological direction, which is critical in essence but can be traced back as early as the late 1760s. Moreover, this dual perspective shapes Kant's overall critique of traditional metaphysics and rational theology, where he pursues the twofold aim of exposing the specific errors in the arguments leading up to certain existential claims as well as explaining why human reason is still drawn to making those errors.

Therefore, in line with this twofold aim, Kant's critique of his own argument, if any, should address two distinct questions: i) Why, or due to what specific error, does the argument fail as an objectively valid demonstration of the existence of God? ii) Why is the argument nevertheless so strongly appealing for human reason, or what causes reason to succumb to the error in it? I will argue that Kant's answer to the first question lies in his critical reinterpretation of the Actualist Principle (AP) as informed by his revolutionary theory of modality. Kant's original argument relies on the ontological interpretation of the AP. However, the critical and epistemological interpretation of the AP, while still compatible with the (subjective) necessity of presupposing the idea of the *ens realissimum* as a conceptual requirement for our abstract thinking about the real possibilities of things in general, does not support the 'hypostatization' of this idea or attribution of an actual object to it. Accordingly, the argument concludes in the claim that the *ens realissimum qua* object does exist, because of a metaphysical error or conflation consisting precisely in mistreating the AP as an ontological principle unrestrictedly applying to the real possibility of things in general. Kant's answer to the second question lies in the doctrine of transcendental illusion. Such conflation, and the consequent hypostatization of the idea of *ens realissimum* are ultimately motivated by the 'supreme principle of reason,' which, as a matter of transcendental illusion, leads reason to presuppose the existence of an unconditioned being as the ground of everything that is given as conditioned, and identify the idea of this being, *ens necessarium*, with that of *ens realissimum*, thereby positing an ultimate common ground of all existence and real possibility at once.

8.1.1. The 'only possible argument' reconstructed

Let us briefly recall the logical construction of the argument as I presented in chapter 4.

(1) Something is really possible if and only if a) its predicates are logically consistent with one another; b) all of its predicates are really possible in virtue of being *grounded* in something existent.

[10] See my (2017) for a more detailed presentation of this account.

(2) AP: Each and every (logically simple) predicate F of any really possible thing must be grounded in something existent that actually instantiates either F itself or a more fundamental predicate F* that grounds F. (1b)
(3) Real possibility gets cancelled when a) there is a logical contradiction between the predicates; b) the predicates are not grounded in actuality. (1, 2)
(4) P is absolutely really impossible if and only if it cancels the entire material ground of all real possibility.
(5) It is absolutely really impossible that nothing at all exists, since in that case no predicate would be grounded and all real possibility would be cancelled. (3, 4)
(6) It is absolutely really necessary that something exists rather than nothing. (5)
(7) Something is absolutely really necessary if and only if its non-existence cancels the entire material ground of all real possibility.
(8) All really possible predicates must be grounded in one existing thing. (Premise)
(9) All really possible predicates must be grounded in one existing thing that instantiates the complete set of the fundamental predicates (8, 2).
(10) There exists an absolutely really necessary being. (6, 7, 9)
(11) There exists an absolutely really necessary being (i.e. *ens necessarium*), and this is a being that instantiates all of the fundamental predicates (i.e. *ens realissimum*). (9, 10)

Let us also recall that the success of the argument depends on whether two of its crucial premises, primarily (2), the AP itself, and secondarily, (8), the claim that all real possibility must be singlehandedly grounded in a single actual being, can be justified. As I claimed in section 4.6., without a justification of (8), Kant cannot rule out the scenario in which multiple beings distributively ground all real possibility, and thus cannot prove the existence of an absolutely really necessary single ground of all real possibility. And as I claimed in section 4.7., though it does not attract as much attention as (8) and the corresponding problem of the singularity of the ground, without a justification of (2) or the AP, Kant's argument cannot even take off the ground in the first place.

8.1.2. Kant's dual perspective on the argument

The most important clue we have about the critical Kant's take on the argument is the dual perspective he adopts towards it. He is still sympathetic to the argument's deductive form and acknowledges its strong rational appeal, but does not accept it as an objectively valid demonstration of God's existence anymore. In his lectures on philosophical theology (1783–84), he is reported to say:

> Even this proof is not apodictically certain; for it cannot establish the objective necessity of an original being, but establishes only the subjective necessity of assuming (*annehmen*) such a being. But this proof can in no way be refuted, because it has its ground in the nature of human reason. For my reason make it absolutely necessary for me to assume a being, which is the ground of everything. (*Th.Pölitz*, Ak. 28:1034)

As I argued in chapter 5, Kant's perspective on the argument as valid merely subjectively, and his conception of the idea of *ens realissimum* as a subjectively necessary hypothesis of human reason can be traced back to the late 1760s. His *Reflexionen* suggest that Kant was already convinced that the argument failed as an objectively valid demonstration of the necessary existence of God, but it still had the merit of subjective validity in demonstrating the (subjective) necessity of the idea of God before he fully mapped out any of his critical doctrines.[11] However, this chronological observation does not warrant the claim, voiced by Schmucker and Logan, that Kant's downgrading of the argument must be altogether philosophically independent of his critical project. The shift in Kant's approach to the argument reflects a broader and genuinely critical shift in his conception of metaphysics in general, and rational theology in particular. From the mid-1760s on, Kant starts drifting away from the traditional conception of metaphysics as an ontology of things in general toward a new conception of metaphysics as an epistemological study of human cognition.[12] Starting around the same period, we also see Kant's repeated emphases on the rehabilitative and critical aspect of this new metaphysics as 'a critique of pure reason,' as opposed to a doctrine *per se*, producing positive knowledge of things.[13] This shift also effects a novel understanding of rational theology as a critical study of the various ways in which pure reason is compelled to assume the existence and attributes of God, instead of a positive theology aiming at demonstrating the existence and attributes of God.[14] Therefore, it is very natural to construe Kant's revaluation of his own precritical argument as part of this critical conception of rational theology, and not as a project that developed independently of the overall critical transformation of Kant's thought. But this general framework does not tell us much about Kant's specific reasons for downgrading the argument and how exactly the downgrading is related to the doctrine of transcendental illusion, which plays a substantial role in the construction of Kant's critical conception of rational theology in the *Critique*.

There are in fact two distinct questions here: i) Why does the argument fail as an objectively valid demonstration of the existence of God, or what is the error involved in it? ii) Why is the argument so strongly appealing for human reason, or what compels reason to commit this error? Let us note here that the first question can be answered on its own.[15] For why something is erroneous can be sufficiently answered

[11] See R 4249 (Ak. 17:481), R 4580 (Ak. 17:600), R 4729 (Ak. 17:689–90), R 4113 (Ak. 17:420–2), R 4244 (Ak. 17:477–9), R 4253 (Ak. 17:482–3), R 4261 (Ak. 17:486), R 4345 (Ak. 17:514), R 4568 (Ak. 17:596–7), R 4585 (Ak. 17:602), R 4587 (Ak. 17:602), R 5492 (Ak. 18:197–8), R 5508 (Ak. 18:203), R 5522 (Ak. 18:206), R 5525 (Ak. 18:207), R 5527 (Ak. 18:208), R 5569 (Ak. 18:235–6).

[12] See, *Inquiry*, (Ak. 2:283); R 3716 (Ak. 17:259), R 3930 (Ak. 17:352), R3946 (Ak. 17:359), R 3952 (Ak. 17:362), R 4146 (Ak. 17:433).

[13] See *Dreams* (Ak. 2:368); R 3964 (Ak. 17:359) (1769), R 4284 (Ak. 17:495) (1770–71).

[14] See R 4113 (Ak. 17:421), R 4117 (Ak. 17:423), R 4589 (Ak.17:603). In the same period between 1769 and 1780, we see Kant also referring to a "moral proof," and a conception of God serving as a necessary hypothesis for the practical use of our reason, as "a *principium* of deeds," necessary "to determine the final ends and hopes of rational beings." See, respectively, R 4253 (Ak. 17:482), R 4118 (Ak. 17:424), R 4255 (Ak. 17:484), R 4589 (Ak. 17:603). These reflections do not have all the details of the moral proof in the *CPrR* (Ak. 5:125–32), but they suggest that the idea of a "moral theology" was not a post-critical afterthought for Kant to but part of his unfolding critical project from very early on.

[15] On the claim that Kant's critique of traditional metaphysics does not only appeal to a general doctrine such as transcendental illusion (2) but also specific logical errors in its major arguments that can be exposed without any appeal to such a general doctrine (1), see Ameriks (1992).

without an explanation of why we are nevertheless drawn to this error. As we will see, Kant himself maintains that this answer "suggests itself on the basis of the discussions of the Transcendental Analytic" (A581/B609), and that it lies in the misuse of a principle of the understanding and cannot be understood through the doctrine of transcendental illusion alone, which rather aims to answer the second question. However, these two questions instantiate two lines of investigation that together constitute the twofold aim of Kant's critique of rational theology as well as his critique of metaphysics in general: exposing the errors resulting in the existential claims that lack objective validity and explaining the subjective basis of these errors in the nature of human reason.[16] Accordingly, Kant presents his answers to two questions in the Ideal as two parts of a single narrative conveying his critical revaluation of the argument. Thus, even though Kant's rejection of the objective validity of the argument has ultimately something to do with the Analytic, and more specifically with the theory of modality conveyed there, in order to understand the specific reference to the latter, we still have to look at the doctrine of transcendental illusion, as the broader context in which Kant presents this unified narrative.

8.1.3. *Transcendental illusion and transcendental ideas of pure reason*

The essence of the doctrine of transcendental illusion is that reason is naturally urged to pursue unity of thought and completeness in explanations, and this pursuit, though subjectively motivated, creates the illusion that it yields objective results. Kant suggests that such pursuit is rooted in the 'logical use' of reason, where reason follows the maxim (P1): "find the unconditioned for conditioned cognitions of the understanding, with which its unity will be completed" (A308/B364).[17] Expressing the nature of reason rather than the nature of objects, P1 is a 'subjective law,' and the necessity it prescribes is a subjective necessity.[18] Anything that is conditioned has a condition that explains it, and logical inferences from given facts to their immediate conditions can be legitimately made. Yet this will lead to an infinite series of inferences and the explanation will remain incomplete unless an unconditioned condition is found. Kant maintains that P1 cannot motivate reason to seek the unconditioned, unless reason also assumes that the unconditioned is in fact available and adopts P2: "when the conditioned is given, then so is the whole series of conditions subordinated one to the other, which is itself unconditioned, also given" (A307–8/B364).[19] Thus, P1, a merely subjective principle expressing the nature of reason, urges reason to adopt another principle, P2, which has the peculiar feature of looking like an objective principle that holds of objects. Reason thus has the inherent propensity of (mis)taking "the subjective necessity of a connection of our concepts"

[16] This twofold aim of Kant's critique of traditional metaphysics in the Dialetic is well established in the literature. See, for instance, Allison (2004), 308; Grier (2001), 3; Guyer (2006), 126–7; Rohlf (2010), 190–1.

[17] I follow here the notations introduced by Grier (2001), 117–30.

[18] This 'subjective necessity' is 'logical' in the broad sense that it issues from a prescriptive maxim of reason in its logical use, but distinct from 'logical necessity' in the narrower sense of a proposition's issuing from the law of contradiction.

[19] Grier (2001), 126 characterizes P2 as the "application condition" of P1. See also Allison (2004), 330.

(i.e. reason's own demand for completeness in the connection of our concepts of the conditioned and its condition) for "an objective necessity in the determination of things themselves" (i.e. the necessity of the existence of the unconditioned) (A297/B354).[20]

Kant argues in the Dialectic that 'soul,' 'world-whole,' and 'God' are in fact different concepts of the unconditioned that pure reason generates in its pursuit of complete explanations with regard to different objects. The problem with these concepts or 'transcendental ideas' is that they cannot refer to possible objects of our theoretical cognition, since anything that is ever given to us must be given as conditioned by the conditions of experience. And yet, reason's commitment to P2 leads to certain metaphysical errors, which, in turn, result in regarding these concepts as having actual objects, that is, 'hypostatizing' them.

Now, it is of great importance to recognize a distinction here. This phenomenon of reason's natural propensity to mistake its own subjective principles, specifically P2, to be objective, Kant calls, "transcendental illusion" (A297). The transcendental illusion is the cause of, but distinct from, the metaphysical errors resulting in the eventual hypostatization of the ideas of reason. The latter belong to the faculty of judgment and consist in the illegitimate or "transcendental" use of the principles and categories of the understanding beyond their legitimate use designated in the Analytic (A296).[21]

Kant also holds the thesis that, unlike standard logical fallacies, which can be removed from an argument upon discovery, transcendental illusions are subjectively necessary in that they are "natural and unavoidable" even upon discovery (A298). Therefore, even if those dialectical inferences of pure reason concluding in the hypostatization of the various concepts of the unconditioned are ultimately fallacious, they nevertheless spring from the nature of human reason via an unavoidable illusion and continue to pose a very strong, even irresistible, rational appeal in their deductive form. The aim of uncovering transcendental illusions cannot thus be to avoid them, but to avoid the metaphysical errors and the consequent erroneous existential judgments caused by them. These errors are what make transcendental illusions harmful, for an illusion cannot deceive if one recognizes that it is an illusion and avoids falling for it! Going back to the aforementioned two types of question, this confirms the primacy of the first over the second for the rehabilitative aspect of Kant's critique of speculative metaphysics, since the real damage of the latter lies in these metaphysical errors and the consequent erroneous knowledge claims.[22] Prevention of this damage requires discovering which specific principle of the understanding gets misused in the hypostatization of each transcendental idea of reason. In the case of the Ideal of Pure Reason, this discovery will reveal where Kant thinks his precritical argument objectively fails.

[20] See Grier (2001), 117.

[21] On the importance of this usually overlooked distinction, see Grier (2001), Introduction, chapters 3 and 4.

[22] Especially in the Appendix to the Dialectic, Kant talks about the positive utility of (the correct, i.e. regulative, use of) the ideas for the unity and systematicity of our scientific inquiries into nature. This specific positive utility of ideas of reason is beyond the scope of this chapter.

8.1.4. *The genealogy of the ideal*

The ideal of pure reason is the type of dialectical inference generating a specific conception of God as the "absolute unity of the condition of all objects of thought in general" (A340/B398). This inference proceeds from possibilities of individual things in general to the idea of a unique being materially grounding all possibility. Kant's strategy is first to examine the generation of this idea and show its subjective necessity, and, then, to provide a critical account of how this idea gets erroneously 'hypostatized,' which, as we will see, contains a tacit revaluation of the 'only possible argument.' This section will focus on Kant's genealogical account.

Kant first lays out two principles that apply to concepts and things, respectively. The "principle of determinability" states that "That of every two contradictorily opposed predicates only one can apply to [a concept]" (A571/B600). The emphasis here is on concepts as incomplete representations of things that are always further determinable with respect to the predicates that they do not already contain.[23] In contrast to concepts, individual things can only exist as thoroughly determined with respect to all possible predicates. "Every thing, *as to its possibility*, further stands under the principle of thoroughgoing determination; according to which, among all possible predicates of things, insofar as they are compared with respect to their opposites, one must apply to it" (A571–2/B600–1).

Unlike Wolff and Baumgarten, who used the 'principle of thoroughgoing determination' as a principle of existence,[24] Kant holds that complete determination does not warrant existence via his T1 and presents the principle as one of real possibility: no individual thing is really possible as incompletely determined.[25] More strikingly, he points out that while the *logical* possibility of a *concept* rests on the *formal* condition of freedom from contradictory predicates, the *real* possibility of a *thing* requires the *material* consideration of that individual thing in relation to the sum total of all possible predicates of things:

> The principle of thoroughgoing determination thus deals with the content and not merely the logical form. It is the principle of the synthesis of all predicates which are to make up the complete concept of a thing... and it contains a transcendental presupposition, namely that of the material of all possibility, which is supposed to contain *a priori* the data for the particular possibility of every thing. (A572–3/B600–1)

So conceiving the real possibility of an individual thing requires conceiving it as thoroughly determined, which in turn requires the complete concept of that thing as determined with respect to all possible predicates. In order for us to form such a concept, the stock of all possible predicates must be available in the first place, i.e. $(F1 \vee -F1) \wedge (F2 \vee -F2) \wedge \ldots (Fn \vee -Fn)$. It is crucial to understand that this sense of 'availability' corresponds exactly to what I formulated in chapter 4 as the

[23] So the more determinate a concept gets, the richer its content and consequently the smaller its extension get. As the principle implies, however, the idea of a concept that is completely determinate and represents a single object is only a negative limit in the logical specification of concepts and thus can never be reached. See *JL* (Ak. 9:97).

[24] *Ontologia* §225; *Metaphysica* §148.

[25] See R 6291 (Ak. 18:560).

weak epistemological interpretation of the AP. That is, the 'availability' of predicates does not refer to their instantiation in actual things but expresses only a conceptual condition of there being data in the logical space for our mere thought of the real possibilities of things in abstraction from the conditions of the actual givenness of any data to our cognition. This means that even though what motivates this search for a complete condition to think real possibilities is reason's demand for completeness (via P1), since the condition in question is a condition of abstract thinking and thus conceptual, what needs to be presupposed here (via P2) as 'available' or 'given' is itself a concept, i.e. the concept of a totality of all possible predicates. Thus, the ultimate conclusion of this inference will have to remain on the level of the necessary presupposition of a concept rather than that of the existence of an object. I will revisit this important point shortly.

Now, if reason is seeking an ultimate ground from which the matter of all possibility can be derived in thinking, this raw aggregate of all possible predicates is subject to a refinement. For, first of all, each negative predicate (−F) derives from, and thus presupposes, its logical contrary (F). This material ground must therefore contain only positive predicates (F1, F2,... Fn,... F∞). Second, when considered "transcendentally," some logically positive predicates turn out to represent "non-being" (*Nichtsein*), as opposed to a being, a "Something" (*Etwas*) or a "reality" (*Realität*) (A574/B602). For instance, 'dark,' though logically positive, represents the non-being of light. Transcendental negations presuppose 'realities': "The person blind from birth cannot form the least representation of darkness because he has no representation of light" (A575/B603). Realities are also limitable by their transcendental negations and thus come in various degrees, e.g., different shades of light can be attained through the gradual introduction of darkness into the light. The intermediate degrees of a reality, therefore, presuppose the highest degree of that reality. It follows that the material ground of all possibility, conceived as providing the 'transcendental content' of all possibility, must include only realities in the transcendental sense. This refined totality or "All of reality" (*omnitudo realitatis*) is the true material ground of all possibility, a "transcendental substratum, which contains as it were the entire storehouse of material from which all possible predicates of things can be taken" (A575/B603). The thoroughgoing determination of the transcendental content of each and every individual thing is in fact the limitation of the *omnitudo realitatis* in a particular way.[26] The idea of *omnitudo realitatis* leaves out no reality with respect to which it can be further determined. Thus, unlike all other concepts that are incomplete and universal representations of things, this concept is thoroughly determinate, and thus is the representation of a unique individual in possession of all realities in the transcendental sense, i.e. the *ens realissimum*.

[26] Kant analogizes this conceptual procedure of thoroughgoing determination to a disjunctive syllogism, as it were the major premise states that the individual thing has either one portion of the *omnitudo realitatis* or the other and the minor premise denies one portion of the realities to the thing (A576–7/B604–5). This analogy would work better if we were talking about merely logical contents of predicates. But on the level of transcendental contents of predicates with various intensive magnitudes, it is hard to see how a disjunction could work toward the determination of things. I believe the analogy is incidental to Kant's overall argument, but for attempts to make sense of it, see Wood (1978), 50–5; Allison (2004), 401–2.

"Only in this one single case is an—in itself universal—concept of one thing thoroughly determined through itself, and cognized as the representation of an individual" (A576/B604).

Kant concludes that the concept of *ens realissimum* is in fact "that of God thought of in a transcendental sense" (A580/B608). However, in this last move he further refines the ideal. In what Kant calls his 'first crude outline,' the *ens realissimum* is characterized as possessing all realities, i.e. logically and transcendentally positive predicates, and so to their highest degree, if they are gradable. Yet this characterization allows the *ens realissimum* to be seen as 'a mere aggregate of derivative beings' out of which the matter of each individual beings is carved. Kant emphasizes that the *ens realissimum* should be construed as grounding rather than instantiating the whole of reality in itself, and the relationship between the possibility of any individual being and the *ens realissimum* is not be one of part and whole, but one of consequence and ground: "The highest reality would ground the possibility of all things as a ground not as a sum total; and the manifoldness of the former rests not on the limitation of the original being itself, but on its complete consequences" (A579/B607). It remains unclear what predicates then the *ens realissimum* is to have,[27] and how all other predicates are to derive from them as their consequences, but it is clear that Kant intends to introduce an ontological distance between the ground of all real possibility and the world.[28]

The bottom line of Kant's genealogical account is that mere rational thinking about the real possibilities of individual things in their thoroughgoing determination necessarily presupposes the idea of *ens realissimum* as the ultimate material ground of all real possibility. This entails only the subjective necessity of a concept with no existential import. Following his genealogical account, Kant unambiguously states, "with this aim . . . reason does not presuppose the existence of a being conforming to the ideal but only the idea of such a being" (A577–8/B605–6); or reason presupposes the concept of *ens realissimum* "without demanding that this reality should be given objectively (*objective gegeben*), and itself constitute a thing" (A580/B608). The relation of the *ens realissimum* to the individuals is therefore not to be regarded as "the objective relation of an actual object to other things," but only as "that of an idea to concepts," and "as to the existence of a being of such preeminent excellence [this relation] leaves us in complete ignorance" (A579/B607). So, the generation of the ideal operates on a merely conceptual and thus existentially neutral level.

One may ask, however, what it means to presuppose a mere idea without presupposing the existence of its object. In the Appendix to the Transcendental Dialectic, Kant points to a subtle distinction between the givenness of something to reason as an "object absolutely" (*Gegenstand schlechthin*) and only as an "object in the idea"

[27] In his reiteration of the ideal in his lectures on theology (*Th.Pölitz*, Ak. 28:1015), Kant refers to perfections, or unlimited and unconditioned realities, e.g. Omniscience, Omnipotence, Eternity, as the divine predicates that ground all other realities.

[28] Allison (2004) 403 argues that Kant introduces this refinement to preclude the Spinozistic implications of his account of the ideal. Boehm (2012), 308, on the other hand, insists that the idea of *ens realissimum* as the ground instead of the sum total or aggregate of all reality is the more genuinely Spinozistic conception of God and brings Kant closer to Spinoza.

(*Gegenstand in der Idee*) (A670/B698). While in the first case the object is regarded as actually existing, in the second case, "there is really only a schema for which no object is given . . . but which serves only to represent other objects to us, in accordance with their systematic unity, by means of the relation to this idea, hence to represent these objects indirectly" (A670/B698).[29] Thus, in presupposing an idea we do not have to commit to the actual existence of an 'object absolutely,' that is, outside that idea, but only to posit the 'object in the idea' as a mere regulative principle, which would bring a systematic unity to all other objects. In the specific case of the ideal, therefore, we only need to posit the *ens realissimum* in the idea, guiding our inquiries into the real possibilities of things 'as if' it were the single object that materially grounds all real possibility.[30] This means both that the regulative use of the ideal is its legitimate and positive use, and that even if the generation of the ideal is due to the illusory principles of reason, and thus unavoidable or subjectively necessary, as long as the ideal is used regulatively and not constitutively, i.e. as having actual existence, the illusion does not turn into an error.

Although Kant does not make any explicit reference to his 'only possible argument' in the Ideal, we can still make two preliminary points at this stage. First, the ideal and the argument share the same type of inference, proceeding from particular real possibilities to the ultimate material ground of all real possibility. A critical period reflection makes it clear that Kant does not discriminate between the two when it comes to the type of modal inference they exemplify:

> Even if through this proof (*Beweis*) the objective necessity of the highest being is not established, the subjective necessity of a hypothesis of this being as the *substrati* of all possibility (the complete determination of everything in general) is established in our reason itself by its speculative use, though this use is not necessary in itself. (R 6293, Ak. 18:561)[31]

Second, both this reflection and the previously cited passage from critical period lectures, stating that the deductive form of the argument, despite failing to prove God's existence, "can in no way be refuted, because it has its ground in the nature of human reason" (*Th.Pölitz*, Ak. 28:1034), suggest that the critical Kant intends to reframe his argument as a procedure of reason that necessarily leads us to the idea of *ens realissimum* once we engage in thinking about individual possibilities. This procedure is nothing but the generation of the ideal itself. Therefore, even though we cannot hope to find a direct formal refutation for the argument in Kant's text, we

[29] Again in search of a formulation of this distinction, in his notes on Eberhard's *Vorbereitung zur natürlichen Theologie*, Kant likens the ideal of the *ens realissimum* to the sensible form of space which is not an object that is actual itself but necessary for the intuition of actual objects, and even calls it "subjectively actual" R 6290 (Ak. 18:559).

[30] This lends support to the argument of Fisher and Watkins (1998), 393 that Kant's downgrading of the existential conclusion of the argument is based on the distinction between the ways reason and the understanding relate to objects. But, as we will see, while this distinction is indeed very important to the downgrading, there is something specifically critical about the latter.

[31] Interestingly enough, even though Kant employs the notion of thoroughgoing determination in the Ideal and not in the formulation of the argument in the *OPA*, this reflection attributes it to the argument itself.

are entitled to expect Kant's critique of the ideal to apply also to the argument and thus read it as an indirect self-criticism about his precritical take on the argument.

It is noteworthy here that while the original argument presents the generation and the 'hypostatization' of the idea of *ens realissimum* as inseparable parts of the same process, the Ideal presents these as two distinct rational procedures, the former, being subjectively necessary and unavoidable, the latter, though still rationally appealing, avoidable. This is in fact how Kant can still endorse the deductive form of the argument despite rejecting its existential conclusion. So Kant adopts the same dual perspective on the ideal as he adopted on the argument, and locates the error regarding the ideal not in its generation but in its hypostatization.

8.1.5. What exactly is wrong with the ideal?

If Kant's critique of the ideal does not target the deductive form of its generation as a mere concept, but its hypostatization, his account of the hypostatization should lead us to the right direction as to why the argument fails to prove the existence of God. But first, I wish to address a view that locates the problem with the ideal in its generation. Both Kemp Smith and England claim that the transition from the idea of *omnitudo relitatis* to that of *ens realissimum* violates Kant's own thesis in the Amphiboly that there can be a real opposition among realities (A264–5/B320–1; A273/B329).[32] While Kant's refinement of the concept of *ens realissimum* as the ground instead of the sum of all realities seems intended to block this objection, one can still argue that there might be a 'real opposition' among the most fundamental predicates that this being is supposed to have.[33]

I argued in chapter 4 that when considering this worry in the *OPA* (Ak. 2:86) Kant maintains that any real conflict among the predicates of the *ens realissimum* would result in the cancellation of some of the predicates of this being and such deprivation would logically contradict its supreme reality. Yet this reasoning by itself does not warrant that there is no real opposition within the set of most fundamental predicates. The precritical Kant seems to simply assume the latter. In the critical context, however, Kant has strong reasons to remain silent on this matter. For as Wood points out, the account in the Amphiboly restricts the knowable instances of real opposition to the realities in the phenomenal world.[34] Moreover, the critical Kant expressly embraces agnosticism regarding the real compatibility of the predicates of the *ens realissimum*: "I have no capacity to judge *a priori* whether the realities combined in the concept of God cancel each other in their effects" (*Th.Pölitz*, Ak. 28:1016). This lends support to Chignell's claim that the critical Kant finds the assumption that the *ens realissimum* is free of real opposition unwarranted. However, the text of the Ideal does not corroborate Chignell's further claim that Kant's reason for 'undoing' his argument is his realization of this unwarranted assumption.[35]

Kant's critique of the ideal comes in a complicated narrative that is distributed over the last two paragraphs of section two and the whole of section three of the Ideal. The first part, contained in section two, argues that the hypostatization of the

32 Kemp Smith (2003), 524; England (1930), 120.
33 See, for instance, Allison (2004), 403–4.
34 See Wood (1978), 59.
35 Chignell (2009), 190.

ideal lacks objective validity because of a series of metaphysical errors involved in the modal inference leading up to it. The second part, contained in section three, argues that reason is still drawn to committing these errors and consequently hypostatizing the ideal because of a transcendental illusion regarding P2, and thus shows that the hypostatization, though lacking objective validity, is not a matter of arbitrary fallacy but is rooted in the very nature of human reason. Here, I will focus on the first part, as it is where Kant's specific critical reasons for downgrading the argument can be found, but will also briefly discuss the second part and its relation to the former.

I propose that the central element in Kant's critique of the ideal and thus in his downgrading of the argument is the critical version of the AP that he lays down in the Postulates. This critical shift in Kant's understanding of the AP comes in the context of a shift in his understanding of modality in general. As I argued in chapter 7, Kant introduces two crucial changes to his approach to real modality in the Postulates. First, he restricts real modality to the domain of experience, that is, to the (relative) real modalities of appearances. Second, he construes modal categories as expressing the modes in which representations of objects are related to the conditions of human experience instead of the modes of being of things themselves. This entails that our assertions of real modality have objective validity only insofar as they are made in reference to the conditions of our experience. Thus, on a fundamental level, Kant's critical and revolutionary conception of modality does not support the conclusion of the argument that posits an absolutely really necessary being as the ground of the absolute real possibilities of things.[36] This, however, is only a framework and too general to capture the critical Kant's specific issues with the argument. That these issues center around the AP can only be discovered through a close scrutiny of Kant's critique of the hypostatization of the ideal.

Following his genealogical account, Kant points out the need for further investigation into the source of the hypostatization of the ideal. For although the latter is unauthorized, human reason is still disposed to it: "Therefore I ask: How does reason come to regard all the possibility of things as derived from a single possibility, and even to presuppose these possibilities as contained in a particular original being?" (A581/B609) Kant distinguishes, albeit sometimes not articulately, between the transcendental illusion, which he takes to be the ultimate ground of the hypostatization of the ideas of reason, and the specific metaphysical errors caused by the transcendental illusion and involved in the hypostatization process. While the transcendental illusion itself is seated in reason and consists in reason's taking its own subjective principles to be objective, the metaphysical errors caused by the illusion belong to the faculty of judgment and consist in the transcendental application of the principles of the understanding. In other words, the latter type of error occurs when the principles of the understanding are forced to live up to reason's demand for the unconditioned, and thus are used beyond their legitimate (i.e. empirical) realm of application. The empirical cognition of the understanding

[36] Both Beck (1969), 435 and Wood (1978), 74 have rightly pointed out, though without providing much explanation, that Kant's account of real modality in the Postulates is what undermines the modal argument of the precritical argument.

is where the 'conditioned' is given to us, and, *ipso facto*, where reason's pursuit of the 'unconditioned' starts.[37] So even though the hypostatization of the ideal is ultimately due to an illusory act of reason, an explanation of the specific errors involved in it must appeal to the understanding and its particular principles that get abused. Without the latter, the normative problem in the hypostatization cannot be disclosed. Why something is wrong and why we nevertheless do it are distinct questions. The illusion itself only explains why the hypostatization is attractive to reason (question 2). But it cannot explain why the hypostatization involves an illegitimate act, and, consequently, why its result, i.e. the judgment that the *ens realissimum* exists, is erroneous (question 1). Therefore, Kant maintains that what is wrong with the hypostatization of the ideal should be sought in "the discussions of the Transcendental Analytic" (A581/B609). He details this connection in the first of the last two paragraphs of section two of the Ideal, which I will divide here into three parts for ease of digestion:

> The possibility of objects of sense is a relation of these objects to our thought, in which something (namely the empirical form) can be thought *a priori*, but what constitutes the material, the reality in appearance (corresponding to sensation) has to be given; without that nothing at all could be thought and hence no possibility could be represented (A581/B609).

This passage directs us back to the Postulates, for the notion of possibility that is at stake here is that of real possibility defined by the first postulate in terms of compatibility with the actual form of our experience (A218/B266). However, Kant's 'elucidation' of this postulate reveals that our cognition of real possibility requires also a material consideration that involves the actual content as well as the form of experience. In order for us to positively verify that an object is really possible, the concept of this object must be 'borrowed from experience' and thus not only the individual predicates of the object but also their synthesis must be empirically exemplified (A222). In this particular sense, our cognition of real possibility is dependent on and bound by our cognition of actuality.

I suggested in chapter 7 that this idea of the material dependence of our cognition of real possibility on our cognition of actuality expresses exactly what I called in chapter 4 the strong epistemological interpretation of the AP. Let us remember that the precritical Kant flirts with this interpretation in the *OPA*, yet since it would not support the desired objective conclusion of the argument, he ends up adopting the ontological interpretation as the central premise (2) of the argument. Now, in accordance with his revolutionary theory of modality, not only is Kant forced to drop the ontological interpretation, which is based on an absolute conception of modality expressing the modes of being of things without reference to the conditions of their givenness to human cognition, he also has an empiricist justification for the strong epistemological interpretation: our verification of real possibility depends on our cognition of actuality, for only through the latter can we positively verify that our logically consistent thoughts are not 'figments of the brain' but representations of things that can exist in empirical reality. Obviously, even the sole fact that this

[37] See A409/B436.

critically informed, epistemological version of the AP applies only to the (relative) real possibility of empirical objects, and not to the absolute real possibility of things in general, substantially undermines the prospects of the argument. But there is more to come:

Now an object of sense can be thoroughly determined only if it is compared with all the predicates of appearance and is represented through them either affirmatively or negatively. But because that which constitutes the thing itself (in appearance), namely the real, has to be given, without which it could not be thought at all, but that in which the real in all appearances is given is the one all-encompassing experience, the material for the possibility of all objects of sense has to be presupposed as given in one sum total; and all possibility of empirical objects, their difference from one another and their thoroughgoing determination, can rest only on the limitation of this sum total. Now in fact no other objects except those of sense can be given to us, and they can be given nowhere except in the context of a possible experience; consequently nothing is an object for us unless it presupposes the sum total of all empirical reality as condition of its possibility (A581–2/B609–10).

One implication of the critical or epistemological version of the AP is that our cognition of the real possibility of a particular empirical object presupposes not just the givenness of its individual material content but the givenness of the sum total of all empirical reality. This is analogous to the first step of the argument as well as that of the ideal, moving from individual real possibilities to the sum total of all material contents. But why does the critical version of the AP lead to the presupposition of the totality of empirical reality? There are a number of points here. First of all, this move is based on a major thesis of the Analytic: "There is only one experience, in which all perceptions are represented as in thoroughgoing and lawlike connection" (A110). Predicates cannot be given to us as isolated atomic contents but only as located in the context of the unity of a single all-encompassing experience, that is, a causally, spatio-temporally structured continuum of all empirical predicates, or what can be called an empirical *omnitudo realitatis*. Accordingly, the complete determination of an empirical object presupposes the givenness of this empirical totality. It also follows that for us to verify that a particular object can be an object of experience, we need to establish that this object is compatible not only with the form of experience but also with the actual totality of empirical reality that constitutes the material content of the whole of experience.

The problem, however, is that the totality of empirical reality is never actually given to us.[38] Because of the spatio-temporal form of our intuition, we perceive only limited regions of space and time, and our picture of empirical reality remains always incomplete.[39] Moreover, even if our intuitions represent objects in their individuality, i.e. in their specific spatio-temporal determinations, since our concepts are always incompletely determinate representations of objects, we cannot cognize the 'complete (*vollständige*)' real possibility of individual objects as determined with respect to

[38] One might note here that Kant has already demonstrated in the First Antinomy that presupposing the whole of all appearances (or the 'world') as given in its completeness leads to contradictions regarding the bounds of the world in space and time (A426–33/B454–61).

[39] See Fisher and Watkins (1998), 392; Guyer (2006), 146.

the totality of all empirical predicates. Our cognition of real possibility, therefore, always remains on the level of incomplete possibilities.

Now, this brings a further critical restriction on the validity of the inference from individual real possibilities to their ultimate material ground. First, the critical version of the AP restricts the validity of the inference to the empirical domain, and, thus, to the real (relative) possibilities of empirical objects. Second, if neither the totality of empirical reality is ever given to us, nor the complete possibility of empirical objects can be cognized by us, then both notions of totality and complete possibility are nothing but ideas of reason and should be construed as merely regulative principles guiding our investigations into the real possibilities of empirical objects. The totality of empirical reality should then be presupposed as a mere schema or as 'given' only as an 'object in the idea' rather than as an 'object absolutely.'[40]

Therefore, the AP can legitimately take us as far as the empirical *omnitudo realitatis* construed as a regulative idea. We can then establish that the argument's failure to demonstrate the existence of the *ens realissimum* must lie in its illegitimate use of the AP:

In accordance with a natural illusion, we regard as a principle that must hold of all things in general that which properly holds only of those which are given as objects of our senses. Consequently, through the omission of this limitation we will take the empirical principle of our concepts of the possibility of things as appearances to be a transcendental principle of the possibility of things in general (A582/B610).

While the precritical argument uses the AP as expressing an ontological condition (*ratio essendi*) of the real possibilities of things in general, the critical Kant interprets this principle as expressing an epistemological condition of our cognition (*ratio cognoscendi*) of the real possibilities of empirical objects. From the critical viewpoint, whether the same material dependence of possibility on actuality holds of things in general regardless of the conditions of their givenness to our cognition, i.e. whether absolute real possibilities (of things) themselves are dependent on the actuality of their matter, is a matter of speculation. This means that the argument conflates two different kinds of conditions and makes a speculative or transcendental use of the AP in its foundational premise (2). This conflation is an instance of a special metaphysical error that Kant calls 'transcendental subreption': what holds of objects of empirical cognition is unwarrantedly assumed to be holding of things in general. A principle about our representation of relative (real) possibilities of appearances gets treated as a transcendental principle about the absolute (real) possibilities of things in general.

Now, Kant assigns a key place to the transcendental subreption in his account of the hypostatization of the ideal. In the last, and, again, a very compact and tortuous

[40] Longuenesse (2005), 222 insightfully argues that the idea of an empirical *totum realitatis* is the only idea of the whole of reality which is critically legitimate to presuppose, and establishes its connection with the precritical argument, albeit unfortunately only in a brief footnote, claiming that "[T]his is the only refutation Kant ever gives (without saying that he is giving it) of his own precritical argument of the existence of God."

paragraph of section two of the Ideal, Kant presents the process of hypostatization as composed of three errors, where subreption plays the final and most decisive part. Here are the first two errors:

> That we subsequently hypostatize this idea of the sum total of all reality, however, comes about because we dialectically transform the distributive unity of the use of the understanding in experience, into the collective unity of the whole of experience; and from this whole of appearance, we think up an individual thing containing in itself all empirical reality.
>
> (A582–3/B610–1)

The critical version of the AP states that our cognition of the complete real possibilities of empirical objects presupposes the givenness of all empirical predicates. But even if the fact that 'complete possibility' is unattainable by our cognition is set aside, the AP still does not necessitate that the predicates be given in a unified whole of appearances. Like the conclusion of the first step of the argument, which necessitates that all possibility be materially grounded but remains neutral as to whether the grounding is done by multiple beings or by one individual being, the AP would be equally satisfied here, whether the stock of all predicates necessary for grounding complete possibilities is distributed over the multiplicity of discrete appearances or instantiated collectively by the whole of appearances at once.

Both space and time, as forms of intuition, and the principles of the understanding, as forms of our empirical thinking, provide the content of our experience with a formal unity. Nevertheless, this is still a unity among distinct appearances. Kant argues that we tend to erroneously treat this distributive unity of our experience, provided by the sensibility and the understanding, as a collective unity of the whole of experience, which in fact is an idea of reason.[41] By thinking this whole as grounding all real possibility singlehandedly, we are in fact also committing the second error of treating it as an individual thing over and above all appearances, 'containing in itself all empirical reality.' It is clear that the equivalent of this error in the argument is embodied in premise (8) warranting that all possibility be singlehandedly grounded in one actual being. Therefore, the critical Kant thinks that the second crucial premise in the argument, motivating the transition from the first to the second step, is also based on a metaphysical error.[42]

[41] In the context of the Paralogisms, the same error amounts to regarding the unity of consciousness among thoughts (i.e. apperception) as an individual substance (i.e. the soul) existing independently of and underlying all thought (A348–50). Perhaps because Kant also refers to this error in the Paralogisms as the "subreption of hypostatized consciousness" (A402), Both Grier (2001), 245–6 and Allison (2004), 408 identify the subreption in the Ideal with the error of transforming the distributive unity of experience into the collective unity of reason. This is quite misleading, because the latter is a conflation between a concept of the understanding and that of reason, and cannot be called 'subreption' in the sense of a transcendental misuse of an empirical principle and/or a category due to a conflation between the sensitive and the intellectual, as described by Kant's original and systematic introduction of the term in the *Dissertation* (Ak. 2:412–13). This sense of subreption more appropriately applies to the transcendental use of the AP and the category of possibility. For a helpful explanation of why Kant's reference to subreption in the Paralogisms (A402) actually deviates from this standard sense of subreption, see Dyck (2014), 86.

[42] Boehm (2012), 313, 316 has also argued that the critical Kant takes issue with premise (8) (or D6 in his reconstruction). However, based on this accurate observation, Boehm claims that Kant rejects the

The third error *en route* to the hypostatization clarifies the decisive role of the transcendental subreption of the AP plays in this process:

[The concept of] an individual thing containing in itself all empirical reality... [is] then—by means of the transcendental subreption we have already thought—... confused with the concept of a thing that stands at the summit of the possibility of all things, providing the real conditions for their thoroughgoing determination. (A582–3/B610–1)

The legitimate use of the AP gets us as far as the presupposition of the distributive whole of empirical reality. This whole, erroneously, gets transformed into a collective whole and then into an individual. We have then, on the one hand, the idea of the empirical world, erroneously posited, *qua* an individual object, as the material ground of the (relative) real possibility of all empirical objects, and the idea of the *ens realissimum*, presupposed, *qua* a mere idea, as the material ground of the (absolute) real possibility of all things in general. Kant's claim here is that we conflate these two ideas due to the transcendental subreption he described earlier, and end up positing the actual givenness of the *ens realissimum* as the single material ground of all (relative and absolute) real possibility. What Kant has in mind, but unfortunately does not articulate seems to be this: the AP is the motor of the whole inference from individual real possibilities to their material ground, and the subreption of the AP, which consists in converting an empirical principle holding only of empirical objects into a transcendental principle holding of all things in general, leads to the very subreption of the conclusion of the inference, that is, the conversion of the idea of an imminent or empirical ground (of relative real possibility of empirical objects) into that of a transcendental ground (of all real possibility). And since the former is posited as actually given, so is the latter. However, one important question still stands. If the inference starts with the real possibilities of empirical objects, it appears that the actuality of the empirical world as a whole would stop the regress and be thought as the ultimate ground of the real possibility of all empirical objects including itself. So it is not clear here what motivates positing a further, transcendental ground, or more precisely, what motivates the transcendental subreption of the AP that results in the positing of

argument as an objective demonstration because he has to give up (8) upon his realization that it results from reason's illusory principle P2, from which he arrives at the further claim that Kant's rejection of the argument is due to his commitment to the doctrine of transcendental illusion. I hold that both claims are problematic. First, as I have shown above, Kant's dismissal of the foundational premise (2) forces him to reject the objective status of the argument on a more fundamental level than the dismissal of any other premise may. For without (2), the modal inference underlying the argument cannot even get started. Second, and more importantly, the direct role of the doctrine of transcendental illusion is to explain not why the argument objectively fails (question 1), but why it is still rationally attractive (question 2). Even if the latter explanation involves a reference to the former, it cannot constitute the substance of the former but can only complement it. Therefore, Kant's commitment to the doctrine of transcendental illusion alone cannot justify his rejection of the argument as an objective demonstration, but only his retention of it as a subjective demonstration of an ideal. Furthermore, the epistemological conception of the AP, as informed by Kant's treatment of modality in the Analytic, sufficiently explains the failure of the argument, without having to depend on the doctrine of transcendental illusion.

this further ground. As we will see, section three of the Ideal will complete the missing link here by grounding the subreption in a transcendental illusion.

At any rate, with this final step, Kant connects what initially appeared to be two distinct stories in section two. Our abstract thinking about the complete (absolute) real possibilities of things necessarily leads us to the presupposition of the idea of the *ens realissimum*. Our empirical thinking about the complete (relative) real possibilities of empirical objects leads us, via the AP, to the presupposition of the actual givenness of the totality of all empirical reality. This totality, a distributive unity of discrete appearances, is, erroneously, first transformed into a collective unity, and then into an individual thing instantiating all empirical reality. Due to the transcendental subreption of the AP, the concept of this individual thing, which is erroneously presupposed as given, is identified with the idea of the *ens realissimum*, whereby the latter gets hypostatized. Of course, neither the whole of empirical reality nor the *ens realissimum* can be actually given to our cognition, and both existential judgments lack objective validity.

8.1.6. What is left of the precritical argument?

If Kant's narrative so far is taken as plausible, it can be said that a critically informed version of the argument, though unable to demonstrate the actual existence of God, would demonstrate the strong rational appeal of two ideas of grounds of real possibility: first, the idea of the whole world of experience, as the immanent ground of the complete real possibilities of empirical objects, and the idea of the *ens realissimum*, as the transcendental ground of the complete real possibilities of everything in general. Yet given the transcendental idealist restrictions on our cognition, we are left in the dark as to how such a transcendental ground, with its noumenal predicates, would ground empirical predicates as it is originally described in the argument, i.e. through either its own noumenal predicates or the consequences of them (*OPA*, Ak. 2:79).

One might, of course, also question the plausibility of Kant's narrative in section two of the Ideal as an account of the way in which the idea of *ens realissimum* is naturally hypostatized by human reason. In fact, somewhat puzzlingly, in section three Kant appears to step back from this narrative and suggests that the drive for an ultimate ground of all real possibility alone is not persuasive enough for reason to hypostatize the ideal, if it were not "urged from another source." Kant thus offers another account of "the natural course taken by human reason, even the most common" (A584/B612). On this account, reason starts not with possibility, but with the existence of something given in common experience, and through the supreme principle (P2), arrives at the presupposition that something exists as the unconditioned ground of all contingent existence. Once reason convinces itself of the existence of a necessary being, it "looks around for the concept of a being suited for such a privileged existence" among all the concepts of possible beings (A585/B613). The idea of *ens realissimum*, which is generated by reason through a separate procedure, is identified with that of a necessary being, *ens necessarium*, for lack of a better choice, since, at least, as the concept of a being grounding all possibility without needing a further ground itself, it does not conflict with that of unconditioned necessity (A586/B614). The ideal thus gets hypostatized into a necessarily existing object.[43]

Despite the appearance that this is an altogether alternative account of the hypostatization, obviating the convoluted account in the last two paragraphs of section two, it should in fact be read as complementing the latter.[44] In his account of the hypostatization of the ideal based on the subreption of the AP, Kant does not offer a sufficient motivation for reason to make a further move from the whole of the empirical world to the *ens realissimum*. We now realize that P2, the ultimate source of transcendental illusions, is at work here. For even though the whole of the empirical world can ground the (relative or conditioned) real possibility of all appearances and of itself, it is still an empirically conditioned totality, and when it is presupposed as actually given, an unconditioned ground even for this whole must also be presupposed as actual via P2. The idea of *ens realissimum*, which is already at the disposal of reason, is conveniently matched up with the concept of this unconditioned ground, *ens necessarium*. With an act of rational economy, the ideal is posited as the single ground of all real possibility and all existence at once.[45]

Therefore, the account in section three both bridges the gap in the account of hypostatization in section two by pointing to P2 as the cause of subreption, and incorporates the argument into the doctrine of transcendental illusion. We understand here why the critical Kant assigns his own argument a status distinct from that of traditional arguments. While he presents (and formally refutes) the three traditional arguments in section four as different ways to argue for the identity of the *ens realissimum* and the *ens necessarium*, he sees the modal inference underlying his own argument as an intrinsic component of the dialectical procedure of human reason which brings about the identification of these two ideas in the first place. He thereby

[43] It should be noted that this is significantly different from the precritical argument. While in the argument the identity of the two distinct notions of God, the *ens realissimum* and the *ens necessarium*, is presented as the conclusion of a single line of modal reasoning, here the existence of the *ens necessarium* is first asserted and its identification with the *ens realissimum* comes as a secondary move. For a discussion of this difference between the argument in the *OPA* and the Ideal, see Henrich (1960), 148 cf., and Pinder (1969), 15 cf. I suggest that here Kant has in mind traditional rationalists, who provide distinct arguments for the existence of the necessary being (cosmological) and the most real being (ontological), instead of a single argument demonstrating the existence of God by way of demonstrating the identity of both notions of God. For respective cosmological and ontological arguments, see, for instance, Leibniz's *Monadology*, §§36–39 and §§40–45; Wolff's *NT I*, §24, and *NT II*, §§6–22; Baumgarten's *Metaphysica* §§307–10, and §§803–11.

[44] See Grier (2001), 233–4 suggests that the accounts in section three and section two are "two stages of a single extended argument." Accordingly, the first stage shows how the ideal is generated and why it is fallacious to hypostatize this idea; and the second stage shows, why, despite the inappropriateness of hypostatizing this idea, we nevertheless do so. For a similar approach, see also Allison (2004), 410. Unfortunately, neither Grier, nor Allison discusses the account in section two in its connection with the precritical argument. Without this connection, it is hard to motivate Kant's account of the fallacious hypostatization of the ideal at the end of section two, since it is indeed the account in section three that explains our ultimate tendency to hypostatize the ideal.

[45] Kant reiterates this move later in his account of the regulative role of the ideal in the Appendix: "[W]e have to consider everything that might ever belong to the context of possible experience as if this experience constituted an absolute unity, but one dependent through and through, and always still conditioned within the world of sense, yet at the same time as if the sum total of all appearances (the world of sense itself) had a single supreme and all-sufficient ground outside its range" (A672/B700).

both naturalizes his own argument by showing how its crux operates in the natural dialectic of reason, and critically reframes it as a manifestation of what happens when reason hijacks our empirical thinking about real possibilities under the spell of its own illusory principles.

I have thus argued that the key to Kant's critical downgrading of his precritical argument lies in the shift in his interpretation of the AP, according to which it no longer expresses an ontological condition of the (absolute) real possibility of things in general to but only an epistemological condition of our cognition of (relative) real possibilities of empirical objects. This shift is in fact an extension of Kant's revolutionary theory of modality, culminating in the Postulates, consisting in redefining modal notions as expressing the modes in which representations of objects are related to our cognition rather than the modes of being of things themselves. Kant's revolutionary theory of modality radicalizes his ambivalence toward ontotheology: it undermines his own precritical attempt to revive ontotheology and leads to its ultimate transformation in the Dialectic as generating the idea of God as a regulative ideal satisfying subjective needs of human reason, though only an illusion of an objective argument of the necessary existence of God.

8.2. Kant's Critical Refutation of the Ontological Argument

Following his implicit revaluation of his own precritical, alternative ontological argument in sections two and three, in the rest of the Ideal of Pure Reason, sections four through seven, Kant goes on to criticize the three traditional arguments of the existence of God., i.e., the ontological, the cosmological, and the teleological arguments. However, his criticism of the ontological argument is central to this endeavour. For he holds that the other two arguments ultimately depend on the ontological argument and thus that a formal refutation of the latter would create a domino effect and undermine the former two arguments as well.

Kant's refutation of the ontological argument has attracted enormous attention in the literature. One widely held view is that the refutation that Kant presents in the *CPR* is a restatement of his objection based on the thesis that existence is not a predicate (T1) in the *OPA*. While there is substantial continuity between the two texts, especially regarding Kant's views on existence, Kant's treatment of the ontological argument in the *CPR* cannot simply be reduced to a restatement of that in the *OPA*. First and foremost, as I emphasized in chapter 3, in the *OPA* Kant seems to take it to be obvious that T1 alone amounts to a knock-down objection to all previous versions of the ontological argument, without much engagement with the specific logical mechanics of those. I suggested that T1 underlines the second line of objection, aiming to block the introduction of existence into the concept of God. In the *CPR*, on the other hand, Kant offers a systematic refutation, based on a multilayered, dialectical strategy, aiming to cover both historical versions of the argument with a number of subtle arguments, all but one of which are usually understudied in the literature because of the aforementioned reduction of Kant's refutation to T1 alone.

Kant's strategy consists of at least four layers. 1) He first questions the intelligibility of the concept of an absolutely necessary being. Of course, one faulty way to make sense of this concept is to construe it in terms of logical necessity, which requires to conceive existence as a predicate contained in the concept of God and 'God exists' as an analytic proposition. 2) He then argues that the ontological argument would still fail even if one were to concede to these conceptions of absolutely necessary being, existence, and the analyticity of 'God exists.' Here Kant presents a version of the first historical line of objection, blocking the inference of actual existence from conceived existence. 3) Thirdly, he argues for why existence is in fact not a predicate that could be contained in the concept of anything (T1), thereby presenting the second line of objection that he first introduced in the *OPA*. Thus, with these second and third steps, Kant provides a combination of the two historical lines of objection, a strategy that he does not espouse in any other text. 4) Finally, Kant utilizes the most mature form of his revolutionary theory of modality, which springs out of the radical core idea embedded in the positive theses on existence initially laid out in the *OPA*. This adds a further, critical layer to his refutation of the argument, solidified through his insistence that all existential (and modal) propositions are synthetic in a peculiar sense. In the following, I will disentangle and go through this multilayered refutation of the ontological argument.

8.2.1. Is the concept of an absolutely necessary being really intelligible?

The conclusion of Kant's revaluation of his precritical argument in section three was that the idea of an absolutely necessary being is posited by reason due to its urge for the unconditioned, though the presence of such an urge is by no means proof of the objective reality of this idea. Reason, for lack of a better candidate, identifies the concept of this being that exists with unconditioned necessity with that of the most real being. I argued in Part I that this identification of the *ens necessarium* with the *ens realissimum* is what both the classical and the modal versions of the ontological argument maintain. In the beginning of his attack on the ontological argument, Kant takes on the concept of absolutely necessary existence, and raises the question of its very intelligibility. If this concept could be shown to be unintelligible, then this alone would be a serious blow to the argument, if not a knock-down refutation.[46]

Kant points out that the definition of the absolutely necessary being as 'something whose non-being is impossible' is merely nominal and not a real definition that instructs us as to the conditions that make the non-being of this being impossible:

By means of the word unconditional to reject all the conditions that the understanding always needs in order to regard something as necessary, is far from enough to make intelligible to myself whether through a concept of an unconditionally necessary being I am still thinking something or perhaps nothing at all. (A593/B621)

[46] The unintelligibility of the concept of an absolutely necessary being is particularly threatening for the modal version of the argument, propounded by Anselm in *Proslogion* 3 and Leibniz in the *Monadology* (§§38–45). This undermines the view defended by Hartshorne (1966) and Malcolm, (1960) that Kant's critique remains irrelevant to the modal version of the argument.

There is something "strange and paradoxical" (A592/B620) in our efforts to conceive this concept: for, on the one hand, the unconditional necessity at stake requires us to remove all conditions upon which the existence of this being may depend, and on the other hand, the understanding requires certain conditions in order to conceive necessity. Kant's elucidation of the postulate of necessity restricts the notion of (real) necessity to that of conditional or relative necessity. The understanding always conceives of necessity (of existence) as relative to or conditioned upon another existence and laws, and thus, always in the context of a lawful connection of things.

This problem of the intelligibility of absolutely necessary existence would have been what Henrich calls the 'critical' objection to the ontological argument, if Kant cared to present it as an independent objection. However, Kant's concern here is not to underscore the problem of intelligibility, which would not have a significant persuasive power on its own anyway. Instead, Kant aims to get at a more compelling objection moving from the very concept of absolutely necessary existence. Thus he levels the more specific charge against the proponents of the ontological argument that they fail to confront the problem of the unintelligibility of the concept of an absolutely necessary existence, because they cash it out in terms of logical necessity. In other words, the proponents of the argument conflate two notions of necessity, the notion of absolute real necessity, which applies to the existence of things, and the notion of absolute logical necessity, which applies to propositions or conceptual connections. Kant insists that "all the alleged examples" of absolute necessity are taken "only from judgments, but not from things and their existence," and "the unconditioned necessity of judgments, however, is not an absolute necessity of things" (A593/B621). In an analytic judgment such as 'A triangle has three sides,' absolute necessity pertains to the logical connection that holds between the concept of the predicate, 'three-sided,' and that of the subject, 'triangle.' The judgment is absolutely necessary in virtue of the meanings of these two concepts, without regard to what exists in reality, or more precisely, in virtue of the fact that the negation of the connection between them is a logical contradiction in itself. However, Kant points out, "the absolute necessity of the judgment is only a conditioned necessity of the thing or the predicate in the judgment" (A593–4/B621–2). Absolute logical necessity does not warrant the existence of the subject or the predicate. The existential import of absolute logical necessity is always conditional. For 'A triangle has three sides' only entails the conditional 'if a triangle exists, then three sides also exist.'

There are two major issues here. First, any proponent of the ontological argument would accept that in this example the actual instantiation of the predicate is conditioned upon the existence of the subject. Again, they would accept that the existence of a non-divine thing like a triangle is conditioned upon other things and cannot be derived from the mere concept of it. Yet the proponent might still insist that there is a unique subject, God, whose existence is not conditioned upon anything else. Second, Kant observes that the proponent can only derive the absolutely necessary existence of a subject from its concept, i.e., from an absolute logical necessity, if they introduce 'existence' into that concept as a first-order predicate in the first place:

Nevertheless the illusion of this logical necessity has proved so powerful that when one has made a concept *a priori* of a thing that was set up so that its existence was comprehended within the range of its meaning, one believed one could infer with certainty that because existence necessarily pertains to the object of this concept (A594/B622).

In other words, only if existence is included in the intension of the concept of God, can one argue that 'God exists' itself is absolutely logically necessary, which, in turn, appears to bypass the problem of the conditional existential import of logical necessity. This seems to be a suitable way to set the stage for Kant's well-known objection: existence is not such a predicate that could be contained in the concept of anything (T1), therefore 'God exists' cannot be an analytical or absolutely logically necessary proposition. Yet the force of this approach relies entirely on the truth of T1, to which a proponent of the argument might not want to concede without further argumentation justifying T1 itself.

8.2.2. Kant's first line of objection

NON-CONTRADICTORINESS OF NON-EXISTENCE

At this point, Kant offers another objection to the argument that does not depend on T1. He aims to demonstrate that even if one were to accept that existence is a predicate contained in the intension of the concept of God, and consequently, that 'God exists' is an analytical proposition that is absolutely logically necessary, the conclusion that God really exists with absolute necessity does not follow and the argument still fails. I understand his objection to be a version of the first historical line of objection as exemplified by Gainulo and Aquinas against Anselm, Caterus against Descartes, and Crusius against Wolff. As I discussed in chapter 3, all of these objectors move from the concession that existence is a predicate. Their objection is therefore not to the first step of the argument introducing existence as a predicate into the concept of God in one way or another. Instead, their objection is to the inference of God's actual existence from the conceived existence of God. Of course, the key to this objection is the distinction between two notions of existence, existence at the level of mere conception, and real or actual existence outside the concept. Gaunilo, Caterus, and Crusius all claim that one cannot infer the latter from the former alone. I noted in chapter 3 that the precritical Kant is ambivalent about criticizing the argument based on the duality of existence: he endorses it in *NE* (1755) and denounces it later in his reflections (R 3706, Ak. 17:240). Yet here in the *CPR*, Kant develops a version of the first line of objection, without employing the dual conception of existence.

Conceding to the proponent of the argument that existence is a predicate of God, Kant goes on to examine the claim that 'God exists' would then be an analytic and absolutely logically necessary proposition. The absolute logical necessity of an analytical proposition stems from the fact that the negation of the conceptual connection between its subject-concept and predicate-concept is or leads to a logical contradiction. For in an analytical proposition the predicate-concept (e.g., 'three-sided') is contained in the subject-concept (e.g., 'triangle' or 'a three-sided closed figure'), and denying their connection amounts to stating that A is not A, e.g., 'A three-sided close figure is not three-sided.' The conceptual connection between the subject-concept and the predicate-concept holds as a matter of logical necessity, regardless of the actual states of affairs in reality, whether these concepts are actually instantiated or not. Yet, as Kant insisted earlier, this only yields a conditional existential import: 'if the subject exists, then the predicate also exists (necessarily).' It is in fact this conditional itself whose negation is a contradiction. Thus, "if I cancel the predicate in an identical judgment and keep the subject, then a contradiction arises; hence I say that the former necessarily pertains to the latter" (A594/B622). Now, Kant argues that negating the antecedent of the conditional does not mean negating the conditional itself, nor should it lead to a logical contradiction. Instead, it leads to a total cancellation of the subject with all its predicates contained in it: "If I cancel the subject together with the predicate, then no contradiction arises; for there is no longer anything that could be contradicted" (A594/B622). For a contradiction only arises when there is something to be posited and cancelled at the same time!

The cancellability of the existence of the subject without contradiction, Kant suggests, should also apply in the case of the proposition 'God exists.' The proponent of the ontological argument claims that the predicate-concept 'exists' is part of the subject-concept 'God,' and thus, there is an uncancellable logical connection between the two concepts. Kant's point, however, is that this only entails the conditional, 'if God exists, then God exists necessarily.' If one denies the antecedent, then all of the predicates that might be contained in the intension of the concept of God, say, omnipotence, omnibenevolence, omniscience, and existence, are cancelled along with the subject, "and in this thought, not the least contradiction shows itself" (A595/B623).

Kant's argument is better understood through the principle that when nothing is posited, no fertile soil is left for logical contradiction. This principle of the non-contradictoriness of non-existence is familiar from the *OPA*, where Kant emphasizes that absolute nothingness is not a state of logical impossibility: "There is no internal contradiction in the negation of all existence. For, in order that there should be an internal contradiction it is necessary that something should be posited and cancelled at the same time" (Ak. 2:78). As I argued in chapter 4, this is exactly why Kant claims in the *OPA* that nothing exists with absolute logical necessity, and the absolutely necessary being, if any, has to be construed as an absolutely really necessary being, whose non-existence does not amount to a logical contradiction but the cancellation of all real possibility. If absolute nothingness in general is logically non-contradictory, then the non-existence of any particular thing must also be non-contradictory.[47] For the same principle of non-contradictoriness of non-existence applies here:

1. If a subject is cancelled, all of the predicates contained in it are also cancelled.
2. A contradiction occurs only when a predicate is posited and negated at the same time.
3. When no predicate is posited, no contradiction can arise.
4. Therefore, the cancellation of a subject is never contradictory.

The proponent of the ontological argument is cornered here. In order to claim that the existence of God is derivable from the very concept of God, they must construe

[47] For some of Kant's early reflections on the same point, see R 3717 (Ak. 17:260), R 3731–33 (Ak. 17:272–5), R 4007 (Ak. 17:383), R 4033 (Ak. 17:391), R 4035 (Ak. 17:392).

God as something whose existence is absolutely logically necessary or whose non-existence is logically impossible or contradictory. The only possible way to construct such a concept of *ens logice necessarium* is to introduce existence into it. Yet this is still a dead end, for there is nothing logically contradictory in the non-existence of any particular thing or the cancellation of all existence.

Kant notes that against his thesis that the existence of a subject is always cancellable, the proponent of the argument might still insist that "there are subjects that cannot be cancelled at all and thus have to remain" (A595/B623). For instance, Malcom claims that the proposition 'God exists' is not translatable into a conditional like a regular analytic proposition such as 'A triangle has three sides' is translatable into the conditional 'If a triangle exists, then three sides exist necessarily.' For, Malcom maintains, the conditional we arrive at, 'If God exists, then He necessarily exists,' entails the proposition 'it is possible that God does not exist,' and yet this very possibility is incompatible with the absolute logical necessity of God's existence stated by the analytic proposition 'God exists.'[48] Thus, even though subjects like triangles can be cancelled, God cannot be cancelled. While this reasoning might serve as a good line of argument for why existential statements should not be conceived as analytic propositions, which Kant would welcome, it misses Kant's point and ends up begging the main question here. For the existence of such an uncancellable subject is exactly what the argument is supposed to prove and thus cannot presuppose: "that would be the same as saying that there are absolutely necessary subjects—just the presupposition whose correctness I have doubted" (A595/B624).

EITHER A CONTRADICTION OR A TAUTOLOGY

The main target of Kant's argument above for the non-contradictoriness of non-existence is the classical version of the ontological argument. One implication of this argument is that the very concept of a being that exists with absolute logical necessity is in fact empty, or worse yet, incoherent. In other words, existence is always absolutely logically contingent. Moving from this implication, Kant also levels an attack against the modal version of the argument. The modal version of the argument is based on the contention that contingent existence or non-existence is not compatible with the concept of an absolutely necessary being, and thus, such a being is either impossible (i.e., necessarily non-existent) or exists necessarily. This exclusive disjunction can be transformed into the conditional: 'If God is at all possible, then He exists necessarily.' Here is a concise formalization of the modal version of the argument:

1. God is the absolutely necessary being.
2. If God is possible, then He exists necessarily.
3. God is the most real being.
4. God is possible, because (positive) realities do not contradict one another.
5. God exists necessarily.

[48] Malcolm (1960), 58.

Kant raises a number of issues here. First, he observes that this argument involves another instance of conflating logical and real modality, specifically, the logical possibility of the concept of the most real being and the real possibility of the most real being itself. While the non-contradictoriness of a concept proves its logical possibility, it "falls short of proving the possibility of its object" (A596/B624), i.e., that the object can exist in reality. Kant explicates this point in a footnote:

> The concept is always possible if it does not contradict itself. That is the logical mark of possibility... Yet it can nonetheless be an empty concept, if the objective reality of the synthesis through which the concept is generated has not been established in particular; but as was shown above this always rests on principles of possible experience and not on the principles of analysis (on the principle of contradiction). This is a warning not to infer immediately from the possibility of the concept (logical possibility) to the possibility of the thing (real possibility).
> (A596/B624)

Obviously, Kant restates here the postulate of possibility. The real possibility of an object or the objective reality of a concept is only warranted when a) the concept is borrowed from actual experience or actually instantiated in experience, and b) the concept is a pure concept of the understanding that is a necessary condition of experience, that is, necessarily instantiated in experience. However, when the concept in question is an idea of reason, we cannot establish whether the synthesis of predicates in that concept is really possible. For a real opposition can obtain between realities that are logically compatible (A273/B329). Kant reiterates this point in relation to the concept of *ens realissimum* at the end of his refutation:

> The analytic mark, which consists in the fact that mere positings (realities) do not generate a contradiction, of course, cannot be denied of this concept; since, however, the connection of all real properties of in a thing is a synthesis about whose possibility we cannot judge *a priori* because the realities are not given to us specifically... the famous Leibniz was far from having achieved... gaining insight *a priori* into the possibility of such a sublime ideal being.
> (A602/B630)

As I argued in chapter 4, in the specific case of the concept of the most real being, real opposition among predicates or realities amounts to the cancellation of the subject. We can thus at best remain agnostic about the real possibility of this being. Yet again, this is not Kant's main concern in the refutation. So, he concedes, for the sake of the argument, that the most real being is really possible.

The second point is about the role of existence. The modal version of the argument too has to introduce existence into the concept of God as a predicate in one way or another in order to support the conditional 'if God is possible, then He exists necessarily,' for only then mere possibility can at least appear to logically entail (necessary) existence. Kant here is still engaging with the argument at the level of the first line of objection, allowing this step in principle. He says, however: "You have already committed a contradiction when you have brought the concept of existence, under whatever disguised name, into the concept of a thing which you would think merely in terms of its possibility" (A597/B625).

This charge is puzzling, for it is not quite clear what exactly Kant takes to be contradictory in this step. Proops proposes that Kant's specific target here is

Baumgarten. He argues that the contradiction Kant mentions must be between Baumgarten's conception of existence as "something external to the concept of a thing" or as complete determination which cannot be found in the mere concept or possibility of a thing, and "the central idea of the ontological argument . . . that the concept of existence is (covertly) contained in the concept of the *most real being*," or, as Kant puts it here, the idea that existence is contained in the very possibility of this unique being.[49] This, however, would not work. I argued in chapter 2 that while Leibniz and Wolff endorse the notion of completely determined mere *possibilia*, Baumgarten holds that mere *possibilia* are indeterminate and existence, as opposed to mere possibility, is the completion of determination. But this conception of existence only applies to non-divine essences. Leibniz, Wolff, and Baumgarten all maintain that God is the unique being whose essence is completely determined and does not need a complement from outside in order to exist. But for Baumgarten, this is because God does not have any contingent determinations, i.e., modes, and his essentials and attributes exhaust the complete set of his determinations. In fact, Baumgarten places such strong emphasis on the complete determinacy of the concept of God that it leads some commentators like Stang to attribute a distinct ontological argument to him based on the very idea that God cannot fail to exist because his essence or possibility is completely determined.[50] Therefore, contrary to what Proops suggests, God is the only case in which Kant's charge of contradiction is off the mark.

So far, then, Kant's contradiction charge does not appear compelling. Once it is accepted that existence is a predicate containable in the concept of a thing, it appears vain to press the contradiction charge. Perhaps Kant should have said that the proponent of the modal argument cannot meaningfully introduce existence into the concept of something whose possibility they are seeking to prove. Since existence already presupposes possibility, this step would be begging the question, even if not contradictory. Yet it is possible to evaluate the contradiction charge in a more charitable fashion once it is located in the context of the broader and potentially much stronger attack Kant levels against the argument. This attack consists in the claim that if the first step of introducing existence into the concept of God does not commit a contradiction, then the second step of inferring the existence of God from this concept is nothing but a tautology:

> If one allows you to [introduce existence into the concept of God], then you have won the illusion of a victory, but in fact you have said nothing; for you have committed a mere tautology. I ask you: is the proposition, This or that thing . . . exists—is this proposition, I say, an analytic or a synthetic proposition? If it is the former, then with existence you add nothing to your thought of the thing; but then either the thought that is in you must be the thing itself, or else you have presupposed an existence belonging to possibility, and then inferred that existence on this pretext from its inner possibility, which is nothing but a miserable tautology. (A597B625)

[49] Proops (2015), 14. [50] See Stang (2016), 56-65.

This is Kant's second, and I think, more original take on the first historical line of objection, and has usually been neglected in the literature.[51] This objection is also both a precursor of Kant's claim that all existential statements are synthetic and a reconfirmation of the previous thesis that all and any existence can be cancelled without contradiction. Here is a reconstruction of Kant's objection in the form of a dilemma:

1. The proposition 'God exists' is either analytical or synthetic.
2. If 'God exists' is analytical, then the predicate-concept 'exists' adds nothing to the subject-concept 'God' that is not already contained in the latter.
3. If existence does not add to the concept 'God,' then either
   a) the concept 'God' in your mind is God himself, which is absurd
      or
   b) the predicate-concept 'exists' repeats what is already contained in the subject-concept 'God' and thus 'God exists' is a mere tautology.
4. If 'God exists' is synthetic, it can be negated without contradiction.

The only premise that might require an explanation here is 3. Accepting 'God exists' as an analytic proposition on the basis of treating existence as an essential predicate contained in the mere concept of God entails two things. First, as stated by (a), the distinction between the object God and its conceptual representation 'God' collapses. The aim of the ontological argument is to demonstrate that the actual existence of God *qua* object outside the mind follows from the concept 'God.' This is what makes the argument both *a priori* and ontological. However, the only possible way in which the actual existence of the object God is part of our conceptual representation 'God' is that the two items, the actual object and the concept are identical. It is this absurdity Kant must have in mind when he frames the introduction of existence into the concept of God as a contradiction. For it is indeed a contradiction to maintain at once that 'God' is a mere concept and that actual existence, which is supposed to apply to the object God outside the mind, is contained in the mere concept 'God.' As I showed in chapter 1, in order to avoid this absurdity, Anselm, for instance, introduces a distinction between merely mental existence (*esse intellectu*), applying to intentional objects in the mind, and real or actual existence (*esse reale*), applying to real objects outside the mind. And as I discussed in chapter 3, the historical representatives of the first line of objection like Gaunilo, Aquinas, Caterus, and Crusius point out that the *a priori* inference from the mental to the actual existence of God is illegitimate, for the former does not in any way logically entail the latter. Here, Kant makes the same point by replacing the language of dual existence, which he never endorsed, with the language of concept and object: if one claims to be licensed to infer the actual existence of an object from the mere concept of that object on the grounds that existence is already contained in that concept, then they are treating the conceptual representation of the object and the object itself as one and the same item, which, of course, is not only absurd but also defeats the purpose of the argument.

[51] With the exception of Proops (2015) and Pasternack (forthcoming). See also my (2008).

Second, if the object God and concept 'God' are indeed distinct items, which implies that actual existence and mental or conceived existence must be distinct, then inferring 'God exists' from the claim that (mental) existence is contained in the concept 'God' is nothing but repeating just the claim that (mental) existence is contained in the concept 'God.'[52] Thus, even if the first step of introducing existence into the concept of God is allowed, the argument cannot run the second step and arrive at the conclusion that God, the object, actually exists outside our mere conceptual representation of this object.

What the proponent of the argument is forced to accept here is that 'God exists' is a synthetic proposition, which, however, comes with a cost: "If you concede, on the contrary, in all fairness as you must, that every existential proposition is synthetic, then how would you assert that the predicate of existence may not be cancelled without contradiction?" (A598/B626) For if the proposition is synthetic, then one cannot claim that the predicate 'exists' is part of the subject 'God' and that 'God exists' is not deniable without a contradiction. Kant therefore connects here his two arguments following the first line of objection, the non-contradictoriness of non-existence and contradiction or tautology. While both stand on their own, the latter presents a further confirmation of the conclusion of the former.

8.2.3. *Kant's second line of objection*

Kant's next attack is an instance of the second line of objection to the ontological argument, consisting in blocking the first step of introducing existence into the concept of God. In chapter 3, I argued that despite the attributions of it to Gassendi in the literature, Kant is the true originator of this line of objection, for T1, the very ground of this objection, is first formulated by Kant. Kant's motivation to appeal to this line of objection in the *CPR* seems to be that he thinks that it reveals a more fundamental fallacy in the argument:

> I would have hoped to annihilate this over-subtle argumentation without any digressions through a precise determination of the concept of existence, if I had not found that the illusion consisting in the confusion of a logical predicate with a real one (i.e., the determination of a thing) nearly precludes all instruction (A598/B626).

I detailed my interpretation of Kant's T1 in chapter 3 and will briefly reiterate it below. But since Kant's formulation of T1 here, "Being is obviously not a real predicate" (A598/B626), employs the notion of 'real predicate' in contradistinction with 'logical predicate,' I wish to dwell on this distinction first. Kant's explanation of the distinction seems clear enough:

> Anything one likes can serve as a logical predicate, even the subject can be predicated of itself; for logic abstracts from every content. But the determination is a predicate, which goes beyond the subject and enlarges it. Thus, it must not be included in it already. (A598/B626)

[52] As Proops (2015), 14 rightly observes, Kant echoes Caterus's objection to Descartes here (AT 7:99, CSM 2:72).

The logical/real predicate distinction should not be construed as a distinction of mutual exclusion. 'Logical predicate' is only a syntactic category expressing that the term occupies the predicate position in a proposition regardless of the intensions of the terms (subject and predicate) of that proposition. The place of a term may qualify it as a logical predicate, but whether the term is also a real predicate requires further consideration regarding the content of the proposition, that is, we are required to establish whether or not the predicate is already contained in the subject-concept. The distinction is then between the predicates that are *merely* logical and those that are also real predicates.

Kant's talk of 'real predicate or determination' is better situated against the background of the idea of complete determination. There are two conditions for a concept P to be a real predicate or determination. The first is the necessary and general condition: For P to be even considered as a candidate for a real predicate, it must take part in the complete determination of at least one concept, or it must be contained in the intension of at least one concept. The second is the sufficient and subject specific condition: P is a real predicate of C iff P adds to C a *further* determination that is not already contained in its intension. If a predicate does not add a further determination to the subject-concept, then it is not a real predicate with respect to that subject-concept, but merely a logical predicate. So merely logical/real predicate distinction is ultimately context sensitive. The predicate 'three-sided' is real with respect to 'my dining table' but merely logical with respect to 'a triangle.' Kant's context-independent thesis that existence is not a real predicate amounts to the radical position that existence does not even satisfy the necessary and general condition of being a real predicate, for it cannot enter into the intension of any other concept and thus be part of its complete determination.

Now, obviously, the predicates of all analytic propositions are merely logical predicates. For a proposition is analytic in virtue of the fact that the predicate-concept does not add any determination to the subject-concept beyond what is already contained in its intension. However, one should be cautious about using the analytic/synthetic distinction as a decisive guide in determining whether a predicate is real or merely logical in a given proposition. Kant appears to allow such use of the analytic/synthetic distinction when he states in his post-critical lectures "We call determinations not analytic predicates <*praedicata analytica*> but rather synthetic predicates <*praedicata synthetica*>" (*Th.Pölitz*, Ak. 28:552). This would be misleading if a 'synthetic predicate' is understood as any predicate in a synthetic proposition. Kant cannot be suggesting this interpretation, nor does he define 'synthetic predicate' anywhere as such.[53] Since, as I will discuss later on, Kant unambiguously categorizes all existential or modal propositions in the form of 'S exists' or 'S is possible/actual/necessary' as synthetic, the syntheticity of a proposition does not warrant that its predicate is a *real, determining, or synthetic* predicate. Kant himself clarifies this point later in his lectures, "we have introduced the categories: possibility, actuality, and necessity, and then we deemed that they are not at all

[53] Stang (2016, 37) produces an "inconsistent triad" based on the interpretation of "synthetic predicate" as the predicate of a synthetic judgment. See 8.2.4 below.

determinations of a thing, or synthetic predicates" (*MM*, Ak. 29:821). Therefore, a 'synthetic predicate' should instead be construed as a predicate that adds a further determination to the subject. But this latter was exactly the sufficient condition of being a real predicate, and does not shed further light on what it means to be a real predicate. More importantly, the analytic/synthetic predicate distinction does not improve our understanding of why existence, though capable of serving as a predicate in synthetic propositions, is never a real predicate in any proposition. For, regardless of the context, existence can be neither an analytic predicate (contained in the subject-concept), nor a synthetic predicate (further determining the subject-concept and enlarging its intension or narrowing its extension). The exceptional status of existence or any modal category, and the special kind of syntheticity of existential/modal propositions, is crucial in understanding the role of Kant's critical theory of modality in his opposition to the ontological argument. I will revisit this point at the end.

Following the introduction of the merely logical/real predicate distinction, Kant states T1: "Being is obviously not a real predicate, i.e., a concept of something that could add to the concept of a thing" (A598/B627). I will examine T1 from the viewpoints of three distinct questions: 1) What exactly does T1 mean? 2) What argument does Kant offer for T1? 3) How does T1 constitute a refutation of the ontological argument?

WHAT T1 MEANS

In chapter 3, I provided a detailed analysis of T1 as part of the following string of theses Kant puts forth regarding the concept of existence.

T1: Existence is not a predicate or a determination of an object.
T2: Existence is a predicate of concepts of objects.
T3: Existence is absolute positing of an object.
T4: The actual contains nothing more than the merely possible.
T5: More is posited through an existent thing than is posited through a merely possible thing.

Let us recall that one prevalent interpretation of T1, defended by Malcolm, Barnes, Plantinga, Van Cleve, and most recently Stang, is that 'exists' does not divide the extension of any concept (of an object) such that there cannot be an object that instantiates a concept C but not 'exists.'[54] Accordingly, existence is a first-order predicate of objects that universally or unrestrictedly applies to all (and not a subset) of them. Stang, quite rightly, further claims that this interpretation implies 'actualism,' the view that there are *no* non-existing objects.[55] My interpretation of T1, on the other hand, suggests that existence is not a first-order predicate of objects and cannot be contained in the intension of the concept of any object. T1 does indeed entail that 'exists' does not divide or narrow the extension of any concept C. However, this is not because 'exists' is a universal first-order predicate of all objects, but because it is

[54] Malcolm (1960), 44; Barnes (1972), 46; Plantinga (1967), 35; Van Cleve (1999), 188; Stang (2016), 39–40.
[55] Stang (2016), 38.

not and cannot ever be part of the intension of the concept of anything, however determinate that concept may be. What cannot be part of the intension of any concept cannot divide or narrow the extension of any concept. For the same reason, adding further determination to a subject-concept that is not already contained in it is out of question for 'exists.' Since existence cannot take part in the complete determination of any concept, it cannot even satisfy the necessary and general condition of being a real predicate or determination, let alone the sufficient and specific condition of adding further determination to a certain subject-concept. As I repeatedly emphasized before, one major implication of Kant's discussions of modality in the *CPR* is the radical separation of modal categories, including existence, from the intensions of the concepts of objects. Accordingly, conceptual representations of things are existentially neutral and existence (and real modality in general) has a strictly extra-conceptual character. The statement of T1 in the *OPA* was Kant's first step toward the separation of modality from intension, and the statement of T1 in the *CPR* enjoys the complete fruition of this idea.

HOW KANT ARGUES FOR T1

Let us first remember the main argument Kant provided for T1 in the *OPA*:

1. If existence were a predicate of a thing (~T1), then the complete concept of a thing would contain existence (or non-existence).
2. No (logically consistent) concept, regardless of its level of determinacy, can contain existence (or non-existence).
3. Therefore, existence is not a predicate of a thing (T1).

In chapter 3, I suggested that this argument relies heavily on the claim that concepts are existentially neutral no matter how determinate they may be. This, however, leads to a circularity, because, on my reading, the ultimate meaning of T1 is the existential neutrality of concepts of objects. Therefore, Kant needs an argument that justifies T1 or the existential neutrality of concepts without using one as a premise for the other. Kant's refutation in the *CPR* follows a different argumentative strategy to avoid this problem.

Contrary to the widespread contention in the literature, Kant's infamous 'hundred dollars' example does not provide an argument directly for T1. Instead, the example aims to justify T4, which, in turn, works as part of a grand argument for T1. I propose that Kant has in the mind the following grand argument:

1. If existence were a real predicate (~T1), then the actual would contain more than the merely possible (~T4).
2. The actual does not contain more than the merely possible (T4).
3. Therefore, existence is not a predicate (T1).

Kant's formulation of T4 in the *CPR* is indeed a little puzzling. His wording, "the actual contains nothing more than the possible" (A599/B627), appears to point toward a comparison between two kinds of objects, the actual and the merely possible, and thereby, unnecessarily commit him to an ontology of mere *possibilia*, on which, as I argued in the previous chapter, Kant deliberately avoids taking a position. However, this is not quite the case. Kant's talk of 'containment' reveals that he is in fact talking about a conceptual content as instantiated and not instantiated. Thus, T4 should more properly be understood as stating that the actual object does not instantiate more content, i.e., positive predicates, than is contained in the concept of it as merely possible (not instantiated). So when we think of an object, respectively, as actual and as merely possible, we do not entertain a larger conceptual content or intension in the former. Based on this conception of T4, it is not hard to see why Kant holds the first premise. If existence were a real predicate applying to objects, it would be the predicate of an actual object, and not contained in the concept of a merely possible object (or what is the same, non-existence would be contained in the latter). In other words, if existence were a real predicate, there would be an intensional difference between the actual and the concept of the merely possible in favour of the latter: more content would be posited if one posited an object as actual than they entertained it as merely possible. Thus, the more questionable premise here is the second one: why is T4 the case?

In support of T4, Kant presents his 'hundred dollars' argument:

> The actual contains nothing more than the merely possible. A hundred actual dollars do not contain the least bit more than a hundred possible ones. For since the latter signifies the concept and the former its object and its positing in itself, then in the case the former contained more than the latter, my concept would not express the entire object and thus would not be the suitable concept of it. (A599/B627)

The argument here can be reconstructed in the following way:

1. If the actual object O contained more (i.e., the real predicate of existence) than the merely possible object O (~T4), then the complete concept of O would not completely represent O.
2. The complete concept of O, by definition, completely represents O.
3. Therefore, the concept of the actual does not contain more than the concept of the merely possible.

First, an important point is in order. 'Complete concept' represents its object as completely determined with respect to all of the predicates it would have were it to actually exist. However, as I showed in the previous chapter, the critical Kant explicitly holds that we cannot form complete concepts of individual things. Our concepts are rather general representations of things, and thus, have only partially or incompletely determinate contents. Even in the beginning of the Ideal, Kant makes it clear that there is always this imperfect congruency between actual things that are always completely determinate and our conceptual representations of them, which are always further determinable and yet never completely determined. The notion of a complete concept of a thing is thus a mere idea of reason, a boundary concept we can never attain. Yet we have also seen that Kant consistently uses the notion of a complete concept, not only in the *OPA* and other precritical texts, but also in the *CPR*, as a rhetorical device to make the same point about modality. The modal status of an object is in principle not a matter of conceptual content or intension, no matter how determined the latter may be. Let us recall Kant's emphasis from the Postulates: "If the concept of a thing is already entirely complete, I can still ask about this object whether it is merely possible, or also actual" (A219/B266). Therefore, Kant's talk of 'complete concepts' in formulating T4 should be understood in line with this argumentative rhetoric, suggesting that if it were possible in principle for us or any mind to form complete concepts of objects, then an actual object would not be instantiating more content or a single more determination than what is contained in the complete concept of that object as merely possible.[56]

This brings us to the crucial first premise of Kant's 'hundred dollars' argument for T4. Kant clarifies that the comparison in T4 is between the positive content a particular actual object would instantiate and the content its mere complete concept would contain. If the actual object were to instantiate more content than the complete concept of that object (i.e., given the grand argument, the excess would be the real predicate of existence), then the complete concept of the object would not be completely congruent with or the complete representation of the actual object, for it would be missing a determination that the actual object instantiates (i.e., existence). But this is absurd because, the complete concept of an object, by definition, is the complete representation of it, with all the determinations it would instantiate if it were to be actual.

There have been conflicting claims regarding the specific target of this argument in recent literature. Proops, for instance, claims that its 'exclusive' target is Leibniz, while Stang holds that this argument specifically targets Baumgarten and his 'possibilist ontology.' Proops' claim is based on a certain reading of the 'hundred dollars' argument, according to which the argument relies on the assumption that the actual does not contain more than the merely possible (T4) and is open to the objection that it begs the question. Proops argues that this objection is not available to Leibniz, since his mature metaphysics (laid out, for instance, in the *Theodicy*), by accepting that the best possible world does not become more real just in virtue of God's actualization of it, commits him to T4, and thus, Kant's argument works specifically against Leibniz. However, this could not be the case. First, the argument, as I reconstructed it above, does not rely on T4 but in fact proves it. Second, given Kant's aim in formulating the 'hundred dollars' is to prove T4, whether or not he succeeds in doing so, it is implausible that he exclusively targets with this specific argument someone who already holds T4. Third, as I showed in chapters 2 and 3, T4 is compatible with the metaphysics not only of Leibniz but also of Wolff, but certainly conflicts with Baumgarten's conception of existence as complete determination.

It is this last point which seems to motivate Stang's claim. Stang rightly singles out Baumgarten as the rationalist figure who explicitly holds that the actual contains more than the mere possible on the grounds that while the former is completely determined, i.e., with respect to all of its essentials, attributes, and modes, the latter remains indeterminate with respect to its modes or contingent predicates. According to Baumgarten, it is this difference in the level of determinacy which defines the modal difference between the actual and the merely possible. Stang is also right that

[56] Thus, the worries such as the following, voiced by Plantinga (1966), 539–40, miss Kant's point: "Of course it will not be true that the concept of an object contains as much content as the object itself. Consider, for example, the concept *horse*. Any real horse will have many properties not contained in that concept; any real horse will be either more than 16 hands high or else 16 hands or less. But neither of these properties is in the content of the concept *horse*."

Baumgarten's conception of existence necessarily commits him to 'possibilism,' the contention that only a subset of objects are completely determinate and thus actual or that there are some objects that are merely possible. However, I disagree with Stang's claim that the specific target of the 'hundred dollars' argument is Baumgarten and his conclusion that Kant's ultimate aim here is to undermine possibilism.

First of all, the explicit aim of the argument is to prove directly T4 and ultimately T1 (via the grand argument). While Baumgarten's conception of existence certainly entails the opposite of T4 (and consequently of T1), it is not the only possible conception of existence that could do so. Second, Kant's concern here is to formulate a knock-down refutation of the ontological argument, which, he holds, in any of its historically available versions, treats existence as a real predicate of God, and thus, commits the proponent of the argument to this conception of existence in their ontotheology, whether or not they espouse it in their general metaphysics. Third, there seems to be no exclusive connection between Baumgarten's conception of existence and possibilism. Both Leibniz and Wolff hold that there are (completely determinate) mere *possibilia*, even though they do not define existence as complete determinacy. Fourth, and more importantly, here as well as in his other discussions of the modal difference, Kant's concern is not to offer a metaphysical position on possibilism. This is true even when he explicitly criticizes the possibilitist contention that "the field of possibility is greater than that contains everything actual" (A230/B283). As I argued in the previous chapter, Kant's actualism is at best rooted in the specific constitution of our cognition and ultimately epistemological, and not metaphysical. Thus, his concern is rather to rule out the general misconception of the modal difference between the actual and the merely possible in terms of intensional difference, that is, the difference between the content that is posited when something is posited as actual and the content through which that thing is thought as merely possible. Kant proposes to define the modal difference in terms of how a content is posited (in relation to our cognition). Sixth and most importantly, T1, as I have interpreted it in this book, does not necessarily commit Kant to actualism at the exclusion of possibilism. T1 does not mean that existence is a predicate of all objects (and that all objects are actual), but that existence is not a predicate of any object, and thus cannot be contained in the concept of any object, including God, as a predicate. Thus, again, as I noted earlier, 'exists' does not divide the extension of any concept, but it is not because 'exists' is a predicate of all objects (hence actualism) but because it is not a predicate of any object and cannot be part of the intension of any concept.

This last point is further clarified by Kant's reformulation of the 'hundred dollars' argument for T4, this time without reference to the modal difference as such:

> Thus, when I think a thing, through whichever and however many predicates I like (even in its thoroughgoing determination), not the least bit gets added to the thing when I posit in addition that this thing is. For otherwise what would exist would not be the same as what I had thought in my concept, but more than that, and I could not say that the very object of my concept exists. (A600/B628)

Now this can be reconstructed through two alternative routes. The first route does not differ too much from the original 'hundred dollars' argument:

1. If the positing of an object as actual were to add (any real predicate) to the content of the complete concept of that object, then the complete concept of that object would not be congruent with or completely representing that object.
2. The complete concept of an object is completely congruent with its object.
3. The positing of an object as actual does not add to the content of the complete concept of an object.

The conclusion of the argument or the antecedent of the first premise is just another statement of T4, employing the talk of 'no addition' instead of 'no more containment.' The point, however, is the same: the difference between what is posited as actual and thought as merely possible should not be sought in the content that is posited. The second route is intended to demonstrate how the negation of T4 leads to a paradox about positing a particular object as actual.

1. If the positing of an object as actual were to add to the content of the complete concept of that object, then the complete concept of that object would not be congruent with that object.
2. If the complete concept of an object is not congruent with that object, then what is posited as actual (O*) is not the same as what is represented by the complete concept (O).
3. Then, we can never posit the exact content of our representation (O) as actual.

One might object that since our concepts are always incomplete anyway, this conclusion should not pose a paradox here. But again, Kant's aim here is to reveal the principle about the very nature of existence and conceptual representation in abstraction from our particular cognitive limitations. We can reconstruct the same argument with a hypothetical appeal to the contents of a divine mind with the ability to form complete concepts. If existence or the positing of something as actual adds (a real predicate) to the content of the complete concept of that thing in God's mind, then God would never be able to actualize or create exactly what he had in mind.

HOW T1 BLOCKS THE ONTOLOGICAL ARGUMENT

Thus, all three versions of the 'hundred dollars' or conceptual congruency argument conclude that T4 must be the case, which, in turn entails T1. But how does T1 constitute a refutation of the ontological argument? Two things can be said here. First, on the more general level, Kant is confident that the proof of T1 is also the proof of the 'impossibility of an ontological proof' as such. For an ontological argument (with the exception of Kant's own admittedly failed precritical attempt at an alternative) is the kind of argument that purports to infer the existence of God from the mere concept of God *a priori* (A590/B618, *OPA*, Ak. 2:156). Such *a priori* inference is only possible if God's existence is somehow contained in or logically entailed by the concept of God. But this latter is only possible if existence is a real predicate of God *qua* object. For only such a first-order or object-level predicate can be contained in the concept of an object or logically entailed by other first-order predicates in that content. Therefore, if existence is not the kind of predicate that could be discoverable in the concept of an object by logical analysis (T1), then the kind of argument that operates on the assumption that it is so is doomed.

Second, on the more specific level, as the survey of the major historical versions of the ontological argument that I provided in chapters 1 and 2 illustrated, all versions do indeed treat existence as a real predicate in their first step of introducing existence into the concept of God, either as a distinct predicate, e.g., perfection (Descartes), reality (Wolff, Baumgarten), or under the guise of another predicate, e.g., greatness (Anselm), self-sufficiency (Leibniz), which they take to entail existence. Then, the thesis that existence is not a real predicate, if correct, blocks all of these specific arguments. Having pointed out the direct relevance of T1 (and thereby T4) to the ontological argument, Kant's refutation in the *CPR* has a further, fourth layer that is based on his positive theses. These theses present a clearer picture of how Kant incorporates his critical and revolutionary theory of modality into his refutation of the ontological argument and thereby substantially develops his precritical attack on the argument.

8.2.4. The kind of predicate existence is and the syntheticity of existential propositions

Although it goes usually unnoticed, Kant presents T3 in the *CPR* as the complement of T1: "Being is obviously not a real predicate... It is merely the positing of a thing or certain determinations in themselves" (A598/B626). Now, while Kant defines 'being' (*Sein*) here, as opposed to his statement of T3 in the *OPA*, which defines 'existence' (*Dasein*) as absolute positing, the two statements do not differ in terms of the semantic aspect of the thesis. For just like he does in the *OPA*, Kant again identifies 'being in general' (*Sein überhaupt*) with 'positing' (*Setzung*), which is asserting a thing, a ground, a state, a proposition or a predicate as being true, real or the case, as opposed to, for instance, cancelling or negating it. And again, Kant distinguishes positing into two kinds: relative and absolute. In merely relative positing, being takes the form of 'is' as copula, serving to posit only a logical relation between the predicate-concept and the subject-concept. For instance, in the proposition 'God is omnipotent,' the predicate-concept 'omnipotent' is posited as belonging to the subject-concept 'God.' In absolute positing, on the other hand, being takes the form of the 'is' of existence or 'exists':

> If I... say God is, or there is a God, then I add no new predicate to the concept of God, but only posit the subject with all its predicates, and indeed posit the object in relation to my concept. Both [the object and my concept] must contain the same... (A599/B627)

Kant formulates absolute positing in two alternative ways here, one from the viewpoint of the concept and the other from the viewpoint of the object, both of which boil down to the positing of the same relation. First, in absolute positing, the subject-concept is posited with all its predicates contained in its intension as related to (or as instantiated by) an actual object outside the concept. This, of course, reiterates the extra-conceptual reference of existence, as is also strongly emphasized by T1. Second, from the viewpoint of the object, the object is posited as related to the very content I conceptually represent. This means that, contrary to the false impression that existence is a predicate of the object caused by our daily usage of 'exists,' absolute positing does not tell us anything further about what the object in question is, but that it just wholly instantiates what I already represent in my mind. Hence, the congruency of the object with the conceptual content that it instantiates. Both of Kant's formulations of absolute positing here entail the thesis (T2) that existence is a second-order predicate or predicate of concepts: existence is not to be found among the predicates that the object instantiates (T1), but expresses that our conceptual representation of the object, with all of the predicates contained in it, is instantiated.

So far, Kant's semantico-logical account of absolute positing does not differ from the one he provides in the *OPA*. In chapter 3, I suggested that it is already possible to observe the core of Kant's radical idea of modality in his positive conception of existence (particularly in T2) in the *OPA*, according to which existence expresses a certain feature of our conceptual representation of the object rather than the object itself. I have demonstrated in the previous chapters of Part III that this core idea culminated in a revolutionary conception of modality in the *CPR*, according to which modalities express the ways in which concepts of objects are related to our cognition rather than the ways objects are. In chapter 7, I argued that through the postulates Kant redefines each modal category as a distinct kind of absolute positing and provides the conditions of positing: (i) possibility is the absolute positing of the concept of the object as agreeing with the formal conditions of empirical cognition; (ii) actuality or existence is the absolute positing of the concept of the object as connected to an actual perception; and (iii) necessity is the absolute positing of the concept of the object as connected to an actual perception through a causal inference.

Thus, Kant's critical account of existence as absolute positing goes beyond the *OPA* exactly in that it provides epistemic conditions of positing and specifies the locus of positing as the realm of experience. This also tells us what exactly more is posited *through* the actual than is posited *through* the merely possible (T5), even though nothing more is contained *in* the former than *in* the latter (T4): the connection with an actual perception and thus with "the content of the entire experience" (A601/B629). Now, this places an important constraint on absolute positing and our cognition of existence in general:

> Whatever and however much our concept of an object may contain, we have to go beyond it in order to provide it with existence. With objects of sense this happens through the connection with some perception of mine in accordance with empirical laws; but for objects of pure thinking there is no means whatever cognizing their existence, because it would have to be cognized entirely *a priori*, but our consciousness of all existence (whether immediately through perception or through inferences connecting something with perception) belongs entirely and without exception to the unity of experience, and though an existence outside this field cannot be declared absolutely impossible, it is a presupposition that we cannot justify through anything. (A601/B629)

Hereby Kant rules out the very possibility of *a priori* cognition of existence, which, of course, is also a major blow to the possibility of an ontological argument in general, purporting to prove the existence of a being through merely *a priori* means. However, Kant's critical account of existence as absolute positing has a further implication regarding the nature of existential propositions, which generates a more specific objection to the ontological argument.

As we saw earlier, in his presentation of his first line of objection, not only does Kant entertain the syntheticity of existential propositions as a hypothesis, he also intimates that this is indeed the case and his interlocutor, the proponent of the ontological argument, must accept it "in all fairness" (A598/B626). Kant repeats the claim in his presentation of the second line of objection: "with actuality the object is not merely included in my concept analytically, but adds synthetically to my concept (which is a determination of my state)" (A599/B627). However, in neither line of objection does Kant present the syntheticity of existential propositions as a direct objection to the argument, nor does he positively explain why he holds it. There is, however, no doubt that Kant holds that all existential propositions are synthetic—let us call this his sixth thesis on existence, T6—and that he takes T6 to be damning for the ontological argument. For if 'God exists' is a synthetic proposition, then it is deniable without a logical contradiction, and cannot obtain with absolute logical necessity. Therefore, T6, if proved, would constitute a direct and distinct objection to the argument.

Kant's commitment to T6 has been seen as problematic by some scholars in the literature, especially against the background of T1. For instance, Schaffer, Everitt, and Goodman claim that T1 and T6 are incompatible, because if existence is not a real predicate that could add to the concept of anything, as Kant insists, then existence cannot be a predicate in a synthetic proposition.[57] Stang, though he concludes that it is ultimately resolvable, problematizes T6 as leading to an 'inconsistent triad':

(1) Existence is not a determination of any concept, i.e. the predicate 'exists' is not synthetic with respect to any concept.
(2) All existential judgments are synthetic.
(3) If a judgment is synthetic, then its predicate is synthetic with respect to its subject.[58]

In both cases, the inconsistency charge is based on the assumption that the syntheticity of a proposition requires its predicate to be a real (or synthetic) predicate in the sense of adding a further determination to its subject-concept. This assumption, however, is just false, and the inconsistency is a non-problem. T6 is not in conflict with T1, but as I have tried to explain above, it follows from the positive theses that complement T1. Taking another look at the peculiar type of syntheticity that existential (and in fact all modal) propositions instantiate will help us realize why exactly existential judgments are synthetic without 'exists' being a real or synthetic predicate.

As I pointed out in chapter 7, Kant distinguishes between two types of syntheticity of judgments: objective and subjective. The objective-synthetic judgments are synthetic in the ordinary sense that the predicate-concept adds a further determination to the subject-concept. Yet, Kant states, since existence and other modal categories cannot make such addition to the intension of the concept of anything, existential or modal judgments cannot be objective-synthetic: "The principles of modality are not... objective-synthetic" (A234/B286). These peculiar judgments must be synthetic

[57] Schaffer (1962), 309, Everitt (2004), 52, and Goodman (1996), 67. [58] Stang (2016), 37.

in a different sense and thus must make a different kind of addition to the subject-concept. Kant argues that these judgments are synthetic "only subjectively, i.e., they add to the concept of a thing... the cognitive power whence it arises and has its seat" (A234/B286). This 'subjective' addition is what makes existential or modal propositions synthetic. Kant specifies this addition in the case of possibility, existence, and necessity assertions, respectively, as that of an agreement with the formal conditions of experience in the understanding, a connection with perception, and a connection with perception through causal inferences of reason. So, unlike the ordinary, objective-synthesis, the subjective-synthesis in a modal proposition does not amplify the content or intension (or narrow the extension) of the subject-concept but adds to it a certain relation with the cognitive subject and the whole of her empirical cognition or experience.

In the specific case of existential propositions, then, the subject-concept is 'synthesized' or connected with a (direct or indirect) perception of the object. An existential proposition does not 'amplify' the concept of the object by adding a further determination to it, but it does amplify our cognition in general by adding to this concept a connection with an actual perception of the object and thus locating its object in the context of experience. This is the ultimate sense in which more is posited *through* the actual than *through* the merely possible (T5) even though the former does not contain more than the latter (T4). This vindicates, once again, Kant's insistence that the modal difference is not to be construed on the level of the intension of the concept of the object, but on the level of the concept's relation to our cognition. T1 and T6 are not at all inconsistent, but are indeed complementary pieces of a novel, critical theory of modality.

9

Absolute Real Modality and Kant's Amodalism Regarding Noumena

I have argued in this book that Kant's critical revolution in modality operates on two major novel ideas: first, modal notions are features of our conceptual representations of things and not of things themselves; second, they express the manners in which our conceptual representations of things are related to our cognition. Kant, thus, relativizes modality to our specific, human mode of representation and cognition in general. This, of course, raises the question of absolute (real) modality or the modality of things in themselves, without regard to our specific mode of representing and cognizing them. Although the precritical Kant utilizes the notions of absolute real possibility and necessity of things in his 'only possible argument,' the Kant of the *CPR* remains either silent or expressly agnostic about the absolute real modality of things in abstraction from their representational and cognitive relation to us, which is one general reason why he gives up on the objective validity of his argument. This appears congruent with Kant's overall critical agnosticism with respect to things in themselves: we know things only insofar as they are given to our cognitive faculty and thus cannot know what they really are in abstraction from this relation. Furthermore, given his critical restriction of the legitimate, cognitive yielding use of the categories (including, of course, modal categories) to appearances, it is not surprising that Kant holds that absolute modality of things in themselves, though thinkable, is beyond our cognition.

However, there is something not quite fitting here. Agnosticism with respect to absolute real modality is too modest a position and does not live up to the radical character of Kant's revolution in the history of modal thought. For if modality is indeed fundamentally a feature of the specific way we represent things and thus is relative to the conditions of our specific mode of cognition of things, then things in themselves just *cannot* have modal properties and absolute (real) modality should be rejected as a category mistake. I suggest that Kant comes to endorse this bolder position and do full justice to his revolutionary theory of modality in a passage in the *Critique of the Power of Judgment* (CJ). In this final chapter, I will offer a reading of Kant's remarks in this passage as both the ultimate logical consequence and the articulation of his revolutionary theory of modality as I reconstructed in this book.

* * *

In §76 of the *CJ* Kant states something striking regarding modal cognition:

It is absolutely necessary for the human understanding to distinguish between the possibility and the actuality of things. The reason for this lies in the subject and the nature of its cognitive faculties. For if two entirely heterogeneous elements were not required for the exercise of these faculties, understanding for concepts and sensible intuition for objects corresponding to them, then there would be no such distinction (between the possible and the actual). That is, if our understanding were intuitive, it would have no objects except what is actual. Concepts (which pertain merely to the possibility of an object) and sensible intuition (which merely gives us something, without thereby allowing us to cognize it as an object) would both disappear (Ak. 5:401–2).

This appears to be an epistemic thesis about the employment of modality in respective acts of representations by two differently constituted minds: while it is an indispensable characteristic of our discursive understanding to distinguish between the merely possible and the actual, an intuitive understanding would not make such modal distinctions between its objects. However, considering that Kant maintains that God would possess an intuitive understanding,[1] one might worry that this epistemic thesis entails a Spinozistic God, who would not represent mere possibilities but only actualities and thus could not have actualized a world otherwise than he actually did.[2] Although the recent studies on §76 have focused on Kant's epistemic thesis, I suggest that Kant also offers a metaphysical thesis in §76, without an account of which the grounds of the epistemic thesis cannot be properly understood.[3] Following the epistemic thesis, Kant adds:

[T]he distinction of the possible from actual things is one that is merely subjectively valid for the human understanding . . . The propositions, therefore, that things can be possible without being actual . . . quite rightly hold for the human understanding without thereby proving that this distinction lies in the things themselves . . . those propositions are certainly valid of objects insofar as our cognitive faculty, as sensibly conditioned, is concerned with objects of these senses, but are not valid of objects in general. (*CJ*, Ak. 5:402)

As I noted above, Kant restricts the validity of all categories to those objects that are given to our sensible intuition. Accordingly, we do not have a license to positively attribute to things in themselves or noumena properties that instantiate the categories (I will call these 'categorial properties'), though we can think them through 'unschematized' categories without making such metaphysical claims about their determinate constitutions. Thus, agnosticism with respect to whether noumena have categorial (and thus modal) properties would trivially follow from the critical restriction of the objective validity of the categories. Yet Kant goes one step further here by singling out the modal categories as "merely subjectively valid for human understanding," which entails a far stronger metaphysical conclusion than agnosticism: noumena do not have modal properties.[4] He seems to hold here that modal categories are nothing but features of our discursive representation, and thus, that the fabric of the mind-independent reality is amodal, perhaps an analog of his claim that space and time are nothing but mere forms of our sensible intuition, and thus, noumena do

[1] On Kant's contention that God would have an intuitive understanding, see for instance, B145, R 4348 (Ak. 17:515); R 4270 (Ak. 17:489), R 6041 (Ak. 18:431); R 6048 (Ak. 18:433); *Th.Pölitz* (Ak. 28:1053). For a rigorous account of why the intuitive understanding Kant refers to in §76 is the divine intuitive understanding, see Winegar (2017, forthcoming).

[2] See Zammito (1992), 255–9; Lord (2011), 98.

[3] I will discuss two such recent studies, Leech (2014) and Winegar (2017), in 9.2.

[4] For another statement of this thesis, see also R6270 (Ak. 18:538).

not have spatio-temporal properties. While the epistemic thesis appears to undermine divine freedom, the metaphysical thesis, as Stang has argued, threatens human freedom since it rules out the contingency of our noumenal volitions.[5] But what argument does Kant have for this metaphysical thesis?

One might want to argue that both of the heterogeneous formal components of our finite cognition are merely subjectively valid: just like space and time are merely subjective forms of our sensible intuition, all categories are merely subjective forms of our discursive understanding. Kohl has recently offered such a blanket argument based on Kant's commitment to the claims that an intuitive or divine understanding would not employ categories in representing things and that the way such an understanding represents things is the ultimate measure of how things really are.[6] For these two claims seem to jointly entail that noumena do not have categorial properties. If this argument succeeds, then it presents a shortcut to Kant's metaphysical thesis: if noumena do not have any categorial properties, they do not have modal properties either. Better yet, Kant's remarks in §76 can then be conveniently interpreted as conveying an instance of this general argument: Kant's epistemic thesis regarding the amodality of divine understanding (given that such an understanding would represent things as they really are) entails his metaphysical thesis regarding the amodality of noumena.

In the first section of this chapter, I will argue that the blanket argument fails in itself and does not offer an explanation as to why Kant's remarks in §76 focus only on the modal categories as opposed to all the categories. In the second section, I will argue that Kant's emphasis on the modal categories in §76 is neither trivial nor an instance of a general thesis about all categories. Instead, the mere subjective validity of modal categories is the culmination of Kant's revolutionary theory of modality and extends from his repeated claim about the 'peculiar' nature of modal categories, distinguishing them from all other categories.

Accordingly, modal categories are exceptional categories in that instead of even purporting to express the most fundamental ways things in general are, they express the various ways in which the representations of objects are related to the cognitive subject. Kant also holds that this peculiarity makes modal distinctions specifically a feature of a discursive mind like ours. Modal properties, therefore, are irreducibly relational, subjective, and discursive. Consequently, neither would noumena have modal properties, nor would an divine understanding represent them as having modal properties. That is to say, Kant's revolutionary theory of modality offers a direct justification for both the metaphysical and epistemic theses. I will finally briefly show that this way of understanding Kant's theses in the framework of his theory of modality offers a remedy for the worries that they undermine divine and human freedom.

9.1. A Blanket Argument for the Mere Subjectivity of All Categories?

Against the metaphysical readings of Kant's transcendental idealism, which attribute some categorial properties to noumena such as substantiality and causality, Kohl

[5] Stang (2016), chapter 10. [6] Kohl (2015).

argues that all categories are merely subjective forms of finite, discursive cognition and that noumena do not have any categorial properties. Kohl's argument turns around two claims that he attributes to Kant: "(1) Noumena have categorial properties . . . only if an intuitive, divine intellect would represent them as having such properties. (2) An intuitive intellect would not represent noumena as having categorial properties."[7] From these two, it follows that noumena do not have any categorial (and modal) properties.

Kohl takes (1) to be an instance of a more general principle (P): "how an intuitive intellect would represent things is the one decisive measure for what things are in themselves."[8] It is *prima facie* plausible that Kant must be committed to something like (P). For Kant has in mind divine intellect as the paradigm of 'intuitive intellect,' and it is a reasonable assumption that divine omniscience would involve the most accurate representation of how things really are.[9] The more questionable premise here is (2). There are two alternative routes one can take to justify (2). First, it can be shown that the only way for an intuitive intellect to represent a thing as having a categorial property is for that intellect to apply the corresponding category to that thing. Second, it can be shown that categorial properties themselves are in fact merely subjective forms of discursive thinking and ontologically dependent on a discursive mind, from which it follows that categorial properties cannot be had by things in abstraction from their relation to such a mind. I will argue below that neither route is ultimately tenable.

9.1.1. *Intuitive intellect and categories*

There is not much doubt that Kant maintains that an intuitive intellect would not employ categories in representing its objects.[10] This appears almost trivial: it should be in the nature of a non-discursive mind not to represent objects discursively, through concepts that pick out the common features of things. The categories, as the concepts of the most general features of objects, would not figure in the representations of a non-discursive mind. Kant's reasoning, however, is in fact less trivial then it seems. A discursive mind like ours depends on the givenness of an object, both logically and temporally prior to its representation, which consists in an act of synthesizing the given manifold data into a unity by means of certain functions of synthesis, i.e., the categories. It is this givenness of the object from elsewhere which requires the kind of intuition we have to be sensible, on the one hand, and requires the kind of understanding we have to be discursive. The discursivity of our understanding and the sensibility of our intuition are then jointly rooted in this kind of relation between our mind and its object. Kant's assumption here is what he offers as a grounding premise of his transcendental deduction of the categories: "combination of a manifold in general can never come to us through senses" and "all combination . . . is an act of the understanding" (B129–30). Thus, what can be given through sensibility is always a manifold, lacking an original

[7] Kohl (2015), 91. [8] Kohl (2015), 91.

[9] See especially R 6041 (Ak. 18:431): "The divine understanding is the highest and pure understanding which cognizes things absolutely, as they are in themselves."

[10] See B93, A256.

synthetic unity in itself, and the synthetizing role of the categories is only applicable to sensibly given objects.

In the case of an intuitive intellect, the object would not be given from elsewhere. For such is an intellect "through whose representation the objects of this representation would at the same time exist" (B138–9) and "which would not represent given objects, but through whose representation the objects would themselves at the same time be given, or produced" (B145). An intuitive intellect would then be purely active or spontaneous and have no receptive component.[11] There would be no logical and temporal distinction between the representation and the object itself, the concept and the intuition, the form and the content of thought, leaving neither any need nor any room for "a special act of synthesis of the manifold" (B139). Consequently, "the categories would have no significance at all with regard to such a cognition" (B145). Yet this negative claim that an intuitive would not employ categories alone does not suffice to prove that it would not represent categorial properties, without the presupposition that the only way for such an intellect to represent categorial properties is to apply categories to its objects. Thus, in order to see whether this presupposition holds, we need to look at how an intuitive intellect would represent properties in general.

9.1.2. *Intuitive intellect and representation*

Based on its non-discursiveness alone, it seems convenient to assume that an intuitive or divine intellect would have a capacity of aconceptual representation. Yet Kant's text is rather baffling on this point. While Kant's remark in §76 that "concepts . . . and sensible intuition . . . would both disappear" (*CJ*, 5:402) suggests that the representations of such an intellect would involve neither concepts nor intuitions, elsewhere he states that the cognitions of the divine intellect are non-sensible sort of intuitions which are, at the same time, ideas. These descriptions are hard to reconcile.

First, it is important to recognize the terminological difficulty in describing the nature of a mind into which we have no real insight. Since an analogy (to our own kind of mind) is the only tool we have in comprehending the divine mind, Kant's attempts are inevitably constrained by the limits of our own (human) mental terminology.[12] Second, when Kant states that the cognitions of the divine intellect are not concepts but ideas and intuitions, he uses 'concept' in the sense of general or discursive representation:

> The originality of the *intellectus originarii* is that it cognizes all parts from the whole and not the whole from the parts, because it cognizes everything and determines *limitando* all things. The cognitions of the *intellectus originarii* are not concepts, but Ideas. Concepts are general discursive representations and general marks of things. Abstraction is required for all concepts, but that is a deficiency; we limit our representation, and thereby we receive clear concepts and representations. But because the *intellectus originarius* is unlimited, it cannot be based on limitation and abstraction. Because human understanding cognizes something by means of general marks, brings it under concepts, and cognizes by means of a rule, the human

[11] *Th.Pölitz* (Ak. 28:1054).

[12] For Kant's explicitly recognition of this linguistic constraint, see *MK3* (Ak. 29:954).

understanding is discursive, but the *intellectus originarius* is intuitive. It does not cognize *per conceptus* but *per intuitus*. Because the original understanding is not bound to limits but discursive cognition is a limitation, the divine cognition is an immediate cognition.

(*MPölitzL1*, Ak. 28:328–9)[13]

Thus, Kant contrasts discursive and intuitive intellects with regard to at least five aspects of cognition of objects.

(i) While the discursive intellect moves from the representations of the parts to the representation of the whole, the intuitive intellect moves from the representation of the whole to the representations of the parts, or, as Kant puts in §77 of the *CJ*, "goes from the synthetically universal (of the intuition of a whole as such to the particular)" (Ak. 5:407). This means that the intuitive intellect cognizes one unified whole in a single act of cognition, which contains the representations of particulars (i.e., objects and their determinations).

(ii) The intuitive intellect has an instantaneous cognition (of everything at once) as opposed to our successive (and therefore, necessarily temporally structured) cognition.[14]

(iii) While our successive cognition from parts to the whole requires us to represent particulars through (general or discursive) concepts of their common determinations, the intuitive intellect's instantaneous cognition of everything at once does not require the employment of general concepts.

(iv) Every general concept picks out a common determination of a set of things and is thus formed through abstracting from other determinations that are not shared by all those things.[15] Because of this dependence on abstraction, the discursive intellect can only form incompletely determinate representations of things. The intuitive intellect, however, is not bound by such a limitation and is capable of forming completely determinate representations of things.

(v) Since general concepts, as abstractions from particulars, are mediate representations of objects, depending on the intuitions' immediate receptivity of data, our discursive intellect is only capable of mediate cognition of objects. In contrast, the intuitive intellect, free from general concepts and productive of its own objects, is capable of immediate cognition.

Therefore, setting aside the conflicts in the mental terminology Kant uses in describing the operation of the intuitive, divine intellect, we should take his point to be that such an intellect would be capable of forming singular, complete, and immediate representations of its objects.[16] More specifically, the intuitive intellect would form a single representation of the whole of everything at once. It is through

[13] I adopted the translation of this passage from Winegar (2017), 314.

[14] On succession, see *Th.Danzig* (Ak. 28:1267); R 4270 (17:489) and *Th.Pölitz* (Ak. 28:1051).

[15] On abstraction, see also *MK3* (Ak. 29:978); *MM* (Ak. 29:888).

[16] An alternative approach here is the suggestion by Westphal (2000), 284 that "concept and intuition would be identical in, and not absent from, and intuitive intellect," and thus, construing the representations of the intuitive intellect still in terms of concepts albeit as those of a non-discursive kind.

the complete representation of the whole that every individual would get completely represented by the intuitive intellect.

9.1.3. Intuitive intellect and representation of categorial properties

We can now raise the question of whether the constitution of an intuitive intellect warrants that it would not represent things as having categorial properties. This is connected to the more general question of whether and how a non-discursive intellect would represent general properties of noumena.

Kant is committed to the claims that (i) an intuitive intellect would represent everything in its complete determination such that no determination of anything would escape from its representation, and (ii) such an intellect would not employ discursive concepts in representing its objects. One might want to move from (ii) to the conclusion that an intuitive intellect would not represent any common properties, which, in turn, entails, via (i), that noumena do not have anything in common and all of their determinations are particular tropes. As I noted earlier, this move could only work with the presupposition that the only way of representing common properties is to employ discursive concepts. An intuitive intellect would have to have a non-discursive way of representing the commonalities among particulars constituting a whole, otherwise its cognition would not be an instantaneous cognition of a synthetic whole, but a successive cognition of an analytical aggregate. Besides, if it is at all meaningful to talk about noumena as things, as a plurality of metaphysically distinct items picked out by the notion of a noumenon, then it would make no sense to claim that they would not have anything in common. For one would think that noumena must instantiate certain common meta-features in virtue of which they all are 'things.' At the very least, being a thing seems to require having the universal meta-feature of having certain properties or at least one property, even if the thing is unique on the level of the latter, first-order properties.[17]

If noumena have common (first- or second-order) properties, then an intuitive, divine intellect would have to represent these common properties as well as the particular ones. But how can common properties can be represented non-discursively? As I suggested before, such an intellect would cognize everything in its complete determinacy and particularity in a single representation of the whole. In a holistic representation of every particular determination of every particular thing, those determinations that are common to multiple things would also get represented, and so without the involvement of general concepts to abstract them away. Marshall has proposed to construe this non-discursive representation of general properties as an 'indirect' representation.[18] An intuitive intellect would directly represent particular properties of things and indirectly represent their general properties. For if general properties are true of things, they are so in virtue of the more particular properties that are true of them, and thus, in representing the particular properties,

[17] One could extend this reasoning to non-modal categorial properties. For instance, it can be argued that noumena, whether there is a plurality of them or a single one as an undifferentiated unity in itself, admit 'quantity.' Again, from the meta-feature of having a constitution, one can infer that a noumenon is a subject of properties, which also entails admitting 'quality.'

[18] Marshall (2018).

an intellect would also be representing the general properties indirectly, without any general concepts to represent them directly.

9.1.4. Categorial properties and the 'neglected alternative'

I showed above that the non-discursive constitution of an intuitive intellect does not rule out the possibility that it (non-discursively) represents general properties. Similarly, the fact that such an intellect would not employ categories does not rule out the possibility of it representing noumena without applying categories to them. Thus, the first route to Kohl's (2), i.e., an intuitive intellect would not represent noumena as having categorial properties does not work. Yet the second route to (2) might still be viable. If the categories are nothing but mere subjective forms of our discursive thought, or more precisely, if the categorial properties signified by the categories are ontologically dependent on a discursive mind's act of representation, then noumena would not have categorical properties and an intuitive intellect would not represent them as having such properties. As Kohl notes, Kant indeed seems to commit to the view that categories are nothing but subjective forms of discursivity, especially when he identifies them as containing "only" (*allein*) "the synthetic unity of apperception" (B148), "*nothing but* the logical function to bring the manifold under a concept" (A245), or "*nothing other than* the mere form of thought" (A567/B595).[19] This impression is further strengthened by the fact that Kant construes the act of synthesis carried out through the categories as applying only to what is given 'from elsewhere' through sensible intuition. This suggests that discursivity is not contingently but necessarily and exclusively attached to sensibility. Thus, the categories, as specific features of discursivity, are dependent on the kind of mind that has not only a discursive understanding but also a sensible intuition.

This need not be the end of story though. Kant consistently makes a distinction between the 'use' (*Gebrauch*) or 'application' (*Anwendung*) or 'function' (*Funktion*) of the categories and their 'significance' (*Bedeutung*) or 'meaning' (*Sinn*), and most of his emphases on the dependence of the categories on sensibility qualify this dependence as pertaining to their use instead of significance.[20] Thus, even if the categories have no cognitively productive function beyond the realm of appearances, their unschematized contents might still signify properties of noumena. This can also be expressed in terms of a distinction between a representation as a mental state and what that mental state is about in reality, and thus, between the categories themselves and the categorial properties. So even if the categories themselves are subjective forms of thinking, ontologically dependent on the specific nature of a discursive mind, the categorial properties such as substantiality or unity can still belong to the fabric of mind-independent reality.[21]

Against this challenge, Kohl points out how the language of Kant's statements about the mere subjectivity of the categories echoes that of his statements about space and time being "nothing but" (A42/B59; A33/B49; A492/B520), "merely" (A48/B66)

[19] Kohl (2015), 99. [20] See, for instance, A156, B305.

[21] Similarly, Marshall (2018) claims that Kant's apparently anti-metaphysical statements can be read as concerning only the natures of the categories themselves and thus as compatible with the possibility of noumena having categorial properties.

"subjective condition[s] of sensibility" (A26/B42), "mere form[s] of... intuition" (A48/B66). Since Kant concludes his statements about space and time with the claim that they are transcendentally ideal and do not represent any property of noumena (A26/B42), Kohl argues that it is fair to assume that his similar expressions in the case of the categories intend to express a similar contention about their mere transcendental ideality and not corresponding to properties in the noumenal realm.

However, this linguistic point does not provide a definitive support for Kohl's conclusion. Kant expressly states that noumena do not have spatio-temporal properties (A26/B41) and provides a direct argument for this metaphysical claim in the Transcendental Aesthetic. The question of its merits aside, Kant's argument is that for propositions regarding the pure structure of space (and time) to be synthetic *a priori*, space (and time) must pertain to the formal properties of only appearances and not of noumena.[22] It is this argument which blocks the so-called 'neglected alternative' objection raised by early critics of Kant that space and time may well be both subjective forms of intuition and objective features of noumena at the same time. Yet Kant does not complement his statements about the mere subjectivity of the categories with an unambiguous metaphysical conclusion that noumena do not have categorial properties, nor does he offer an argument for this position anywhere in the *CPR*.[23] Without such an argument, the alternative that categories are both subjective forms of discursive thought and representations of genuine (categorial) properties of noumena remains viable.

9.1.5. Practical cognition of noumena and the categories

Kant's account of the categories in the *CPR* does not rule out the possibility of noumena having categorial properties. Yet Kant's explicit commitment to the application of the category of causality to our noumenal freedom in his practical philosophy suggests that he holds, at least, that some noumena do have some categorial properties. Kohl maintains that this would not poke a hole in his argument for the inapplicability of the categories to noumena, for, he claims, causality gains a new, 'non-discursive' meaning in the practical context: "since this practical category no longer represents a logical function of synthesis, what it represents might also enter into the representation of a non-synthesizing understanding."[24] However, this claim about the new, practical meaning of causality is dubious.

As I emphasized earlier, the discursivity of a representation is about its 'function' or 'use,' i.e., a representation is discursive if it is used to pick out a general property of an object while abstracting from its other properties, as opposed to its 'meaning,' content, or what properties it represents. Thus, the claim that the categories have a different, 'non-discursive' function or way of representing objects in the domain of practical reason does not entail that they acquire different meanings and come to represent different general properties of things. Besides, if the categories do have a

[22] B41; A48/B66. See Shabel (2004).

[23] For a thorough defense of the claim that here is no such argument in the *CPR* that the properties represent through the categories (i.e., categorial properties) are 'subject-dependent' and does not belong to things in themselves, see Watkins (2002).

[24] Kohl (2015) 107.

'non-discursive' meaning, this only shows that they are in fact more than their discursive functionality and are not 'nothing but mere subjective forms of discursive thought' after all.

The problem that Kant's practical philosophy raises for Kohl's or any anti-metaphysical reading of Kant is in fact far deeper.[25] First of all, identifying the enigma of all critical philosophy as "how one can *deny* objective *reality* to the supersensible *use of the categories* in speculation and yet *grant* them this *reality* with respect to the objects of pure practical reason" in the beginning of the *Critique of Practical Reason* (*CPrR*) (Ak. 5:5), Kant himself takes his practical philosophy as offering an explanation of how the categories (and not just the category of causality) gain (not new meanings or contents but) a legitimate use or application with regard to noumena so long as they can be considered as objects of pure practical reason. Kant argues that this application yields a "practical cognition," through which we cognize cardinal noumenal objects, (not just) freedom, (but also) God, and immortality, "the supersensible *in* us, *above* us, and *after* us" (*Progress*, Ak. 20:295–300). What is "merely thought indeterminately and problematically" in the theoretical domain thus gets "determined" and "cognized assertorically" "from a practical perspective" (*CPrR*, Ak. 5:105, 134).

The key here is to recognize that 'practical cognition' is a genuine kind of cognition, albeit different from theoretical cognition. While both practical and theoretical cognition are instances of conscious representation of an object (A320/B377), they differ in at least four important ways: i) the use of reason; ii) its purpose; iii) the causality of the representation and its object; iv) its objects.

i) While in its theoretical use reason is merely a cognitive faculty, in its practical use reason is both a cognitive and a desiderative faculty, "concerned with the determining grounds of the will, which is a faculty either of producing objects corresponding to representations or of determining itself to effect such objects" (*CPrR*, Ak. 5:15).
ii) While the purpose of theoretical cognition is merely to yield a determinate representation of the object, practical cognition aims to also actualize the object that is represented (Bix–x; *CPrR*, Ak. 5:57, 89–90).
iii) In theoretical cognition, the object, as given to us 'from elsewhere,' is existentially independent of the representation and is the causal ground of the latter. In practical cognition, the very act of representation is the causal ground of the object (*CPrR*, Ak. 5:46).
iv) While theoretical cognition is "that through which I cognize what exists," practical cognition is "that through which I represent what ought to exist" (A633/B661). So as opposed to the mere representation of what exists in the world, practical cognition involves the representation of an action "as an effect possible through freedom" (*CPrR*, Ak. 5:57). Practical reason can be efficacious of its object, exactly because its object is either the action (where the action itself is the end) or the consequence of the action (where the action is a means to a distinct end), toward which reason determines the will to

[25] I provide my more detailed discussion of Kant's theory of practical cognition, in my (forthcoming).

> move. Of course, the action here is not represented descriptively, but normatively, as an action that is good to perform (either in itself or as a means to another end). In this sense, the ultimate aim of practical cognition is not the action *per se*, but the valuable 'end,' 'the good,' which is to be produced through this action.[26] Practical judgments are then those that employ the concepts of good and evil, attributing them as predicates to actions (i.e., 'A is good'), but (at least in the case of finite rational beings like us) take the form of imperatives or obligations prescribing one to perform (or omit) an action.[27] Kant calls the set of most fundamental normative concepts of obligation, expressing various determinations of the concepts good and evil, the "categories of freedom" (*CprR*, Ak. 5:66).

Now, given the nature of its object as action, normatively construed, one might hold that our practical cognition does not employ the categories of the understanding (or of 'nature') but only the categories of freedom. This is why Kohl claims that Kant's account of practical cognition does not threaten his reading (in the specific case of causality and noumenal freedom): "once the concept of causality has acquired this practical meaning, it can function as a 'category of freedom' as opposed to a 'category of nature'."[28] This line of thinking, however, is based on an incomplete picture of practical cognition.

What has been said so far concerns actions as imperatives, which are only the primary objects of practical cognition. Yet Kant also defines a second kind of objects of practical cognition, which are not imperatives but connected to imperatives through a 'grounding' relation: "Practical cognitions are, namely, either 1. imperatives, and are to this extent opposed to theoretical cognitions; or they contain 2. the grounds for possible imperatives and are to this extent opposed to speculative cognitions" (*JL*, Ak. 9:86).[29] So the objects of the second kind are not defined under 'what ought to exist' but under 'what exists (or can exist)' with a special relation to the former. Abstracted from this practical 'grounding' relation, these objects are mere articles of speculation expressible only in formally theoretical propositions. Thus, if noumena are to be objects of practical cognition, they will be objects of the second kind, and the practical cognition of noumena, as objects of the second kind, must employ the categories of understanding instead of that of freedom.

The relevant conception of freedom here is that of noumenal or transcendental freedom of the will, a special kind of causality in rational agents that is capable of initiating actions without being conditioned by any determining cause of nature.[30] Kant argues that we can cognize the actuality of our noumenal freedom through its connection with the moral law: "[F]reedom is . . . the only [speculative idea] the (real) possibility of which we know *a priori* . . . because it is the condition of the moral law, which we do know" (*CPrR*, Ak. 5:4). Thus, once we cognize that

[26] For Kant's formulation of the '*Gegenstand*' of practical cognition not as the '*Handlung*' itself, but as that which will be made actual through it, see, for instance, R 2796 (Ak. 16:517), R 3316 (Ak. 16:666), *Corr* (Ak. 10:131).

[27] See R 2798 (Ak. 16:518), R 3115 (Ak. 16:665), *LV* (Ak. 24:900–1), *JL* (Ak. 9:86).

[28] Kohl (2015), 104.

[29] Cf. *LDW* (Ak. 24:751), *LV* (Ak. 24:901), R 2798 (Ak. 16:518).

[30] See A446/B474, A803/B381.

the moral law actually applies to us, which, Kant claims, is a 'fact of reason' that we become immediately conscious of, we can also derive that we are actually free.[31] Thereby not only is noumenal freedom "*cognized* assertorically" (*CPrR*, Ak. 5:105), but the category of causality contained in the concept of noumenal freedom also gains a transcendental use.

While freedom is the necessary condition of the application of the moral law itself, the existence of God and the immortality of the soul are the necessary conditions of the realizability of the ultimate end of the moral law, the highest good, the ideal of a moral world where virtue and happiness are proportionately maximized in rational beings.[32] The immortality of the soul is the necessary condition of the real possibility of the complete conformity with the moral law or holiness, and the existence of God is the necessary condition of the real possibility of the distribution of happiness proportionate to that morality.[33] If we hold that the highest good is really possible, we also have to hold that the necessary conditions of its real possibility, God and immortality, are actual.

Even though the epistemic ground of cognition here is practical, i.e., a necessary relation to the moral law, the form of cognition is still theoretical. Kant calls a proposition that posits noumenal freedom, God, or immortality based on a practical relation a 'postulate of practical reason' and characterizes it as "a *theoretical* proposition, though one not demonstrable as such, insofar as it is attached inseparably to an *a priori* unconditionally valid *practical* law" (*CPrR*, Ak. 5:122). These postulates are "theoretical in themselves," that is, they are propositions that purport to report what exists or can exist, but "practical *in potentia*," in that they ground possible imperatives for our actions.[34] So the judgments that express the practical cognitions of noumena (i.e., postulates) are not different from theoretical judgments in their conceptual form or meaning, but only in the way they are justified. Accordingly, the practical cognition of the objects of the second kind (i.e., noumenal freedom, God, and immortality) employs the same stock of categories as theoretical cognition. For while practical reason provides the ideas of freedom, God, and immortality with objective reality, the contents of these ideas are in fact still given by theoretical reason. Hence, Kant writes:

> Every use of reason with respect to an object requires pure concepts of the understanding *(categories)*, without which no object can be thought.... here *ideas* of reason, which cannot be given in any experience at all, are what I would have to think by means of categories in order to cognize an object. Here, however, our concern with these ideas is not for the sake of theoretical cognition of their objects but only with whether they have objects at all. Pure practical reason provides this reality, and theoretical reason has nothing further to do in this than merely *to think* those objects through categories. (*CPrR*, Ak. 5:136)

Practical cognition employs categories because it consists in providing a concept, whose content is already constructed in accordance with the categories, with an object through a practical relation. So in practically cognizing God, we posit, on a

[31] See *CPrR* (Ak. 5:6, 31, 42–3, 47–8, 91–4, 104–8); *CJ* (Ak. 5:468, 474).
[32] See A814/B842; *CPrR* (Ak. 5:110); *CJ* (Ak. 5:451).
[33] *CPrR* (Ak. 5:122, 124).
[34] *LV* (Ak. 24:901); *LDW* (Ak. 24:751); *JL* (Ak. 9:86).

practical ground, that the idea of a just distributor of virtue and happiness in the world, and thus, an idea determined with respect to the categories such as substance, causality, unity, as having an object. It is in this way that the transcendent use of categories, which is "no use at all" (A248) in theoretical cognition, turns into an immanent use in practical cognition.

Therefore, the blanket argument for the claim that noumena do not have any categorial properties does not succeed. For neither of the two routes to secure its premise that an intuitive understanding would not represent noumena as having categorial properties works: a) the representation of categorial properties does not necessarily require the employment of discursive categories; b) without an additional argument ruling out the 'neglected alternative,' the discursivity of the categories does not warrant the conclusion that the categorial properties are ontologically dependent on the discursive subject. Moreover, Kant's account of practical cognition commits him to the attribution of certain categorial properties to noumenal freedom or will, God, and the immortal soul.

9.2. Kant's Revolution: Modality as Irreducibly Relational, Subjective, and Discursive

The blanket argument, even if it did not fail on its own terms, would have to take Kant's metaphysical thesis regarding the amodality of noumena in §76 as a specific illustration of the general position that noumena do not have categorial properties, and thus would not provide any explanation of why Kant limits his remarks to modality. I suggest that this is a major issue, for the missing explanation points exactly to why these remarks should not be taken to offer a specific instance of a general thesis regarding all categories. Kant's remarks are limited to modal categories because he assigns them an exceptional status among all classes of categories. The claim that Kant holds that all categories are merely subjective features of discursive representation and thus noumena would not have categorial properties is not warranted. Yet Kant's revolutionary theory of modality commits him exactly to the view that modal categories are special categories in that they are merely subjective features of our discursive intellect, expressing only the various ways in which representations of objects are related to our cognitive faculty. This view of modality unambiguously warrants the metaphysical thesis that noumena would not have modal properties signified by modal categories.

9.2.1. *The peculiarity of modality and the metaphysical thesis*

Let us recall from chapters 6 and 7 that in his discussions of modality in the *CPR* Kant consistently emphasizes that there is something 'peculiar' about modal categories, distinguishing them from all the other categories. For instance, in his exposition of the modal functions of judgments, Kant states: "The modality of judgments is a quite special (*besondere*) function of them, which is distinctive in that it contributes nothing to the content of the judgment" (A74/B100). For "besides quantity, quality, and relation there is nothing more that constitutes the content of a judgment," and modality "rather concerns only the value of the copula in relation to thinking in general" (A74/B100). So while all other functions exhaustively construct the logical structure of a judgment through which an objective content is represented, the modality of a judgment expresses the way in which this logically fully structured content is related to our cognitive faculty.

Again, in his exposition of the schematized modal categories in the Postulates, Kant first underscores the 'peculiarity' of modality: "The categories of modality have this peculiarity (*Besondere*): as a determination of the object they do not augment the concept to which they are ascribed in the least, but rather express only the relation to the faculty of cognition" (A219). As opposed to all the other categories which can function as predicates or determinations of objects in judgments and thus contribute to the content of the conceptual representation of an object, modal categories do not add any further determination to this representation but only express the manner in which this representation is related to our faculty of cognition. In other words, the constitution of an object itself is expressed only by the other three classes of categories. The schematized categories of quantity and quality express the properties of an object in terms of its extensive (A162/B202) and intensive magnitudes (A167/B207), and the schematized categories of relation express the relational properties of an object in connection with other objects (A176/B218). However, the schematized categories of modality express merely the different ways in which the concept of an object, whose intension is already constructed by the other categories, is related to our faculty of cognition and the whole of its empirical cognition.

Let us also recall that the special character of modality as strictly separated from the intension of the concept of an object also leads Kant to hold that modal assertion constitute a peculiar, "subjective," type of synthetic judgment (A234/B286). While in objectively synthetic judgments the synthesis operates on the level of conceptual representation of the object and thus contributes to the determination of that representation, in modal judgments neither an inner determination of the object nor an (external) relation to another object is directly asserted. Kant correlates the three kinds of modal assertion with three distinct cognitive faculties and their specific cognitive acts (A75/B100). In a possibility assertion, e.g., S is possible, the *understanding* posits the *compatibility* between the concept of S and the form of experience. In an existence assertion, e.g., S exists, the *power of judgment* posits a *connection* between the concept of S and an actual perception (and thus an actual object). In a necessity assertion, e.g., S exists necessarily, *reason* posits the *connection* between the concept of S and an actual perception *through a causal inference*. What is added to or synthesized with the subject-concept in a modal assertion is then not a predicate, but the positing of a specific kind of relation with the subject's empirical cognition.[35]

The upshot of Kant's recurrent emphasis on the peculiarity of modal categories is that they do not even purport to represent the properties or ways of being of things, but the properties of our conceptual representations of things, or more specifically, the properties of the relations of those representations to our cognition. Modal categories, when they apply to objects, are not first-order predicates of objects but second-order or concept-level predicates, and signify a relation to the cognitive

[35] See R 5558 (Ak. 18:232); A601/B629.

subject. This peculiarity is what underlines the merely subjective character of modality in distinction from other categories: while the possibility of other categories corresponding to the properties of things in themselves is not excluded, modal categories cannot possibly represent any properties of things as they are in themselves and independent of any relation to our cognitive faculty. For modal categories purport to represent only and precisely the mode in which the relation between the representation of an object and our cognitive faculty holds. In abstraction from that relation, modality does not signify anything. So the peculiar nature of modality makes not only the modal categories but the modal properties they signify irreducibly second-order, relational, and subjective. The 'neglected alternative' that is not eliminated in the case of other categories is thereby eliminated in the case of modality. Thus, this account both substantiates Kant's metaphysical thesis in §76 of the *CJ*, i.e., noumenal reality is amodal, and offers an explanation of why Kant's remarks there are limited to modality.

9.2.2. The epistemic thesis: The necessary discursivity of modality

The mere subjectivity of modal categories directly entails the metaphysical thesis that noumena would not have modal properties. Combined with the claim that an intuitive, divine intellect would have the most accurate representations of things, this metaphysical thesis entails that modality should have no place in the divine cognition of noumena. However, one might still wonder whether Kant's epistemic thesis about modality is justifiable in itself. Why is modality only a feature of a discursive mind like ours? What is it about the way a discursive mind represents things that requires the modalization of its representations? In fact, Kant's remarks in §76 involves a direct, albeit brief, epistemological explanation of why a discursive intellect necessarily employs modal distinctions:

> It is absolutely necessary for the human understanding to distinguish between the possibility and the actuality of things. The reason for this lies in the subject and the nature of its cognitive faculties. For if two entirely heterogeneous elements were not required for the exercise of these faculties, understanding for concepts and sensible intuition for objects corresponding to them, then there would be no such distinction (between the possible and the actual). That is, if our understanding were intuitive, it would have no objects except what is actual. Concepts (which pertain merely to the possibility of an object) and sensible intuition (which merely gives us something, without thereby allowing us to cognize it as an object) would both disappear
> (*CJ*, 5:401–2).

Two interpretations of this passage are available in the current literature. According to the more common interpretation, Kant has in mind here the creative aspect of the divine intellect: since such an intellect creates its own object in its act of representation, whatever it represents is always actual.[36] Or as Leech formulates it, what excludes mere *possibilia* from divine representation is the "epistemic guarantee" that its objects exist.[37] However, even with Leech's metaphysically less presumptuous formulation, this interpretation has two problems. First, the idea of 'epistemic

[36] Lord (2011), 98; Zammito (1992), 257. [37] Leech (2014), 347.

guarantee' of the existence of the object of representation, whether via creation or some other way, is neither found in §76, nor even implied by it. Kant talks about this guarantee in the *CPR* (B138–9; B145), but clearly takes it to lead to the conclusion that an intuitive intellect would not employ (any and not just modal) categories in representing its objects. Second, this interpretation does not espouse the metaphysical thesis and consequently retains the possibility of modal distinctions at the level of noumenal reality in principle. It is as though there may be merely possible noumena in the metaphysical space and yet because of the 'epistemic guarantee' that whatever God represents is actual, these mere possibilities remain unrepresented by God. This seems to impose a constraint not only on divine representative powers, but also on divine free will.

As an alternative to the common interpretation, Winegar has offered another explanation of the actuality of all objects of divine cognition:

> Because God's intuitive understanding cognizes finite things in their thoroughgoing determination, the only possible things that God cognizes are actual things. But, unlike God's intuitive understanding, human understanding employs general concepts that omit particular determinations and, thus, represent things in an indeterminate way. Consequently, humans can represent merely possible things.[38]

However, this interpretation relies on a misunderstanding of Kant's view of the relation between modality and determinacy. On Winegar's reading, Kant holds that what is represented indeterminately is merely possible and what is represented determinately is actual, and thus the modal distinction between possibility and actuality consists in whether or not something is represented with complete determinacy. Of course, then, since a discursive mind represents objects through general concepts that are incompletely determinate, albeit always further determinable, while a divine mind represents things in their complete determinacy, the former employs modal distinctions, the latter does not. A few of Kant's reflections seem to give credence to this view.[39] However, this cannot be Kant's considered view. For, as I showed in chapters 7 and 8, this construal of modal distinctions in terms of the determinacy of representation is something Kant strongly rejects on the basis of the aforementioned peculiarity of modality. Kant insists that even if we were able to form the complete concept of a thing, like a divine mind presumably is, the actuality of that thing would not be warranted or there would not be an 'epistemic guarantee' that our concept has an existent object (A219). Unlike, for instance, Baumgarten, who takes thoroughgoing determinacy as a sufficient condition of actuality, Kant holds that complete determinacy is neither a sufficient condition, nor a distinguishing mark of existence (as opposed to mere possibility). The "serious error" in Baumgarten's contention, Kant points out, is the failure to see that there is a certain sense in which indeterminacy cannot belong to the merely possible as much as it cannot belong to the actual! (*OPA*, Ak. 2:76).

[38] Winegar (2017), 318–19.
[39] See R 4017 (Ak. 17:387), 4297; (Ak. 17:499); R 5723 (Ak. 18:335). See also Schneeberger (1952), 6–11.

For Kant the difference between the actual and the merely possible does not lie in how determinate the representation of an object is or what determinations are posited in the intension of the representation but how that representation as a whole is posited in relation to the subject's cognition. This is precisely why Kant repeatedly complements his T1 that existence is not a predicate or determination of anything with T4 that "the actual contains nothing more than the possible" (A599/B627; B284; *OPA*, Ak. 2:75). He argues that the intension of our concept of an actual hundred dollars would not be different (or 'larger') than that of our concept of a merely possible hundred dollars, even if they were complete concepts. Kant also sees the misconception of modal distinctions in terms of the intensions of concepts as what motivates the contention that modal distinctions reflect extensional differences, i.e., the extension of possibility is greater than that of actuality, which, in turn, is greater than that of necessity (A230–2/B282–4).

Therefore, Kant would not embrace the view that the things God cognizes are always actual because God represents things in their complete determinacy, since the very conception of modality that grounds this view is explicitly rejected by him. To be fair, Winegar adds an important detail here, and suggests that "In cognizing an actual thing as thoroughly determined, God cognizes the thing's connection to his own will and, thus, cognizes the thing's actuality."[40] Accordingly, a thing's complete concept includes its connection with God's will, and thus, in a quite rigorous sense, its actuality. This, however, is again in stark conflict with Kant's separation of modality and intension. The modality of a thing is strictly external to the mere representation of that thing, and the mere representation of the thing is modally and existentially neutral: "In the mere concept of a thing no characteristic of its existence can be encountered at all" (A225/B272). As I showed in chapters 3 and 8, Kant's main issue with the traditional ontological argument is that even God's complete concept is not immune from this separation of modality and intension. He goes as far as to say that "God himself cannot know his own existence through concepts" (*MK2*, Ak. 28:784). Moreover, if the complete concepts of things were to contain their connection to God's will, they could also contain the lack of that connection. For being determined with respect to a predicate can be negative as well as positive, and not having a causal connection to God's will would be a determination as much as having it. In that case, mere *possibilia* would also be completely determinate, and thus, God would be representing them as much as actual things, which contradicts what Winegar sets out to explain: all objects of God's representations are actual.

Both interpretations of Kant's epistemic thesis in §76 that I sketched above place emphasis on the claim that an intuitive intellect would only have actual objects (or would not modalize its objects) and aim to explain this claim by recourse to the positive features of such an intellect's representational powers, i.e., 'epistemic guarantee' or complete determinacy. This, I think, is the wrong interpretative course to take. For just like his metaphysical thesis, Kant's epistemic thesis is also rooted in his conception of modality as an irreducibly subjective feature of a specifically human, discursive intellect, and his claim about the amodality of the intuitive intellect is only

[40] Winegar (2017), 320.

a negative consequence of this conception of modality. Kant's brief explanation of the epistemic thesis, "the reason for this lies in the subject and the nature of its cognitive faculties" (*CJ*, Ak. 5:401), confirms exactly this. The distinction between possibility and actuality is not to be construed in terms of the intension of the representation of a thing, but how that representation is posited in relation to distinct cognitive faculties of a discursive subject: "For if two entirely heterogeneous elements were not required for the exercise of these faculties, understanding for concepts and sensible intuition for objects corresponding to them, then there would be no such distinction" (*CJ*, Ak. 5:402). Our discursive understanding is the faculty of conceptual thought, through which we can represent objects that are not but can be given to us, and sensible intuition is the faculty of receptivity, through which objects that actually instantiate our concepts are given to us. This excess between what we can conceptually represent and what is given to us is what requires us to employ the distinction between the merely possible and the actual.[41]

As we saw earlier, in his more elaborate discussions of modality, Kant correlates the three kinds of modal assertion (possibility, actuality, and necessity) with three distinct cognitive faculties (understanding, power of judgment, and reason) and how these faculties connect the representation of an object to our cognition in general (A75/B100). Kant also observes that modality, thus understood, reflects the progressive nature of our cognition. Mere possibility expresses that the object in question can be part of our cognition; actuality expresses that it is indeed a part of it, and necessity expresses that it is an inseparable part of it due to its very laws. Thus, modalities express various levels of the "gradual incorporation" of the representation of an object into the whole of our cognition (A76/B101). This also means that modalization of an individual representation always occurs against the background of a set of representations to which the former can be related in various ways (i.e., as merely compatible with that set, as part of it, as part of it through the laws that unify the set), and thus is necessarily attached to the progressive type of cognition that operates through establishing various modes or stages of relation between the individual and the background.

Therefore, on Kant's account, modality is a mere feature of a specifically constituted mind. For modal distinctions are only valid for a discursive mind whose cognition involves the operation of distinct and heterogeneous faculties and thus is inevitably progressive. The divine understanding, due to its unity, simplicity, and the complete homogeneity between its representations and its objects, would have neither a need nor a ground to employ modal distinctions. Its operation would not display a progressive, gradual incorporation of individuals into a whole, but the cognition of the whole at one stroke, leaving no room for different modes of relations between the representation of an individual and the cognition of the whole. If different modal notions express different ways in which the representation of an object is related to the cognitive faculty and the different stages of progress in the cognition of an object, then the objects of the divine intellect would not be modally

[41] For a similar account, See R 4298 (Ak. 17:499): "[Possibility] is thought, without being given. [Actuality] is given, without being thought."

determined. This is why it is misguided to focus on explaining why all objects of God's cognition are actual, as though possibility, actuality, and necessity are all applicable notions for God but, due to the special nature of his intellect, God cognizes only actual objects. Kant's position in §76 is not that modal notions retain their meanings and conceptual distinctness in divine cognition and yet are co-extensive with respect to the objects of such cognition, but that modal notions are not valid at all in the case of divine cognition. Thus, the lack of the representation of the merely possible (or necessarily actual) is not in fact a cognitive limitation on the divine intellect. It is not as if God is missing certain modalities (possibility and necessity) among all, but modality as an expression of the limitations of the discursive intellect (i.e., the heterogeneity of its cognitive faculties and the progressive structure of its cognition) does not enter into the composition of such a perfect, simple, and holistic cognition.

9.2.3. *Amodal actuality*

One objection to the idea that the objects of divine cognition would not be modally determined might be that 'actuality' is itself a modal notion. After all, Kant himself states that "if our understanding were intuitive, it would have no objects except what is actual" (*CJ*, Ak. 5:402). First of all, Kant's main point in this statement is that the removal of the heterogeneity between sensibility and understanding results in the removal of the modal distinctions. Second, even if we concede that 'actuality' is meaningful for the divine intellect for the sake of the argument, it cannot have the same sense for the divine intellect as it has for the human intellect. For while actuality is one of the three modes in which the representation of an object can be related to the discursive cognition, it would be, if anything, the only mode of relation between the representation of an object and the divine cognition. But the reason for modalizing objects is precisely that we have multiple heterogeneous cognitive faculties and consequently, there are multiple ways in which our representations can be related to our cognition. If there is only one possible mode of relation, then that relation is not genuinely modal. Thus, a more accurate description of the objects of divine cognition should exclude any talk of modality, since those objects would be represented by the divine intellect as just existing or being without any modal determination. In the more elucidatory part of his remarks in §76, Kant does exactly so:

> For an understanding to which this [thinking and intuiting] distinction did not apply, all objects I cognize would be (exist), and the possibility of some that did not exist, i.e., their contingency if they did exist, as well as the necessity that is to be distinguished from that, would not enter into the representation of such a being. (*CJ*, Ak. 5:403)

Being or existence can here be understood as a case of what Kant calls "absolute positing" (*OPA*, Ak. 2:73, A598/B626). But while for a human subject absolute positing is a mere cognitive act, positing the representation as having an object distinct from that representation, for God absolute positing would be a creative as well as a cognitive act. Interestingly enough, in his earliest presentation of existence as 'absolute positing' (T3) in the *OPA*, Kant refers to God's creation as an act of absolute positing: "If I imagine God uttering His almighty '*Let there be*' over a possible world... He adds no new predicate to it. Rather, He posits the series of things absolutely" (Ak. 2:74). However, while the Kant of 1763 still seems to embrace the Leibnizian idea of God creating one among the many possible worlds, Kant's critical theory of modality redefines modality as valid only for human discursive intellect and thus rules out the idea of God representing things as modally determined. Thus, the Kant of 1790 would understand God's absolute positing as a joint act of intellect and will through which God both represents and creates the whole world at once, without thereby modalizing it.

9.2.4. 'Noumenal amodalism' and noumenal freedom

This brings us to the question of divine freedom in light of Kant's amodalism regarding the objects of divine cognition. In the absence of mere *possibilia*, how can we construe divine creation as free? If the actual world is the only world God represents and creates at once, wouldn't that mean that God does not have freedom to create another possible world but is necessitated to create this, actual one? This is the worry that Kant's position in §76 leads to a Spinozistic or necessitarian conception of the relationship between God and the world. This worry, however, is not well founded. First, as we saw above, Kant is clear that not only mere possibility but also necessity (and thus, contingency) would be missing in the divine picture of the world. Second, this worry about creative freedom is based on the fallacy of looking at the divine from our human standpoint. For only if we neglect the fact that modalizing objects is a specific feature of our limited cognitive make-up and take modality to belong to the texture of mind-independent reality, do we fall for seeing lack of mere *possibilia* as a cognitive and creative constraint. Thus, divine freedom is not to be understood in modal terms, or more specifically, in terms of the availability of alternate (unactualized) possibilities in the mind-independent metaphysical space.

Kant himself defines the kind of freedom, i.e., transcendental or noumenal, that God would possess in non-modal terms, as rational agents' capacity to initiate actions without being determined by any temporally antecedent (sensible) causes, "an absolute causal spontaneity" or self-determination (A446/B474; A803/B381).[42] He emphatically states that this kind of freedom "does not consist in the contingency of an action" or possibility of doing otherwise (*Religion*, Ak. 6:50n). God's will is thus free in taking acts of absolute positing in the sense that it is not determined (to posit or not to posit) by any antecedent conditions. This also means that such a will is neither informed nor bound by any antecedently set modal constraints. Divine volition occurs in an amodal or premodal context of absolute positing.[43] We comprehend free divine volitions or actions as modalized, i.e., as contingent or compatible with alternate possibilities, only due to our discursive mode of representation of positing in comparison with a set of background conditions. The remedy for this fallacy lies in the correct application of the actualist principle (AP) here. Our representation of God's free, original act of absolute positing, which in fact does

[42] On divine freedom being a transcendental kind of freedom, see *Th.Pölitz* (Ak. 28:1067–8).

[43] Kant appears to reiterate Descartes' idea of God's will being the ultimate ground of modal truths here. But that is not really the case, because Kant's position is that modal distinctions don't apply to God's will. So for Kant, it does not make sense to say that God can bring about what is impossible.

not take place in relation to any background conditions and thus is in itself strictly amodal, is what grounds our representation of a space of alternate possibilities (even regarding God's original act itself). This modal space is thus not the ground of God's original act, but only a function of our retrospective representation of this act through our discursive intellects.

Admittedly, the case for human freedom is far more complicated. Yet it can still be seen that the non-modal conception of transcendental freedom should also offer a framework of a remedy for the problematic implication of Kant's metaphsyical thesis concerning human freedom. Stang has offered a formulation of this problem in terms of an apparent inconsistency between Kant's "noumenal amodalism" and his theory of freedom, naming it, rather strongly, "the antinomy of Kant's modal metaphysics."[44] Stang argues that given his sustained commitment to the principle of 'ought implies can' from the *CPR* on, Kant's theory of freedom entails that our noumenal selves have modal properties. Accordingly, if the moral law is indeed binding for me such that I *ought* to refrain from doing something wrong, then I can refrain from doing it or if I actually didn't, I could have. In the *CPrR*, Kant seems to tie this possibility of doing otherwise to the transcendental freedom of the subject considered as a noumenal being: "So considered, a rational being can now rightly say of every unlawful action he performed that he could have omitted it even though as appearance it is sufficiently determined in the past, and so far, is inevitably necessary" (Ak. 5:98). From this connection, Stang infers that Kant's theory of freedom "*rests* on attributing a modal property to my noumenal will: as a noumenon, I could have determined my will otherwise, so my actions are contingent."[45]

However, this "antinomy" need not arise in the first place. First of all, even if Kant holds that humans do have the capacity to do otherwise than they actually do, he does not take this to be the same as or essential to transcendental freedom, which he formulates rather non-modally as the absolute spontaneity of the noumenal will or its self-determination.[46] It is not the former, but the latter which is fundamental to Kant's conception of human freedom. For the real ground of the application of the moral law to us and thus of what it implies, our capacity to do otherwise, is our transcendental freedom. Second, it is important to see that in the above quote from the *CPrR* (Ak. 5:98) Kant's principle of 'ought implies can' attributes contingency to the experienced, actual moral failure or success (unlawful or lawful action) of the moral agent, and thus to the phenomenal effect of the free noumenal will, and not to the free noumenal will that is the ground of that effect. The phenomenal effect or the empirical action in question is necessary in relation to empirical causal grounds and yet contingent in relation to its noumenal ground. Yet, the freedom of the noumenal ground of action cannot consist in the contingency of the empirical action, but in its capacity to determine itself to initiate a causal chain without being determined by any antecedent causes.

Of course, one might still ask: doesn't my capacity or possibility to act otherwise entail that my noumenal will has the modal property that it can determine itself

[44] Stang (2016), 298. [45] Stang (2016), 298. [46] Pereboom (2006), 542.

otherwise than it does? Isn't the presupposition of the latter, the only way to make sense of the former? Kant's account of 'radical evil' in the *Religion* further motivates this line of questions. For the idea of an agent's freely choosing between subordinating the moral law to self-love and subordinating self-love to the moral law, between evil and good, seems to assign a modal property to her noumenal will: its radical choice is contingent and thus it can determine itself otherwise than it actually does (for better or worse) (Ak. 6:25, 36). Kant's talk of the possibility of moral reform (or of overcoming radical evil) also seems to commit him to the contingency of one's noumenal character (Ak. 6:29, 37, 40, 43).[47]

However, it should still be kept in mind that only if the radical choice is independent of any antecedent determining causes, can it be genuinely free. This strongly suggests, again, that the contingency in question has non-modal grounds, and that the noumenal will is not modally determined in itself. The issue here is that the only way we can represent absolute spontaneity from a normative standpoint is to modalize it. For conceiving ourselves as morally obligated to will in one way or another requires us to represent our failure or success to will so as contingent. More specifically, we retrospectively modalize the actual (good or evil) volition of an absolutely spontaneous will as a contingent 'choice' between alternatives once we compare them with its non-actual volitions in the abstract. Such comparison and abstraction, as prerequisites of modalization, are merely features of the operation of a discursive mind. In fact, there are no relevant background conditions that our free noumenal volitions can be related to (compatible with or entailed by). Thus, contingency is not a property of our noumenal volitions themselves but of our representations of them. The actualist principle should be at work here again as in the case of our modalization of free divine volitions. The alternate possibilities that we represent with respect to a certain noumenal volition of a human agent are grounded on our representation of the (amodal) actuality or the positing of her noumenal volition. But the noumenal volition itself is neither grounded on nor constrained by those represented possibilities. Therefore, both the fact that we represent the moral law as an 'ought' or an imperative and the fact that we represent our freedom as a 'can,' a capacity to do or will otherwise are due to our own subjective, sensibly conditioned, and discursive human viewpoint.[48] Beyond this particular subjective viewpoint, neither 'ought' nor 'can' or modality in general is valid.

[47] In fact, Kant's doctrine of moral reform in the *Religion* leads to a parallel tension between his metaphysical commitment to the atemporality of noumena and our inevitably temporal representation of reform or change in general. I am thankful to Ben Randolph for bringing this point to my attention.

[48] Stang (2016), 316 rightly recognizes that the representation of freedom in modal terms "(e.g., the possibility of acting otherwise than they do) is due to us." However, the resolution he offers for his "antinomy" essentially consists in replacing the "possibility of acting otherwise" with "power to act (or choose) otherwise." This seems to me to be reintroducing modality into noumenal volition under the disguise of not so apparently modal sounding language. I suggest here, however, that any reference to an alternative to or a state 'otherwise' than the actual noumenal volition itself, whether expressed in an obviously modal language or in disguise, is due to our modalization of an original act of absolute positing, which itself takes place prior to and thus in fact grounds our representation of any alternatives to itself.

Bibliography and Key to Abbreviations

References to primary historical sources in these texts are either by complete title or by abbreviation or short title, keyed to this bibliography. Works cited by abbreviation are cited by page number unless otherwise noted below. I provide here a full list of complete titles paired with abbreviations or short titles, when applicable, and the English translations from which I quoted. Where no published English translation is listed for a work in German or Latin, either translations are my own or the work is cited but not quoted.

Works of Kant

All citations from Kant's works use the pagination of the Akademie edition (Ak.), *Kants gesammelte Schriften* (Berlin: Walter de Gruyter, 1902–), except those to the *Critique of Pure Reason*, which use the standard (A/B) format to refer to the page numbers of the first edition of 1781 (A) and the second edition of 1787 (B), respectively.

Ak. (1902–). *Kants gesammelte Schriften* (vols. 1–29). Hrsg. (formerly Königlichen Preussichen) Akademie der Wissenschaften. Berlin: Walter de Gruyter.

CJ (2000 [1790]). *Critique of the Power of Judgment.* Ak. 5: 165–485. Paul Guyer and Eric Matthews (trans. and ed.). Cambridge: Cambridge University Press.

Corr. (1999). *Correspondence.* Ak. 10–13. Arnulf Zweig (trans. and ed.). Cambridge: Cambridge University Press.

CPrR *Critique of Practical Reason* [1788]. Ak. 5: 1–164. In Kant. (1996a). *Practical Philosophy*, Marry J. Gregor (trans. and ed.). Cambridge: Cambridge University Press, 137–271.

CPR (1998 [1781/1787]). *Critique of Pure Reason*, Paul Guyer and Allen W. Wood (trans. and ed.). Cambridge: Cambridge University Press.

Dissertation *On the Form and Principles of the Sensible and Intelligible World* [1770]. Ak. 2:385–420. In Kant (1992a). *Theoretical Philosophy 1755–1770*, David Walford and Ralf Meerbote (trans. ad ed.). Cambridge and New York: Cambridge University Press, 107-201373-416.

Dreams *Dreams of a Spirit-seer elucidated by Dreams of Metaphysics* [1766]. Ak. 2: 315–73. In Kant (1992a), 301–59.

Inquiry *Inquiry Concerning the Distinctness of the Priniciples of Natural Theology and Morality* [1764]. Ak. 2: 165–204. In Kant (1992a), 243–75.

JL *The Jäsche Logic* [1800]. Ak. 9: 1–150. In Kant (1992b). *Lectures on Logic*, Michael Young (trans. and ed.). Cambridge: Cambridge University Press, 521–640.

LB *The Blomberg Logic* [1771?]. Ak. 24: 7–301. In Kant (1992b), 1–246.

LDW *The Dohna-Wundlacken Logic* [1792]. Ak. 24: 697–784. In Kant (1992b), 425–516.

LV *The Vienna Logic* [1780–82]. Ak. 24: 787–940. In Kant (1992b), pp. 249–377.

MK2 *Metaphysik K2* [early 1790s]. Ak. 28: 705–816. Selections in Kant (1997). *Lectures on Metaphysics*, Karl Ameriks and Steve Naragon (trans. and ed.). Cambridge: Cambridge University Press, 393–413.

MK3 *Metaphysik* Vigilantius (K3) [1794–5]. Ak. 29: 943–1040. In Kant (1997), 415–506.

MPölitzL1 *Metaphysik L1 (Pölitz)* [1777–80]. Ak. 28:185–91, 195–350.

ML2 *Metaphysik L2* [1790–91?] Ak. 28: 531–94. In Kant (1997), 297–354.

MM *Metaphysik Mrongovius* [1782/3]. Ak. 29: 747–940. In Kant (1997), 107–286.

NE *New Elucidation of the First Principles of Metaphysical Cognition* [1755]. Ak. 1: 385–416. In Kant (1992a), 1–45.

NM *An Attempt to Introduce the Concept of Negative Magnitudes into Philosophy* [1763]. Ak. 2:165–204. In Kant (1992a), 203–241.

OPA *The Only Possible Argument in Support of a Demonstration of the Existence of God* [1763]. Ak. 2: 63–164. In Kant. (1992a), 107–201.

Optimism *An attempt at some reflections on Optimism* [1759]. Ak. 2:27–35. In Kant (1992a), 67–76.

Prolegomena (2004 [1791/1804]). *Prolegomena to Any Future Metaphysics*. Ak. 4: 255–383. Gary Hatfield (trans. and ed.). Cambridge: Cambridge University Press.

Th.Danzig *Danzig Rational Theology According to Baumbach* [1783/4]. Ak. 28: 1230–319. In Eberhard and Kant. (2016). *Preparation for Natural Theology*, Courtney D. Fugate and John Hymers (trans. and eds.). London: Bloomsbury, 131–218.

Th.Pölitz *Lectures on the Philosophical Doctrine of Religion* [1783-6?]. Ak. 28: 989–1126. In Kant (1996b). *Religion and Rational Theology*, Allen Wood and George Di Giovanni (trans. and ed.). Cambridge: Cambridge University Press, 335–451.

Progress *What Real Progress has Metaphysics Made in Germany since the Time of Leibniz and Wolff?* [1791/1804]. Ak. 20: 253–332. In Kant (2002). *Theoretical Philosophy after 1781*, Henry Allison and Peter Heath (trans. and ed.). Cambridge: Cambridge University Press, 337–424.

Religion *Religion within the Boundaries of Mere Reason* [1793]. Ak. 6: 1–202. In Kant (1996b), 55–215.

R *Kants handschriftlicher Nachlaß* [*Reflexionen*], Ak. 14–23. Cited by four-digit number and volume and page number. I have also consulted the English translations of selected reflections in Kant (2005). *Notes and Fragments*, Paul Guyer (ed.), Curtis Bowman, Paul Guyer, Frederick Rauscher (trans.). Cambridge: Cambridge University Press.

(1992). *Reflexionen Kants zur kritischen Philosophie*, Aus Kants handschriftlichen Aufzeichnungen hrsg. von Benno Erdmann, (reprint) Norbert Hinske, Fromman-Holzboog (eds.). Stuttgart: Bad Cannstatt.

Works of Baumgarten

All citations from Baumgarten are by title or short title, followed by the appropriate paragraph number (e.g., *Metaphysica* §226).

Logica (1773). *Acroasis Logica* (2nd Edition), Johann Gottlieb Toellner (ed.). Halle Magdeburg: C. H. Hemmerde.

Metaphysica (1757). *Metaphysica* (4th edition). Halle. In Ak. 17.

(2004). *Alexander Gottlieb Baumgartens Metaphysik*, George Friedrich Meier (trans.). Jena: Dietrich Scheglmann Reprints.

(2013). *Metaphysics: A Critical Translation with Kant's Elucidations, Selected Notes and Related Materials*, Courtney Fugate and John Hymers (trans. and ed.). London: Bloomsbury.

Works of Descartes

All citations from Descartes are by abbreviation, followed by the appropriate volume and page number.

AT (1964–76). Oeuvres (vols. 1–12), C. Adam and P Tannery (eds.). Paris:Vrin/ CNRS.

CSM/CSMK (1984–1991). *The Philosophical Writings of Descartes* (vols. 1–3), John Cottingham, Robert Stoothoff, Dugald Murdoch (trans. and ed.). Cambridge: Cambridge University Press. Cited by volume and page number. Volume 3 includes Anthony Kenny as a translator, and is abbreviated here as CSMK.

Works of Leibniz

A (1923–). *Sämtliche Schriften und Briefe*. Darmstadt and Berlin: Berlin Academy.

AG (1989). *Philosophical Essays*, Roger Ariew and Daniel Garber (trans. and ed.). Indianapolis: Hackett.

C (1966 [1903]). *Opuscules et fragments inédits de Leibniz*, Louis Couturat (ed.). Paris: Félix Alcan. Reprint, Hildesheim: Georg Olms.

G (1978 [1875-90]). Die *Philosophischen Schriften* von Gottfried Wilhelm Leibniz, C. I. Gerhardt (ed.). Berlin: Weidemann. Reprint, Hildesheim: Georg Olms.

Gr (1948). *Textes inédits*, Gaston Grua (ed.). Paris: Presses Universitaires de France.

L (1969). *Philosophical Papers and Letters*, 2nd ed. L. Loemker (ed.). Dordrecht: D. Reidel.

New Essays (1996). *New Essays on Human Understanding*, Peter Remnant & Jonathan Bennett (trans. and ed.). Cambridge: Cambridge University Press.

Theodicy (1985). *Theodicy: Essays on the Goodness of God, the Freedom of Man and the Origin of Evil*, E. M. Huggard (trans.). La Salle: Open Court. Cited by paragraph number.

Works of Wolff

All citations from Wolff are by abbreviation or short title, followed by the appropriate paragraph number.

(1965–). *Gesammelte Werke*, J. École, H. W. Arndt, Ch. A. Corr, J. E. Hoffmann, and M. Thomann (eds). Hildesheim, Zurich, and New York: Georg Olms.

Cosmologia Wolff, Christian. (2009 [1737]). *Cosmologia generalis, method scientifica pertractata*. J. École (ed.). Hildesheim: Georg Olms.

DM (1983b [1740]).*Vernünftige Gedanken von Gott, der Welt und der Seele des Menschen, auch allen Dingen überhaupt*, Charles A. Corr (ed.). Hildesheim: Georg Olms Verlag.

Logica Wolff, Christian. (1983a [1740]). *Philosophia rationalis sive Logica*, J. École (ed.). Hildesheim: Georg Olms.

NT I (1978 [1738]) *Theologiae Naturalis, Pars I*, J. École (ed.). Hildesheim: Georg Olms Verlag.

NT II (1981 [1738]) *Theologiae Naturalis, Pars II*, J. École (ed.). Hildesheim: Georg Olms Verlag.

Ontologia (1977 [1736]). *Philosophia prima, sive ontologia*, J. École (ed.). Hildesheim: Georg Olms Verlag.

Other Primary Sources

Danish nama Avicenna (Ibn Sina). (1973). *The Metaphysica of Avicenna*, Parviz Morewedge (trans. and commentary). New York: Columbia University Press.

Disputationes Suárez, Francis. (1983). *On the Essence of Finite Being As Such On the Existence of That Essence and Their Distinction [Disputationes Metaphysicae:*

Discussion XXXI], Norman J. Wells (trans.). Milwaukee: Marquette University Press.

Entwurf — Crusius, Christian August. (1964 [1766]). *Entwurf der Nothwendigen Vernunft-Wahrheiten.* In G. Tonelli (ed.). *Die philosophischen Hauptwerke.* Hildesheim: Georg Olms. Cited by paragraph number.

QuodQs. — Duns Scotus, John. (1975). *God and Creatures: The Quodlibetal Questions,* trans. Felix Alluntis and Allan B. Wolter (trans. and ed.). Princeton and London: Princeton University Press, 1975.

Shifa — Avicenna (Ibn Sina). (2005). *The Metaphysics of The Healing (al-Ilahiyat min al Shifa),* Michael E. Marmura (trans.). Provo UT: Brigham Young University Press.

Vernunftlehre — Meier, Georg Friedrich. (1752). *Auszug aus der Vernunftlehre,* Johan Justinus Gebauer (ed.). Halle. Cited by paragraph number.

Anselm. (2007). *Proslogion.* In Thomas Williams (trans. and ed.). *Basic Writings.* Indianapolis: Hackett.

Aquinas, Thomas. (1981). *Summa Theologica.* In Anton C. Pegis (ed.). *Basic Writings of St. Thomas Aquinas.* New York: Random House. Cited by part, question, and article.

Aristotle. (1984). *The Complete Works of Aristotle,* The Revised Oxford Translation, Jonathan Barnes (ed.). Princeton: Princeton University Press.

Hume, David. (1980). *Dialogues Concerning Natural Religion.* Indianapolis: Hackett.

Lambert, Johann Heinrich. (1965 [1764]). *Neues Organon,* Hans Werder Arndt (ed.). Hildesheim: Georg Olms Verlag. Cited by paragraph number.

Secondary Sources

Abacı, Uygar. (2008). 'Kant's Theses on Existence', *British Journal for the History of Philosophy,* 16(3): 559–93.

Abacı, Uygar. (2013). 'Leibniz and Kant on the Syntheticity of Existential Statements'. In Stefano Bacin, Alfredo Ferrarin, Claudio La Rocca, and Margi Ruffing (eds.). *Kant und die Philosophie in weltbürgerlicher Absicht. Akten des XI. Kant-Kongresses.* Berlin: de Gruyter, vol. 5, 297–308.

Abacı, Uygar. (2014). 'Kant's Only Possible Argument and Chignell's Real Harmony', *Kantian Review,* 19(1): 1–25.

Abacı, Uygar. (2016). 'The Coextensiveness Thesis and Kant's Modal Agnosticism in the "Postulates"', *European Journal of Philosophy,* 24(1): 129–58.

Abacı, Uygar. (2017). 'Kant, the Actualist Principle, and the Fate of the Only Possible Proof', *Journal of the History of Philosophy,* 55(2): 261–91.

Abacı, Uygar. (2018). What does Existence add to Possibility? Wolff and Kant on Complementum Possibilitatis. In Waibel Violetta and Ruffing Margit (eds.). *Nature and Freedom: Proceedings of the 12. International Kant Congress.* Berlin: De Gruyter.

Abacı, Uygar, (forthcoming). 'Kant's Enigmatic Transition: Practical Cognition of the Supersensible'. In Nicholas Stang and Karl Schafer (eds.). *On the Sensible and Intelligible Worlds: New Essays on Kant's Metaphysics and Epistemology.* Oxford University Press.

Adams, Robert M. (1994). *Leibniz: Determinist, Theist, Idealist.* New York: Oxford University Press.

Adams, Robert M. (2000). 'God, Possibility and Kant', *Faith and Philosophy,* 17(4): 425–40.

Allison, Henry E. (1983). *Kant's Transcendental Idealism: An Interpretation and Defense.* New Haven and London: Yale University Press.

Allison, Henry E. (2004). *Kant's Transcendental Idealism*, revised and enlarged ed. New Haven and London: Yale University Press.
Ameriks, Karl. (1992). 'The Critique of Metaphysics: Kant and Traditional Ontology'. In Paul Guyer (ed.). *The Cambridge Companion to Kant*. Cambridge: Cambridge University Press, 249–79.
Barnes, Jonathan. (1972). *The Ontological Argument*. London: Macmillan Press.
Beck, Lewis White. (1969). *Early German Philosophy: Kant and His Predecessors*. Cambridge, MA: Harvard University Press.
Bennett, Jonathan. (1966). *Kant's Analytic*. Cambridge: Cambridge University Press.
Bennett, Jonathan. (1974). *Kant's Dialectic*. Cambridge: Cambridge University Press.
Bissinger, Anton. (1970). *Die Struktur der Gotteserkenntnis:Studien zur Philosophie Christian Wolffs*. Bonn: H. Bouvier und Co. Verlag.
Bissinger, Anton. (1975). 'Das ontologische Argument bei Christian Wolff'. In Helmut Kohlenberger (ed.). *Analecta Anselmiana: Untersuchungen über Person und Werk Anselms von Canterbury*. Frankfurt/Main: Minerva, 243–7.
Blackwell, Richard J. (1963). 'Translator's Introduction'. In Wolff, Christian: *Preliminary Discourse on Philosophy In General*, Richard J. Blackwell (trans.). New York: Bobbs-Merrill.
Blecher, Ian. (2013). 'Kanton Formal Modality', *Kant-Studien*, 104(1):44–62.
Blumenfeld, David. (1973). 'Leibniz's Theory of Striving Possibles', *Studia Leibnitiana*, 5(2): 163–77.
Blumenfeld, David. (1995). 'Leibniz's ontological and cosmological arguments'. In Nicholas Jolley (ed.). *The Cambridge Companion to Leibniz*. Cambridge: Cambridge University Press, 353–81.
Boehm, Omri. (2012). 'Kant's Regulative Spinozism', *Kant-Studien* 10(3): 292–317.
Boehm, Omri, (2014). *Kant's Critique of Spinozism*. New York: Oxford University Press.
Brandt, Reinhard. (1995). *The Table of Judgments: Critique of Pure Reason A67–76; B92–101*, Eric Watkins (trans. and ed.). Atascadero: Ridgeview Press.
Burns, John V. (1966). *Dynamism in the Cosmology of Christian Wolff: A Study in Pre-Critical Rationalism*. New York: Exposition Press.
Buroker, Jill Vance. (2006). *Kant's Critique of Pure Reason: An Introduction*. Cambridge: Cambridge University Press.
Byrne, Peter. (2007). *Kant on God*. Burlington: Ashgate Pub. Co.
Campbell, Richard. (1976). *From Belief to Understanding—A Study of Anselm's Proslogion Argument on the Existence of God*. Canberra: Australian National University Press.
Camposampiero, Matteo Favaretti. (2012). 'Dividing Fiction from Reality: Existence and Nature in Christian Wolff's Metaphysics'. In Matteo Favaretti Camposampiero and Matteo Plebani (eds.). *Existence and Nature: New Perspectives*. Piscataway, Ontos Verlag: 65–98.
Cassirer, H. W. (1954). *Kant's First Critique: An Appraisal of the Permanent Significance of Kant's Critique of Pure Reason*. London: MacMillan.
Chignell, Andrew. (2009). 'Kant, Modality, and the Most Real Being', *Archiv für Geschichte der Philosophie*, 91(2): 157–92.
Chignell, Andrew. (2012). 'Kant, Real Possibility and the Threat of Spinoza', *Mind*, 121(483): 635–75.
Chignell, Andrew. (2014). 'Kant and the 'Monstrous' Ground of Possibility', *Kantian Review*, 19(1): 53–69.
Corr, Charles A. (1973). 'The Existence of God, Natural Theology, and Christian Wolff', *International Journal for Philosophy of Religion*, 4(2): 105–18.
Corr, Charles A. (1975). 'Christian Wolff and Leibniz', *Journal of the History of Ideas*, 36(2): 241–62.
Couturat, Louis. (1972). On Leibniz's Metaphysics'. In Harry Frankfurt (ed.). *Leibniz: A Collection of Critical Essays*. New York: University of Notre Dame Press, 19–45.

Cunning, David. 'Descartes' Modal Metaphysics'. In Edward N. Zalta (ed.). *The Stanford Encyclopedia of Philosophy*, Spring 2014 Edition. https://plato.stanford.edu/archives/spr2014/entries/descartes-modal/.

Curley, Edmund. (1972). 'The Root of Contingency'. In Harry Frankfurt (ed.). *Leibniz: A Collection of Critical Essays*. New York: University of Notre Dame Press, 69–97.

Dell'Oro, Regina O. M. (1994). *From Existence to the Ideal: Continuity and Development in Kant's Theology*. New York: P. Lang.

Dicker, Georges. (2004). *Kant's Theory of Knowledge: An Analytical Introduction*. New York: Oxford University Press.

Dyck, Cory W. (2014). *Kant and Rational Psychology*. New York: Oxford University Press.

England, Frederick Ernest. (1930). *Kant's Conception of God*. New York: Dial Press.

Everitt, Nicholas. (2004). *The Non-Existence Of God*. London and New York: Routledge.

Fisher, Mark and Watkins, Eric. (1998). 'Kant on the Material Ground of Possibility: From *The Only Possible Argument* to *The Critique of Pure Reason*', *Review of Metaphysics*, 52(2): 369–95.

Forgie, Wiliam. (2000). 'Kant and Frege: Existence as a Second-Level Property', *Kant-Studien*, 91(2): 165–77.

Forgie, Wiliam. (2007). 'Gassendi and Kant on Existence', *Journal of the History of Philosophy*, 45(4): 511–23.

Forgie, Wiliam. (2008). 'How Is the Question 'Is Existence a Predicate?' Relevant to the Ontological Argument?', *International Journal for Philosophy of Religion*, 64(3): 117–33.

Frege, Friedrich Ludwig Gottlob (1979). *Posthumous Writings*, Hans Hermes, Friedrich Kambartel, Friedrich Kaulbach (eds). Chicago: The University of Chicago Press.

Gardner, Sebastian. (1999). *Kant and the Critique of Pure Reason*. London: Routledge.

Gilson, Etienne. (1963). *Modern Philosophy: Descartes to Kant*. New York: Random House.

Gilson, Etienne. (2005 [1949]). *Being and Some Philosophers*, 2nd edition. Toronto: PIMS.

Goichon, A. M., (1969). *The Philosophy of Avicenna and its Influence on Medieval Europe*, M. S. Khan (trans.). Delhi: Motilat Banarsidass.

Goodman, Lenn. (1996). *God of Abraham*. New York: Oxford University Press.

Grier, Michelle. (2001). *Kant's Doctrine of Transcendental Illusion*. Cambridge: Cambridge University Press.

Grünewald, Bernward. (1986). *Modalität und empirisches Denken: eine kritische Auseinandersetzung mit der Kantischen Modaltheorie*. Hamburg: F. Meiner.

Guyer, Paul. (1987). *Kant and the Claims of Knowledge*. New York: Cambridge University Press.

Guyer, Paul. (1998). 'The Postulates of Empirical Thinking in General and the Refutation of Ideralism'. In Georg Mohr and Marcus Willaschek (eds.). *Immanuel Kant: Kritik der reinen Vernunft*. Berlin: Akademie Verlag, 297–324.

Guyer, Paul. (2000). *Kant on Freedom, Law, and Happiness*, (New York: Cambridge University Press, 2000).

Guyer, Paul. (2006). *Kant*. New York: Routledge.

Hanna, Robert. (2001). *Kant and the Foundations of Analytic Philosophy*. Oxford: Clarendon Press.

Hanna, Robert. 'Kant's Theory of Judgment'. In Edward N. Zalta. *The Stanford Encyclopedia of Philosophy*, Winter 2016 Edition. https://plato.stanford.edu/archives/win2016/entries/kant-judgment/.

Hartnack, J. (1967). *Kant's Theory of Knowledge*, M. Holmes Hartshorne (trans.). New York: Harcourt.

Hartshorne, Charles. (1965). *Anselm's Discovery: A Re-Examination of the Ontological Proof for God's Existence*. La Salle: Open Court.

Hartshorne, Charles. (1966). 'Is the Denial of Existence Ever Contradictory', *The Journal of Philosophy*, 63(4): 85–93.
Hebbeler, James. (2015). 'Kant on Necessity, Insight, and A Priori Knowledge', *Archiv für Geschichte der Philosophie*, 97(1): 34–65.
Heidegger, Martin. (1982). *The Basic Problems of Phenomenology*, Albert Hofstadter (trans.). Bloomington and Indianapolis: Indiana University Press.
Henrich, Dieter. (1960). *Der Ontologische Gottesbeweis: Sein Problem und seine Geschichte in der Neuzeit*. Tübingen: J. C. B. Mohr.
Hettche, Matt. 'Christian Wolff'. In Edward N. Zalta (ed.). *The Stanford Encyclopedia of Philosophy*, Spring 2014 Edition. http://plato.stanford.edu/archives/spr2014/entries/wolff-christian/.
Hindrichs, Gunnar. (2008). *Das Absolute und das Subject: Untersuchungen zum Verhältnis von Metaphysik und Nachmetaphysik*. Frankfurt am Main: Vittorio Klostermann.
Honnefelder, Ludger. (1990). *Scientia transcendens: Die formale Bestimmung der Seindheit und Realität in der Metaphysik des Mittelalters und der Neuzeit (Duns Scotus– Suárez–Wolff-Kant-Peirce)*. Hamburg: Felix Meiner Verlag.
Kahn, Charles. (1976). 'Why Existence does not Emerge as a Distinct Concept in Greek Philosophy', *Archiv für Geschichte der Philosophie*, 58(4): 323–34.
Kahn, Charles. (2003). *The Verb 'Be' in Ancient Greek*. Indianapolis: Hackett.
Kannisto, Toni. (2012). *From Thinking to Being: Kant's Modal Critique of Metaphysics* (PhD Thesis). University of Oslo.
Kannisto, Toni (2016). 'Positio Contra Complementum Possibilitatis—Kant and Baumgarten on Existence', *Kant-Studien*, 107(2): 291–313.
Kannisto, Toni. (2018). 'Kant and Frege on Existence', *Synthese*, 195(8): 3407–3432.
Kohl, Markus. (2015). 'Kant on the Inapplicability of the Categories to Things in Themselves', *British Journal for the History of Philosophy*, 23(1): 90–114.
Korte, Tapio and Maunu, Ari and Aho, Tuomo. (2009). "Modal Logic from Kant to Possible Worlds Semantics". In Leila Haaparanta (ed.). *The Development of Modern Logic*. Oxford: Oxford University Press
La Croix, Richard. (1972). *Proslogion II and III: A Third Interpretation of Anselm's Argument*. Leiden: E. J. Brill.
Laberge, Pierre. (1973). *La théologie kantienne précritique*. Ottawa: Editions de l'Université d'Ottawa.
Lamacchia, Ada. (1969). *La Filosofia Della Religione in Kant*. Manduria: Lacaita.
Leech, Jessica. (2010). 'Kant's Modalities of Judgment', *European Journal of Philosophy*, 18(1): 1–25.
Leech, Jessica. (2014). 'Making Modal Distinctions: Kant on the Possible, the Actual, and the Intuitive Understanding', *Kantian Review*, 19(3): 339–65.
Logan, Ian. (2007). 'Whatever Happened to Kant's Ontological Argument?', *Philosophy and Phenomenological Research*, 74(2): 346–63.
Longuenesse, Béatrice, (1998). *Kant and the Capacity to Judge* (Princeton: Princeton UP, 1998).
Longuenesse, Béatrice, (2005). *Kant on the Human Standpoint*. New York: Cambridge University Press.
Look, Brandon C. 'Leibniz's Modal Metaphysics'. In Edward N. Zalta (ed.). *The Stanford Encyclopedia of Philosophy*, Spring 2013 Edition. https://plato.stanford.edu/archives/spr2013/entries/leibniz-modal/.
Lord, Beth. (2011). *Kant and Spinozism: Transcendental Idealism and Immanence from Jacobi to Deleuze*. Houndmills: Palgrave Macmillan.

Lovejoy, Arthur. (1907). 'Kant's Classification of the Forms of Judgment', *Philosophical Review*, 16(6): 588–603.

Lovejoy, Arthur. (1936). *The Great Chain of Being*. Cambridge MA: Harvard University Press.

Malcolm, Norman. (1960). 'Anselm's Ontological Arguments', *The Philosophical Review*, 69(1): 41–62.

Marshall, Colin. (2018). 'Never Mind the Intuitive Intellect: Applying Kant's Categories to Noumena', *Kantian Review*, 23(1): 24–40.

Mates, Benson. (1972). 'Individuals and Modality in the Philosophy of Leibniz', *Studia Leibnitiana*, 4(2): 81–118.

Mattey, G. J. (1986). 'Kant's Theory of Propositional Attitudes', *Kant-Studien*, 77(1–4): 423–40.

Mensch, Jennifer. (2007). 'The Key to All Metaphysics: Kant's Letter to Herz, 1772', *Kantian Review*, 12(2): 109–27.

Michaelis, Anneliese. (1937). *Der Ontologische Sinn des Complementum Possibilitatis bei Christian Wolff*. Berlin: Kalbfleisch.

Morewedge, Parviz. (1973). *The Metaphysica of Avicenna (Ibn Sina)*. New York: Columbia University Press.

Motta, Giuseppe. (2007). *Kants Philosophie der Notwendigkeit*. Berlin: Peter Lang.

Nachtomy, Ohad. (2012). 'Leibniz and Kant on Possibility and Existence', *British Journal for the History of Philosophy*, 20(5): 953–72.

Newlands, Samuel. (2013). 'Leibniz and the Ground of Possibility', *The Philosophical Review*, 122(2): 157–87.

Pape, I. (1966). *Tradition und Transformation der Modalität. Band I: Möglichkeit-Unmöglichkeit*. Hamburg: Felix Meiner.

Pasternack, Lawrence. (forthcoming). 'Kant's Four-fold Critique of the Ontological Argument: Conceptual Containment, Predication, and the Portents of Free Logic'. In Graham Oppy (ed.). *The Ontological Argument*. Cambridge: Cambridge University Press.

Paton, H. J. (1936). *Kant's Metaphysic of Experience*. London: Georg Allen and Unwin.

Patzig, Günther.(1976). "Immanuel Kant: Wie sind synthetische Urteile a priori möglich?" In J. G. Speck (ed.). *Grundprobleme der großen Philosophen: Philosophie der Neuzeit II*. Göttingen: Vandenhoeck & Ruprecht.

Pereboom, Derk. (2006). 'Kant on Transcendental Freedom', *Philosophy and Phenomenological Research*, 73(3):537-567.

Pichler, Hans. (1910). *Über Christian Wolff's Ontologie*. Leipzig: Durr'schen Buchhandlung.

Pinder, Tillmann. (1969). *Kants Gedanke vom Grund aller Möglichkeit: Untersuchungen zur Vorgeschichte der "transzendentalen Theologie"*, Berlin: Freien Universität Berlin.

Plantinga, Alvin. (1966). 'Kant's Objection to the Ontological Argument', *The Journal of Philosophy*, 63(19): 537–46.

Plantinga, Alvin. (1967). *God and Other Minds*. Ithaca: Cornell University Press.

Plantinga, Alvin. (1974). *The Nature of Necessity*. Oxford: Clarendon Press

Poser, Hans. (1983). 'Mögliche Erkenntnis und Erkenntnis der Möglichkeit. Die Transformation der Modalkategorien der Wolffschen Schule in Kants kritischer Philosophie', *Grazer Philosophische Studien*, 20: 129–47.

Poser, Hans. (1989). 'Kants absolute Modalitäten'. In Giovanni Corsi, Corrado Magnione, and Massimo Mugnai (eds.). *Atti del Convegno Internazionale di Storia della Logica: Le Teorie delle Modalita*. Bologna: CLUEB.

Prichard, H. A. (1909). *Kant's Theory of Knowledge*. Oxford: Oxford Clarendon Press.

Proops, Ian. (2015). 'Kant on the Ontological Argument', *Nous*, 49(1): 1–27.

Reich, Klaus. (1992). *The Completeness of Kant's Table of Judgments*. Stanford: Stanford University Press.

Rescher, Nicholas. (1967). *The Philosophy of Leibniz*. New Jersey: Prentice Hall.

Reuscher, John. (1977). 'A Clarification and Critique of Kant's *Principiorum Primorum Cognitionis Metaphysicae Nova Dilucidatio*', *Kant-Studien*, 68(1–4): 18–32.

Rohlf, Michael. (2010). 'The Ideas of Pure Reason'. In Paul Guyer (ed.). *The Cambridge Companion to Kant's Critique of Pure Reason*. New York: Cambridge University Press, 190–209.

Rosefeldt, Tobias. (2011). *Frege, Pünjer, and Kant on Existence, Grazer Philosophische Studien*, 82(1): 329–51.

Rosenkoetter, Timothy. (2010). 'Absolute Positing, the Frege Anticipation Thesis, and Kant's Definitions of Judgment', *European Journal of Philosophy*, 18(4): 539–66.

Rosenkoetter, Timothy. (2013). 'A Non-embarrassing Account of the Modal Functions of Judgment'. In Stefano Bacin, Alfredo Ferrarin, Claudio La Rocca, and Margi Ruffing (eds.). *Kant und die Philosophie in weltbürgerlicher Absicht. Akten des XI. Kant-Kongresses*. Berlin: de Gruyter, 383–442.

Russell, Bertrand. (1937). *The Critical Exposition of the Philosophy of Leibniz, with an Appendix of Leading Passages*, 2nd edition. London: Allen & Unwin.

Sala, Giovanni. (1990). *Kant und die Frage nach Gott: Gottesbeweise und Gottesbeweiskritik in den Schriften Kants*. Berlin and New York: Walter de Gruyter.

Schaffer, Jerome. (1962). 'Existence, Predication, and the Ontological Argument', *Mind*, 71(283): 307–25.

Schmucker, Joseph. (1972). 'On the Development of Kant's Transcendental Theology'. In Lewis White Beck (ed.). *Proceedings of the Third International Kant Congress*. Dordrecht: D. Reidel Publishing Co., 495–500.

Schmucker, Joseph. (1980). *Die Ontotheologie des vorkritischen Kant*. Berlin and New York: Walter de Gruyter.

Schneeberger, Guido. (1952). *Kants Konzeption der Modalbegriffe*. Basel: Verlag für Recht und Gesselschaft.

Schönfeld, Martin. (2000). *The Philosophy of the Young Kant: The Precritical Project*. Oxford: Oxford University Press.

Seigfried, Hans. (1972). 'Kant's Thesis about Being Anticipated by Suárez?'. In Lewis White Beck (ed.). *Proceedings of the Third International Kant Congress*. Dordrecht: D. Reidel Publishing Co., 510–20.

Shabel, Lisa. (2004). 'Kant's "Argument from Geometry"', *Journal of the History of Philosophy*, 42(2): 195–215.

Shields, Christopher. (1986). 'Leibniz's Doctrine of the Striving Possibles', *Journal of the History of Philosophy*, 24(3): 343–57.

Smith, A. D. (2014). *Anselm's Other Argument*. Cambridge MA: Harvard University Press.

Smith, Norman Kemp. (2003 [1918]). *A Commentary to Kant's Critique of Pure Reason*. New York: Macmillan.

Stang, Nicholas. (2010). 'Kant's Possibility Proof', *History of Philosophy Quarterly*, 27(3): 275–99.

Stang, Nicholas. (2011). 'Did Kant Conflate the Necessary and the a Priori?', *Nous*, 45(3): 443–71.

Stang, Nicholas. (2016). *Kant's Modal Metaphysics*. New York: Oxford University Press.

Swing, Thomas K. (1969). *Kant's Transcendental Logic*. New Haven and London, Yale University Press.

Van Cleve, James. (1999). *Problems from Kant*. New York: Oxford University Press.

Van Peursen, Cornelius Anthonie. (1987). 'Christian Wolff's Philosophy of Contingent Reality', *Journal of the History of Philosophy*, 25(1): 69–82.

Warnock, G. J. (1948). 'Concepts and Schematism', *Analysis*, 9(5): 77–82.

Watkins, Eric. (1998). 'From Pre-established Harmony to Physical Influx: Leibniz's Reception in Early 18th Century Germany'. *Perspectives on Science* 6 (Special Issue "Leibniz and the Sciences" ed. D. Garber), 136–203.

Watkins, Eric. (2002). 'Transcendental Idealism and the Categories', *History of Philosophy Quarterly*, 19(2): 191–215.

Westphal, Kenneth. (2000). 'Kant, Hegel, and the Fate of "the" Intuitive Intellect'. In S. Sedgewick (ed.). *The Reception of Kant's Critical Philosophy*. Cambridge: Cambridge University Press, 283–305.

Williams, Thomas. 'Saint Anselm'. In Edward N. Zalta (ed.). *The Stanford Encyclopedia of Philosophy*, Spring 2016 Edition. http://plato.stanford.edu/archives/spr2016/entries/anselm/.

Wilson, Kirk Dallas. (1978). 'Studies in the Formal Logic of Kant's Modal Functions of Judgment', *Kant-Studien*, 69(1–4): 252–72.

Winegar, Reed. (2017). 'Kant on God's Intuitive Understanding', *Kantian Review*, 22(2): 305–29.

Winegar, Reed. (forthcoming). 'Kant on Intuitive Understanding and Things in Themselves'. *European Journal of Philosophy*.

Wippel, John F. (1982a). 'Essence and Existence'. In Norman Kretzmann, Anthony Kenny, and Jan Pinborg (eds.). *The Cambridge History of Later Medieval Philosophy*. Cambridge: Cambridge University Press, 385–410.

Wippel, John F. (1982b). 'The Relationship Between Essence and Existence in Late-Thirteenth Century Thought: Giles of Rome, Henry of Ghent, Godfrey of Fontaines, and James of Viterbo'. In Parwiz Morewedge (ed.). *Philosophies of Existence: Ancient and Medieval*. New York: Fordham University Press, 131–64.

Wippel, John F. (2000). *The Metaphysical Thought of Thomas Aquinas: From Finite Being to Uncreated Being*. Washington DC: The Catholic University of America Press.

Wolff. Michael. (1995). *Die Vollständigkeit der kantischen Urteilstafel: Mit einem Essay über Freges Begriffsschrift*. Frankfurt am Main: Vittorio Klostermann.

Wolff, Robert Paul. (1963). *Kant's Theory of Mental Activity: A Commentary on the Transcendental Analytic of the Critique of Pure Reason*. Cambridge MA: Harvard University Press.

Wood, Allen. (1978). *Kant's Rational Theology*. Ithaca and London: Cornell University Press.

Wood, Allen. (2005). *Kant*. Oxford: Blackwell.

Yong, Peter. (2014). 'God, Totality and Possibility in Kant's Only Possible Argument', *Kantian Review*, 19(1): 27–51.

Zammito, John H. (1992). *The Genesis of Kant's Critique of Judgment*. Chicago and London: The University of Chicago Press.

Index